Topics in
Current Physics

19

W0106084

Topics in Current Physics   Founded by Helmut K. V. Lotsch

Theory of
Chemisorption

Edited by J. R. Smith

With Contributions by

J. A. Appelbaum   F. J. Arlinghaus   T. L. Einstein
Ş. Ellialtıoğlu   J. G. Gay   D. R. Hamann   J. A. Hertz
A. B. Kunz   J. R. Schrieffer   J. R. Smith   T. Wolfram
S. C. Ying

With 116 Figures

Springer-Verlag Berlin Heidelberg New York 1980

John R. Smith, Ph.D.

Physics Department, General Motors Research Laboratory,
Warren, MI 48090, USA

ISBN-13: 978-3-642-81433-4   e-ISBN-13: 978-3-642-81431-0
DOI: 10.1007/ 978-3-642-81431-0

Library of Congress Cataloging in Publication Data. Main entry under title: Theory of chemi-
sorption. (Topics in current physics; v. 19.) Bibliography: p. Includes index. 1. Chemisorption.
I. Smith, John Robert, 1940–. II. Appelbaum, Joel A. III. Series.
QD547.T45 541.3′453 79-27542

This work is subject to copyright. All rights are reserved, whether the whole or part of the
material is concerned, specifically those of translation, reprinting, reuse of illustrations, broad-
casting, reproduction by photocopying machine or similar means, and storage in data banks.
Under § 54 of the German Copyright Law where copies are made for other than private use,
a fee is payable to the publisher, the amount of the fee to be determined by agreement with
the publisher.

© by Springer-Verlag Berlin Heidelberg 1980
Softcover reprint of the hardcover 1st edition 1980

The use of registered names, trademarks, etc. in this publication does not imply, even in the
absence of a specific statement, that such names are exempt from the relevant protective laws
and regulations and therefore free for general use.

2153/3130-543210

Preface

The theory of the chemical interaction of molecules with surfaces has advanced handsomely in the last few years. This is due in part to the application of the entire arsenal of bulk solid-state theory and molecular quantum chemistry methods. This considerable activity was stimulated by an outpouring of experimental data, particularly of photoemission spectra. In many cases the theoretical techniques are now such that accurate, atomistic pictures of chemisorption phenomena are computed from first principles. This level of capability has been reached only recently, and has not been described anywhere in a comprehensive manner.

The purpose of this monograph is to review these recent advances and, at the same time, to indicate a number of important questions which have not been answered. We discuss chemisorption on oxides, semiconductors, and both simple and transition metals. Solid surfaces as well as clusters are considered. While the review should be valuable to workers in the field, care has been taken to make the chapters understandable to the nonspecialist.

Warren, Michigan   *John R. Smith*
December, 1979

Contents

List of Contributors

Appelbaum, Joel Alan
    Bell Telephone Laboratories, Murray Hill, NJ 07974, USA

Arlinghaus, Frank J.
    Physics Department, General Motors Research Laboratories, Warren, MI 48090, USA

Einstein, Theodore L.
    Department of Physics and Astronomy, University of Maryland,
    College Park, MD 20742, USA

Ellialtıoğlu, Şinasi
    Department of Physics, University of Missouri-Columbia, 223 Physics Building,
    Columbia, MO 65211, USA
    Permanent address: Physics Department, Middle East Technical University,
    Ankara, Turkey

Gay, Jackson Gilbert
    Physics Department, General Motors Research Laboratories, Warren MI 48090, USA

Hamann, Donald Robert
    Bell Telephone Laboratories, Murray Hill, NJ 07974, USA

Hertz, John A.
    James Franck Institute, University of Chicago, Chicago, IL 60637, USA

Kunz, Albert Barry
    Department of Physics, University of Illinois, Urbana, IL 61801, USA

Schrieffer, John Robert
    Department of Physics, University of Pennsylvania, Philadelphia, PA 19104, USA

Smith, John Robert
    Physics Department, General Motors Research Laboratories, Warren, MI 48090, USA

Wolfram, Thomas
    Department of Physics, University of Missouri-Columbia, Columbia, MO 65211, USA

Ying, See-Chen
    Department of Physics, Brown University, Providence, RI 02912, USA

1. Introduction

J. R. Smith

When a gas or liquid molecule forms chemical bonds with the surface of a solid or cluster, we say it is chemisorbed. Chemisorption plays an important role in, for example, catalysis, corrosion, and electrolysis.

In fundamental research of the chemisorption bond, experiment is generally well ahead of theory. The advent of ultra high vacuum systems and particularly their use in angular photoemission experiments has yielded an extensive amount of experimental data. Recently, however, there has been a rapidly expanding theoretical effort which is beginning to give densities of states results which agree in detail with angular photoemission spectra. There are now surface band structure calculations for chemisorbed layers which are of a quality comparable to modern bulk solid state calculations. The techniques of quantum chemistry have been applied to chemisorption on clusters, again to a sophistication comparable to modern molecular calculations. However, the chemisorbed state is intermediate between the solid state and the gaseous or molecular state. Thus there are properties which have features peculiar to the chemisorption process, such as electron correlation effects, which are not well understood.

Because chemisorption theory is being brought into expanded contact with experiment, a review of theoretical techniques, successes and limitations appears timely. This critical review will hopefully be of value to experimentalists or theorists who would like to acquaint themselves with the state of the art in chemisorption theory, whether or not they have experience in the chemisorption field. For this reason, Chaps.2-6 are devoted to theoretical descriptions, while Chap.7 gives a critique of the theoretical methods used in the preceding chapters and discusses research in improved model Hamiltonians, with emphasis on electron correlation effects.

Chapter 2 contains the most complete description of the physical properties of chemisorption as well as an introduction to the density functional theory used in many of the other chapters. This part of Chap.2 serves nicely as a detailed introduction to the remainder of the book. The bulk of Chap.2 is devoted to a description of chemisorption on simple metals such as the alkalies and aluminum, with some properties of transition metals included. Chemisorption on semiconductors is treated in Chap.3. This is perhaps the most mature field in chemisorption theory.

It is fair to say, however, that more experimental data has been taken on transition and noble metal surfaces than on any other class of materials. The theory
of chemisorption on these metals is discussed in Chap.4. Cluster chemisorption
is the subject of Chap.5. Since clusters usually contain a relatively small number
of atoms they can be treated via the modern techniques of quantum chemistry. Chapter
5 contains a good review of these theoretical techniques of molecular chemistry as
well as their application to cluster chemisorption. Chemisorption on transition
metal oxides is dealt with in Chap.6. The theoretical and experimental state of
the art for these materials is not as advanced as it is for the other materials
described. However, an unusual amount of physical insight relative to oxides of
cubic perovskite structure is contained in this chapter. While Chaps.2-7 contain
results which are often impressive upon comparison with experiment, in Chap.7 a
number of deficiencies in these commonly used methods are exhibited. A rather complete discussion of model Hamiltonian approaches is also given.

It is interesting to make some comparisons between the various approaches discussed in the chapters. In Chaps.2-5 the importance of self-consistency between
the potential used and the wave functions computed is emphasized. However, self-
consistency plays only a minor role in Chaps.6 and 7. The calculations discussed
in Chaps.2-5 often involve extensive numerical work, whereas Chaps.6 and 7 rely
heavily on analytical techniques. A number of treatments of electron correlation
are described. Chapters 2-4 employ primarily the local density approximation. That
is, many of the calculations discussed in those chapters are similar in approach
to bulk band-structure calculations. In Chap.5 the quantum chemist's approach to
correlation, namely configuration interaction, is presented. In Chap.7 a second
solid state approach to correlation is given in terms of model Hamiltonians. Calculations in Chaps.2-4 have as their principal variable the three-dimensional
charge density, while those of Chap.5 key more on electron wave functions. A few
parameters in model Hamiltonians are the key variables in many of the calculations
of Chaps.6 and 7. In Chaps.2 and 3 local pseudopotentials are used successfully to
represent the core electrons, although the adequacy of such pseudopotentials has
been the subject of considerable research of late. Ab initio treatments of the core
electron potentials are presented in Chaps.4 and 5. While the role of s and p electrons in chemisorption on transition metals is emphasized in Chap.5, strong d electron effects are discussed in Chaps.4 and 6. Single adsorbate atom effects are
treated in Chaps.2 and 5-7, whereas Chaps.3 and 4 are primarily limited to fractional monolayer adsorption. Surface states play an important part in chemisorption
on semiconductors and transition metal oxides in Chaps.3 and 6. They are beginning
to appear in a surprisingly important way in chemisorption on transition metal surfaces, as discussed in Chap.4.

The variety of techniques discussed in this book forms the basis for an exciting
period of growth in the theory of chemisorption.

2. Density Functional Theory of Chemisorption on Simple Metals

S. C. Ying

With 12 Figures

In this chapter we discuss the density functional theory of chemisorption on simple metals. The center of attention here is on the ground state properties such as binding energy versus separation from the substrate, charge density, and induced dipole moment. This is in accordance with the very nature of the density functional approach, which is a rigorous formulation only for the ground state of the inhomogeneous electron gas. Although we shall also have occasion to discuss such quantities as the adsorbate induced change in the density of states, this kind of single-particle property in the present density functional approach bears no simple relation with the usual quasi-particle description of the low-lying excited states and is less amenable to theoretical interpretation. Our emphasis will be on general concepts and trends. We shall discuss the chemisorption of adsorbates such as H, Li, Na, K, Rb, Cs, O, Si, and Cl, etc., on various substrates like Al and W. Although the theory outlined in this chapter is designed primarily for adsorption on simple metals, it can be used to understand some qualitative aspects of adsorption on more complicated substrates such as the transition metals. The choice of topics here is by no means exhaustive. For detailed calculations of specific systems, the reader is referred to other recent review articles [2.1,2].

2.1 Background Information

2.1.1 Importance of Self-Consistency

The reason for choosing simple metals for our discussions is that to date, most of the self-consistent calculations for chemisorption on metals have been performed for very simple models of the substrate. See Chap.4 for a review of theory of chemisorption on transition metals. The simplest of these, known as the jellium model, is one in which ions are replaced by a uniform positive background occupying a half space. In more sophisticated models, the ions are represented by pseudopotentials which are usually treated perturbatively. The simplicity of these models allows one to make a detailed study of the various properties of the system while maintaining full self-consistency between the charge density and the effective potential. The satisfaction of self-consistency is of the utmost importance for an inhomogeneous system as can be seen from the following arguments. Consider a semi-infinite solid and surround each lattice plane of ions by a flat slab. This corresponds to a particular choice of unit cell. In a bulk slab, because of the requirements of symmetry, the total electronic charge exactly balances the ionic charge even when the electronic wave functions are not consistent with the potential. Thus charge neutrality is always satisfied within each slab and one does not pay a large price in the Coulomb energy for violating self-consistency. However, the situation is completely different in the surface region. For a non-self-consistent study, one can have deviation from charge neutrality in any number of surface slabs, depending on the approximations involved [2.3]. These configurations usually involve a high Coulomb energy and are hence far removed from the true ground state. Even when the net charge of the surface region is correct, the chemisorption energy still depends sensitively on the amount and shape of the charge transfer between the adsorbate and the substrate, which can only be accurately obtained through a detailed self-consistent study.

The simple models for the substrate described above are good representations only for a limited number of metals such as the alkalies and other s-p bonded metals. Unfortunately, few experimental data exist for chemisorption on these simple metals. In our discussions below, we shall often compare the theoretical results obtained from the simple models with experimental data obtained from chemisorption on transition metals. This should produce useful qualitative information and trends when there are no strong directional bonding or resonant states situated in a rapidly varying region of the density of states of the substrate. Many of the chemisorption quantities are then expected to depend mainly on the tail distribution of the surface charge density, which is reasonably described by the simple models.

2.1.2 Density Functional Theory

The underlying basis of our discussion is the density functional theory developed by KOHN et al. [2.4,5]. We shall only present an outline of the basic theory here. The reader is referred to the original articles for a more detailed description of the formalism. Atomic units are used below, with e (the magnitude of the charge on an electron), m (the mass of the electron, and $\hbar$ all set equal to unity.

Consider the class of systems comprised of interacting electrons moving in various static external potentials. According to the density functional theory there exists a one to one correspondence between the ground state wave function Ψ and the ground state density $n(\underline{r})$. One can therefore write the total ground state energy as a functional of the density $n(r)$.

$\hbar$ = h/2π (normalized Planck's constant)

$$E[n(\underline{r})] = \int v_0(\underline{r})n(\underline{r})d\underline{r} + F[n] \quad , \tag{2.1}$$

where

$$F[n] = <\Psi|T + U|\Psi> \quad , \tag{2.2}$$

T and U corresponding, respectively, to the operator representing electron kinetic energy and Coulomb interaction in the Hamiltonian. The important point here is that F[n] is a universal functional of the density $n(\underline{r})$, its functional form independent of the external potential $v_0(\underline{r})$.

KOHN and SHAM [2.5] pointed out that it is useful to separate the functional F[n] into the following components:

$$F[n] = T_s[n] + \frac{1}{2} \int [n(\underline{r})n(\underline{r}')/|\underline{r} - \underline{r}'|]d\underline{r} \, d\underline{r}' + G_{xc}[n] \quad . \tag{2.3}$$

The first term $T_s[n]$ is the kinetic energy of a system of noninteracting electrons with density $n(\underline{r})$, the second term is the classical Coulomb energy, and the remainder $G_{xc}[n]$ is by definition the exchange correlation energy functional.

The correct ground state density minimizes the total energy. Thus it can be determined through a variational procedure. The Euler's equation reads as

$$\delta\{E[n] - \mu \int n(\underline{r})d\underline{r}\} = 0 \quad . \tag{2.4}$$

For large N, the Lagrange multiplier μ is just equal to the chemical potential.

Equation (2.4) is of practical use only when the explicit form of the energy functional is known. GUNNARSON and LUNDQUIST [2.6] have shown that the exchange correlation functional $G_{xc}[n]$ defined above can be expressed in terms of the pair distribution function $g(\underline{r},\underline{r}';\lambda)$ for a value of the coupling constant λ between 0 and 1, as

$$G_{xc}[n] = \frac{1}{2} \int d\underline{r}d\underline{r}'|\underline{r} - \underline{r}'|^{-1} \int_0^{e^2} d\lambda n^2(\underline{r})[g(\underline{r},\underline{r}';\lambda) - 1] \quad , \tag{2.5}$$

and that the function $g(\underline{r},\underline{r}';\lambda)$ satisfies the sum rule

$$\frac{1}{\sqrt{4\pi}} \int d\underline{r}'[g(\underline{r},\underline{r}';\lambda) - 1] = -1 \quad . \tag{2.6}$$

The most commonly used approximation for $G_{xc}[n]$ is the so-called local density approximation

$$G_{xc}[n] = \int \varepsilon_{xc}[n(\underline{r})]n(\underline{r})d\underline{r} \quad , \tag{2.7}$$

where $\varepsilon_{xc}(n)$ is the exchange correlation energy per electron of a homogeneous electron liquid with constant density n. This corresponds to replacing the exact $g(\underline{r},\underline{r}';\lambda)$ in (2.5) by the pair distribution for a homogeneous electron liquid $g_h(\underline{r},\underline{r}')$ with density equal to $n(\underline{r})$. This approximation satisfies the sum rule (2.6). Moreover, as pointed out in (2.6), the exact functional $G_{xc}[n]$ depends only on the integrated value of $g(\underline{r},\underline{r}';\lambda)$ and there is a systematic cancellation of errors. Thus the local approximation is expected to work well even in the presence of rapidly varying density. This has been borne out by numerous applications to such diverse systems as atoms and molecules, bulk solids, clean metal and semiconductor surfaces, and chemisorption systems [2.6]. Attempts have been made to improve the local approximation by addition of gradient terms [2.7] or by explicit inclusion of nonlocal density dependence [2.8].

For the kinetic energy functional $T_s[n]$, there are two approaches used in the literature. The first is to expand in a gradient series, starting with the local approximation just as in our discussion of $G_{xc}[n]$:

$$T_s[n] = \int \frac{3}{10} (3\pi^2)^{2/3} n^{5/3}(\underline{r}) d\underline{r} + c \int |\underline{\nabla}n(\underline{r})|^2 n^{-1} d\underline{r} \quad . \tag{2.8}$$

The coefficient c in the first gradient correction has been shown to have the value 1/72. HOHENBERG and KOHN [2.4] have argued on dimensional grounds that the gradient expansion requires $|\underline{\nabla}n|/n \ll k_f(n)$ and $|\underline{\nabla}_i\underline{\nabla}_j n|/|\underline{\nabla}n| \ll k_f(n)$. With the explicit form for $T_s[n]$ and $G_{xc}[n]$ as in (2.8) and (2.7), one can work with the Euler's equation (2.4) directly and avoid calculating wave functions. This is the so-called statistical approach. The second approach, developed by KOHN and SHAM [2.5], treats $T_s[n]$ exactly. They showed that the ground state density $n(\underline{r})$ can be obtained from the self-consistent solution of the effective single electron Schrödinger equation

$$\left[-\frac{\nabla^2}{2} + v_{eff}(\underline{r}) \right] \phi_i(\underline{r}) = \varepsilon_i \phi_i(\underline{r}) \quad , \tag{2.9}$$

where

$$v_{eff}(\underline{r}) = v_0(\underline{r}) + \int [n(\underline{r}')/|\underline{r} - \underline{r}'|] d\underline{r}' + \frac{\delta G_{xc}[n]}{\delta n(\underline{r})} \tag{2.10}$$

is the effective one-electron potential, and the ground state density $n(\underline{r})$ is

$$n(\underline{r}) = \sum_{i=1}^{N} |\phi_i(\underline{r})|^2 \tag{2.11}$$

with $\{\phi_i(\underline{r})\}$ the N lowest lying orthonormal solutions of (2.9). The exact kinetic energy functional $T_s[n]$ is evaluated as

$$T_s[n] = \sum_i \varepsilon_i - \int V_{eff}[\underline{r};n]n(\underline{r})d\underline{r} \quad . \qquad (2.12)$$

As mentioned earlier, the eigenvalues ε_i from (2.9) are not the true quasiparticle energies. SLATER [2.9] showed that they are just the derivative of the total energy with respect to orbital occupation numbers. Thus relaxation effects associated with a true excited state are not properly accounted for. The eigenvalues ε_i are expected to be close to the true excitation energies only when the orbitals involved are quite delocalized, such as the states belonging to the s-p valence band of simple metals [2.9].

2.2 Statistical Approach

In this section we describe theoretical studies of chemisorption on simple metals based on the statistical approach. The first subsection concerns the development of a linear response formalism. Subsequent sections describe the application of this formalism to the study of hydrogen and alkali chemisorption. In the last subsection, a study based on the direct variation of the energy functional, including nonlinear effects, is discussed.

2.2.1 Linear Response Function

YING et al. [2.10] have developed a general linear response formalism within the density functional approach for the purpose of studying chemisorption. We outline the basic steps here. Consider the ground state of a metallic system corresponding to an external potential $v_0(\underline{r})$ and density $n_0(\underline{r})$. An additional source change of strength Z is now introduced at point $\underline{r}'$. Let the change in the charge density at a general point $\underline{r}$ by denoted by $ZR(\underline{r},\underline{r}')$ and that in the electrostatic potential be $ZL(\underline{r},\underline{r}')$. To first order in Z (linear approximation), one can determine $R(\underline{r},\underline{r}')$ and $L(\underline{r},\underline{r}')$ through the linearized Euler's equation (2.4),

$$\int \left. \frac{\delta^2(T_s+G_{xc})}{\delta n(\underline{r})\delta n(\underline{r}'')} \right|_{n=n_0} R(\underline{r}'',\underline{r}')d\underline{r}'' - L(\underline{r},\underline{r}') - \delta\mu = 0 \quad , \qquad (2.13)$$

and the linearized Poisson's equation

$$\nabla^2 L(\underline{r},\underline{r}') = 4\pi\{R(\underline{r},\underline{r}') - \delta(\underline{r} - \underline{r}')\} \quad . \qquad (2.14)$$

Equations (2.13,14) are completely general.

Within the linear approximation, the response functions $L(\underline{r},\underline{r}')$ and $R(\underline{r},\underline{r}')$ are characteristic of the substrate and independent of the strength of the perturbing charge Z. Armed with the knowledge of these response functions, one can calculate various ground state properties of chemisorption on a given substrate, treating the various adsorbates as distributions of perturbing charges.

We now restrict our attention on the jellium model description of the metal surface, in which the positive ions of the metal are replaced by the uniform background

$$n_+(\underline{r}) = n_+\theta(-x) \quad . \tag{2.15}$$

In this model, the unperturbed electronic density $n_0(x)$ is a function of the coordinate x normal to the surface only, and the response functions $L(\underline{r},\underline{r}')$ and $R(\underline{r},\underline{r}')$ depend directly on x and x', but only on the differences in the y and z coordinates $(y - y'; z - z')$. It is therefore convenient to perform a Fourier transform in these coordinates parallel to the surface, e.g.,

$$R(\underline{r},\underline{r}') = (2\pi)^{-1} \int d\underline{Q}\, R(Q;x,x')\exp(i\,\underline{Q}\cdot\underline{u}) \quad , \tag{2.16}$$

with

$$\underline{u} = (0,\, y - y',\, z - z') \quad , \qquad \underline{Q} = (0,\, Q_y,\, Q_z) \quad . \tag{2.17}$$

Introducing $J(\underline{r},\underline{r}')$ as the inverse of the kernel $\delta\{T_s[n] + G_{xc}[n]\}/\delta n(\underline{r})\delta n(\underline{r}')$ (2.13) and (2.14) can be combined into a single equation

$$\left(\frac{d^2}{dx^2} - Q^2\right)L(Q;x,x') - \int 4\pi J(Q;x,x'')L(Q;x'',x')dx'' = -4\pi\delta(x - x') \quad , \tag{2.18}$$

where $Q \equiv |\underline{Q}|$. Equation (2.18) is valid for $Q \neq 0$ for which YING et al. [2.11,12] showed that $\delta\mu = 0$. The $Q = 0$ case can be obtained by a careful limiting procedure. The explicit form of (2.18) depends of course on the choice of the energy functional. YING et al. [2.11,12] have solved this equation choosing the gradient expansion (2.8) for $T_s[n]$ and the local approximation for exchange,

$$\varepsilon_x(n) = -\frac{3}{4}\left(\frac{3}{\pi}\right)^{1/3}n^{1/3}(\underline{r}) \quad . \tag{2.19}$$

These authors have found that the inclusion of an additional local correlation energy such as the Wigner interpolation form [2.13] has few effects on the responses. This is not to say that correlation effects are not important, since the cancellation in the higher order gradient terms depends crucially on the inclusion of both exchange and correlation effects [2.7,17]. These authors choose a value for n_+ corresponding to $r_s = 1.5[n_+ = (4\pi/3r_s^3)^{-1}]$, and a form for $n_0(x)$ obtained earlier by LANG and KOHN [2.14]. This was motivated by the fact that for $r_s < 4$, the solution in [2.14] for $n_0(x)$ is very close to the variational solution that SMITH [2.15] obtained as the zeroth order solution of the Euler's equation (2.4). Figure 2.1 shows the Q = 0 component of the density response thus obtained for two different positions of the source charge. Notice that when the source charge is inside the jellium, the screening charge is centered at the source charge and rather symmetric. As the source charge moves away from the surface, the screening charge lags behind and becomes asymmetric. Ultimately with the source charge at infinity, the screening charge is left behind completely in the surface region. Asymptotically, for $x \gg 1$, and $x' \gg 1$, the potential response takes the form

$$L(Q;x,x') = -2\pi/Q\{(1 + 2dQ)\exp[-Q(x + x')] - \exp(-Q|x - x'|)\} \quad . \tag{2.20}$$

The first term in (2.20) is the induced potential, and the second term is the contribution from the bare charge. In the limit $Q \to 0$ and $x \to \infty$, one has

$$\lim_{x \to \infty} \lim_{Q \to 0} L(Q;x,x') \equiv L(Q;\infty,x') = 4\pi(x' - d) \quad . \tag{2.21}$$

From Poisson's equation, it can be shown that

$$L(Q;\infty,x') = 4\pi\left[x' - \int_{-\infty}^{\infty} R(Q;x,\infty)x \, dx\right] \quad , \quad \text{for } x' \gg 1 \quad . \tag{2.22}$$

Comparing (2.21) and (2.22), we see that the parameter d is just the position of the center of mass of the induced charge with the source charge at infinity. By back Fourier transform into real space, the asymptotic form of the induced potential $V_1(x = x', u = 0)$ at the position of the source charge is found to be

$$V_1(x = x' , u = 0) = - Z[2(x' - d)]^{-1} + O(1/x'^3) \quad . \tag{2.23}$$

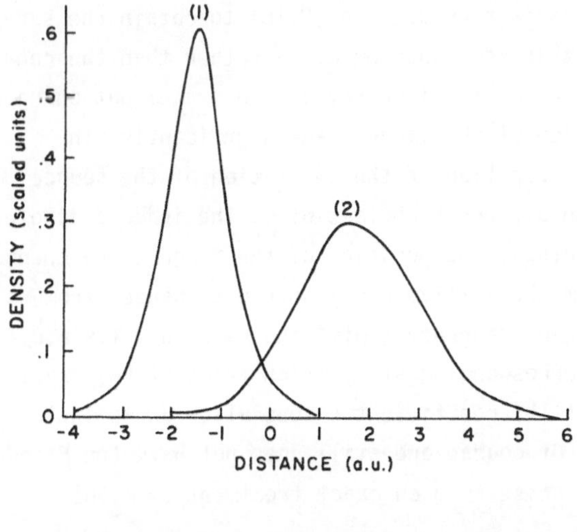

Fig.2.1. Density response of a jellium ($r_s = 1.5$) to a point charge perturbation located at x = -1.5 a.u. (1) and x = 3.0 a.u. (2)

This has the classical image potential form, with the image plane at a distance
d away from the jellium edge. This result was also independently obtained by LANG
and KOHN [2.16]. YING et al. [2.12] have also examined the effect of the gradient
expansion for the functional $T_s[n]$ on the linear response function. Consider the
response of a uniform electron gas to a perturbation with wave vector $\underline{k}$. The susceptibility $f(\underline{k}) = \delta n(\underline{k})/\delta v(\underline{k})$ in the random phase approximation is well known to
be

$$f(k) = - (k_f/2\pi^2) \left[1 + (1 - \eta^2)/(2\eta)\ln\left|\frac{1+\eta}{1-\eta}\right|\right] , \quad \eta = k/2k_f \quad . \tag{2.24}$$

On the other hand, f can also be derived from the general response function in the
density functional approach from (2.18). If one takes the expansion in (2.8) for
$T_s[n]$ and neglects $G_{xc}[n]$, the corresponding f(k) is

$$f(k) = - k_f[\pi^2(1 + 3c\eta^2)]^{-1} \quad . \tag{2.25}$$

Comparison of (2.24) and (2.25) shows for small η, the value of c = 1/72 mentioned
earlier gives the correct leading term for f(k) in powers of η, whereas for large
k, the value c = 1/9 (the VON WEIZÄCKER value [2.17]) gives the correct leading
term in powers of $1/\eta$. Thus, for the response to be correct at all wavelengths,
c has to be a function of η. The surface response to an external charge is comprised of many different wavelengths, and the inclusion of the first gradient term
is only a good approximation for those components corresponding to $\eta < 1$. KAHN and
YING [2.18] have devised the following scheme to include in an approximate way the
effect of higher order gradient terms. In the region around the spatial point x,
the typical wavelength of the response is taken to be equal to the local Thomas-
Fermi wave vector $k_{Tf}[n_0(x)]$. The parameter $\eta = k_{Tf}/2k_f$ is hence a function of
$n_0(x)$, the unperturbed electron density distribution. Substitution of this
$\eta[n_0(x)]$ into the function $c(\eta)$ obtained through the comparison of (2.24) and (2.25)
converts c into a function of $n_0(x)$. This is then used in (2.18) to obtain the surface response functions. The choice of this new function $c(n_0)$ rather than the constant value c = 1/72 has little effect on the total energy of the system but changes
the induced dipole moment and the position of the image plane signficantly. In
Fig.2.2 is shown the induced dipole as a function of the separation of the source
charge from the jellium edge for the two different choices of c. The induced dipole
is larger for the modified coefficient $c(n_0)$. The position of the image plane changes
from 3.5 a.u. to 2.0 a.u. (measured from the jellium edge) when one changes from
c = 1/72 to $c(n_0)$. This brings it into close agreement with the value d = 1.9 a.u.
obtained by LANG and KOHN [2.16] in a corresponding study which treats $T_s[n]$ exactly.
Thus it is seen that the modification of the coefficient c does allow us to include
some higher order gradient corrections. Of course one still does not have the Friedel
oscillations in the response which only arise from an exact treatment of $T_s[n]$.

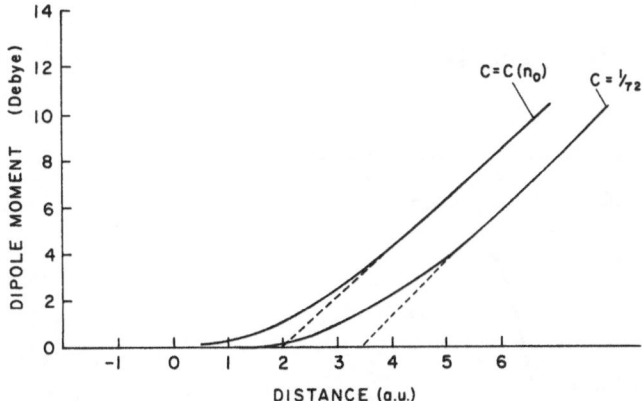

Fig.2.2. Induced dipole moment as a function of distance of a point charge from a jellium surface. The curves for different c values correspond to different approximations to the kinetic energy functional

2.2.2 Chemisorption on Jellium: Hydrogen and the Alkalies

The first application of the linear response formalism described in the previous section is to a single hydrogen atom chemisorbed on a jellium model substrate corresponding to r_s = 1.5 [2.11]. The zeroth order configuration of the system was taken as a proton situated in the surface region plus an extra electron in the conduction band of the metal. Then, in the linear approximation, the change in the electronic density and the electrostatic potential due to the adatom can be directly obtained from the response function by scaling the result for an infinitesimal source charge to that for a point charge of +1 unit. The results are summarized below. First, the interaction energy between the H atom and the metal substrate is determined as a function of the coordinates of the proton (u' = 0,x'). This is given, according to the Hellmann-Feynman theorem, by

$$W(x') = V_0(x') + V_1(x = x', u = 0)/2 \quad , \tag{2.26}$$

where $V_0(x')$ is the electrostatic potential of the bare surface and $V_1(x = x', u = 0)$ is the potential due to the screening charge evaluated at the proton location. The latter can be obtained from the total potential response $L(x,x',u)$ through the subtraction of the self-potential term

$$V_1(x,x',u) = L(x,x',u) - [(x - x')^2 + u^2]^{-\frac{1}{2}} \quad . \tag{2.27}$$

The results are shown in Fig.2.3. According to (2.23), for large x';$W(x')$ tends to the image potential, i.e.,

$$\lim_{x' \to \infty} W(x') = 1/[4(x' - d)] \quad . \tag{2.28}$$

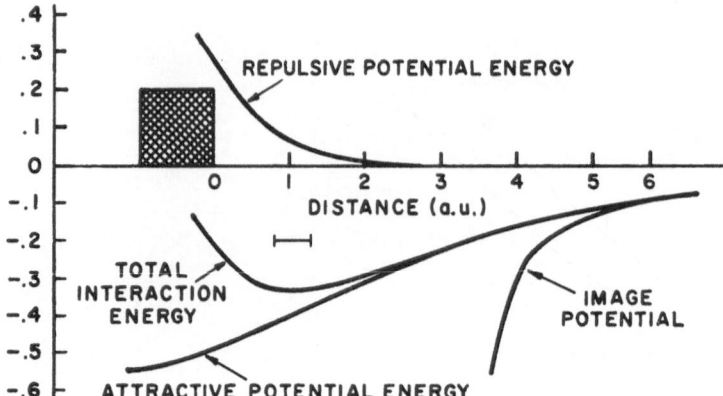

Fig.2.3. Interaction energy between a proton and a jellium as a function of separation distance. [2.11]

As x' approaches the surface region (x' < 5 a.u.), the interaction energy begins
to deviate significantly from the image potential as shown in Fig.2.3. Finally a
minimum in W(x') is reached at x' ~1.08 a.u. The minimum results from a competition
between the repulsive terms $V_0(x')$ and the attractive term $V_1(x = x', u = 0)/2$. The
curve is rather broad near the minimum. The depth of the minimum relative to a
proton at infinity is 9 eV. The adsorbed hydrogen will exhibit vibrational modes
in the potential well displayed in Fig.2.3. The vibrational frequency is

$$\omega = \frac{1}{m} \left(\frac{d^2W}{dx^2} \right)_{x_m}^{\frac{1}{2}} \quad , \tag{2.29}$$

where x_m is the coordinate of the hydrogen at its energy minimum (1.08 a.u.), and
m is the proton mass.

    As mentioned earlier, even though the calculation is for a jellium model ap-
propriate for simple metals, it is useful to make a comparison with experimental
results obtained for H on transition metals, as a number of chemisorption properties do not depend sensitively on the detail band structure of the substrate but
rather on the density distribution $n_0(x)$.

    For gauging the theoretical value of the binding energy, one should note that
while experimentally hydrogen desorbs in atomic or molecular form, the theoretical
configuration obtained in this calculation for the hydrogen at infinite separation
from the substrate is a bare proton. This is because of the adoption of linear response formalism. The atomic configuration at infinite separation requires the inclusion of nonlinear effects. Thus for testing the results at equilibrium separation, one should compare the depth of the minimum in Fig.2.3 with the ionic
desorption energy E_I. This can be derived from the experimentally measured atomic
desorption energy E_a through a Born-Haber cycle relation

$$E_I = E_a + I - \phi \quad , \tag{2.30}$$

where I is the hydrogen ionization potential and ϕ the work function of the substrate metal. If one now takes the system of H on W (electronic density of W
corresponds to $r_s \sim 1.5$) and inserts the experimental value of 5.3 eV for ϕ, 13.6 eV
for I, and 3 eV for E_a[2.19], one obtains from (2.30) a value for $E_I = 11.3$ eV.
The theoretical value of 9 eV compares favorably with this result. The vibrational
frequency for hydrogen adsorbed on tungsten has also been measured through field
emission studies [2.20] and inelastic electron diffraction [2.21,22]. The theoretical value of 200 MeV is in fair agreement with the experimental results ranging
between 130 and 150 MeV.

The induced dipole moment can be read directly from the curve in Fig.3.2 by
locating the equilibrium distance at 1.08 a.u. It is seen that the theoretical
value of induced dipole moment at this separation is rather sensitive to the coefficient of the gradient term in $T_s[n]$, indicating that higher order gradient corrections are important for this quantity. Moreover, LANG and WILLIAMS [2.23] have
also shown that at this separation the induced dipole does not scale linearly
with the magnitude Z of the perturbing charge. Thus the only conclusion that can
be drawn from the linear response calculation is that the induced dipole moment
is very small — the absolute value being less than one Debye. This is due to the
rather symmetric distribution of the screening charge around the proton and little
net charge transfer between the H atom and the substrate metal. Experimentally,
the induced dipole for a single adatom can be deduced from the change in work function in the zero coverage limit through the relation

$$\Delta\phi = 4\pi\mu\theta \tag{2.31}$$

where θ is the coverage. For H on W, the experimental value for μ is indeed very
small. The absolute value of measured μ is less than 0.2 Debye.

At the equilibrium separation, the screening charge obtained in [2.11] is rather
widespread and the peak density at the position of the proton is only 0.36 of that
of an isolated hydrogen atom. This leads to a relatively larger forward scattering
cross section for electrons scattering off the chemisorbed hydrogen as compared
to an isolated H atom [2.12,24]. However, as shall be shown by the results in later
sections, the linear approximation underestimates the magnitude of the screening
charge near the proton. Thus this conclusion about the scattering cross section may
be peculiar to the linear response formalism.

Although the statistical approach focuses directly on the ground state properties,
one can also derive information about possible resonance levels associated with the
adatom through the self-consistent potential. YING et al. [2.12,24] wrote the total
effective potential $V(\underline{r})$ of the system (W + H) as the potential of bare W plus a
remainder: $V(\underline{r}) = V_0(\underline{r}) + V_1(\underline{r})$. In keeping with the small dipole moment it is found

that $V_1(r)$ is nearly spherical symmetric and may be replaced by its spherical average. One bound state is found to exist in this symmetrized $V_1(r)$. The position of the resonance level ϵ_1 is found by shifting this level downward by the value $V_0(x_m)$, x_m being the equilibrium position. The result is that ϵ_1 lies at 5.6 eV below the Fermi level. The width of the resonance level has also been calculated in the spirit of the Hartree-Fock approximation to the Anderson Hamiltonian [2.25]. In this model, the width of the resonance level Γ is given by

$$\Gamma = \pi N(\epsilon_1) \left\langle |V_{1k}|^2 \right\rangle_{\epsilon_1} \, , \tag{2.32}$$

where $N(\epsilon_1)$ is the substrate density of states, and

$$\left\langle |V_{1k}|^2 \right\rangle = \; < \; | \int \psi_k^*(r) V_1(r) \psi_1(r) dr |^2 > \tag{2.33}$$

is the mean square coupling matrix element of the spherically symmeterized perturbing potential $V_1(r)$ between substrate wave functions $\psi_k(r)$ and bound state wave function $\psi_1(r)$. The halfwidth Γ of the level is thus evaluated to be 2.7 eV. It is difficult to gauge these results by comparison with experimental results for hydrogen on transition metals. Besides the fact that theoretical values are obtained for a jellium model substrate, (2.32) does not properly account for the lack of orthogonality between adatom and metal wave functions. We only note here that a peak at about the same position has been found in the difference spectrum of H on W [2.26].

One can also investigate the question of molecular versus atomic adsorption within this formalism. Again, by applying the Hellmann-Feynman theorem, it is seen that the response function $L(x,x',u)$ is also the interaction energy between two adsorbed hydrogen atoms whose nuclei are located at (x,u) and $(x',0)$, respectively. For representative x and x' values, the short-range interaction is found to be repulsive. This is in agreement with the apparent dissociative adsorption of hydrogen on tungsten and certain other high density metals [2.27].

KAHN and YING [2.28,29] subsequently applied the linear response formalism to study the adsorption of alkali atoms on a jellium substrate. We have seen that even for hydrogen, the linear approximation and the gradient expansion for $T_s[n]$ are not quite satisfactory for the region just around the proton. Thus at first sight, the strong potential around the alkali nuclei would present a more serious problem. This is indeed so if one treats the bare nuclear potential as the perturbation. However, because of the strength of the nuclear potential, the core electrons of the alkali atom are very tightly bound. Upon chemisorption, they are only slightly perturbed. Thus an appropriate zeroth order configuration of the chemisorption system is an unperturbed alkali ion plus an extra electron in the conduction band of the metal. Still, the core electron distribution cannot simply be treated as external charge perturbation. The valence electrons experience an additional effective repulsive interaction around the nuclei because of the requirement of orthogonality to the core level wave functions. In the study of bulk metals, this effect is often included approximately through the replacement of the ionic potential by an approximate local pseudopotential [2.30]. For example, in the study of band structure and cohesive energies of alkali metals, the empty core potential modelled by ASHCROFT [2.31] has been very successful. For computational convenience, KAHN and YING represented the alkali ion by a pseudopotential which is similar in physical content but simpler to handle mathematically. They took the potential as arising from a Gaussian pseudocharge distribution

$$\rho_{ad}(r) = (R_0^3 \pi^{3/2})^{-1} \exp(-r^2/R_0^2) \quad . \tag{2.34}$$

R_0 in (2.34) is chosen to be equal to 2 R_c, R_c being the core radius parameter tabulated by ASHCROFT [2.31]. For this choice, the Fourier component of the resulting pseudopotential is identical to that of the empty core potential to order k^2. The calculated values of the bound state position in this potential compare favorably with experimental values of the ionization energies of isolated alkali atoms.

The change in the total energy of the system as a function of the position r_0 of the alkali nuclei is given as

$$W(r_0) = \frac{1}{2} \int dr \, dr' \, \rho_{ad}(r - r_0)L(r,r')\rho_{ad}(r' - r_0)$$

$$\tag{2.35}$$

$$+ \int \rho_{ad}(r - r_0)V_0(r)dr$$

where $L(r,r')$ is the response function defined earlier. Equation (2.35) includes an additive constant corresponding to the self-energy of the pseudocharge distribution which describes the adatom. In writing down the expression in (2.35), a correction term of the form

$$\delta W(r_0) = \int [Z\delta(r - r_0) - \rho_{ad}(r - r_0)][n_+(r')/|r-r'|]drdr' \tag{2.36}$$

has been omitted. This term arises from the fact that the positive background interacts with the true Coulomb potential of the ion rather than the pseudopotential appropriate for the electrons. For the systems studied in [2.28,29], the equilibrium distance of the adatom turns out to be sufficiently far away from the substrate that the effect of this correction term is negligible. The results described below are derived from (2.35). The first application was to a series of alkali atoms adsorbed on a jellium model substrate of $r_s = 1.5$. The result for $W(x)$ as a function of x obtained in [2.28] is shown in Fig.2.4. The main feature

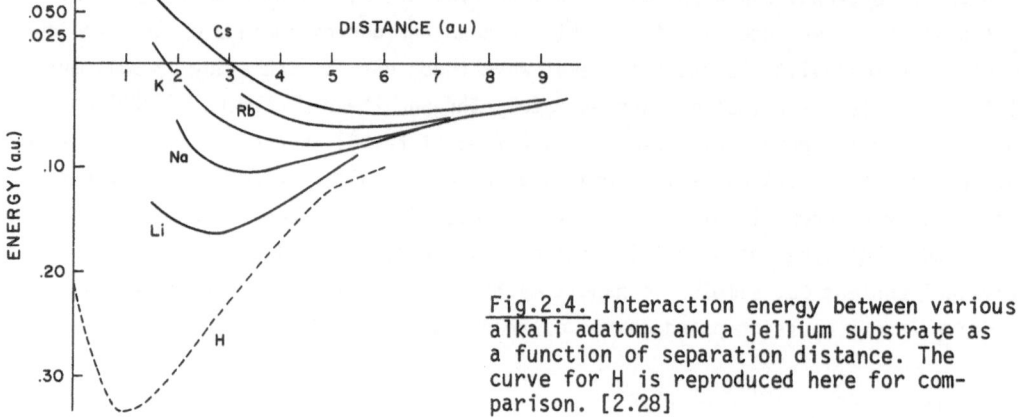

Fig.2.4. Interaction energy between various alkali adatoms and a jellium substrate as a function of separation distance. The curve for H is reproduced here for comparison. [2.28]

here is that the equilibrium distance of the adion is mainly determined by the size of the ionic core and increases monotonically in the alkali series from Li to Cs. The interaction energy is weaker the further the adion is from the substrate. Thus there is a systematic decrease in the binding energy going from Li to Cs. On the same figure we have also indicated the corresponding curve for hydrogen described earlier for comparison. The induced dipole due to the adsorption of the alkali atom is given by the expression

$$\mu_0 = \int \mu(x)\rho_{ad}(x - x_0, \underline{u})d\underline{u} \, dx \tag{2.37}$$

where $\mu(x)$ is the induced dipole due to a point charge of unit magnitude at a distance x away from the jellium edge as shown in Fig.2.2. The coordinate x_0 is the equilibrium separation of the adion from the jellium edge. Since the induced dipole increases monotonically with the distance of the adion from the substrate, it is larger for the heavier alkalis.

In Table 2.1 we summarize the theoretical results obtained by KAHN and YING for various chemisorption properties of alkalies on jellium. Since the only available experimental data are for alkalies on transition metals, we list in Table 2.1 the results obtained from alkalies chemisorbed on tungsten ($r_s \sim 1.5$) wherever available. The experimental values of binding energies listed in Table 2.1 are extracted from field desorption studies by TODD and RHODIN [2.32]. The actual quantities measured are the desorption fields, and the binding energies derived depend on the theoretical model used for interpretation. We shall discuss this in more detail in the next section. Comparing the theoretical values with experimental results, we see that the agreement is in general quite good. In particular, the trends of the binding energy and the induced dipole through the series of alkalies are correctly described by the theory. In nearly every case the dipole moments calculated with $c = 1/72$ and $c(n_0)$ bracket the experimental value. A similar calculation was also

Table 2.1. The ionic binding energy E_I, the induced dipole moment μ, vibrational frequency ω and equilibrium distance r_0 of alkali chemisorption. Theoretical results are from [2.29]. Experimental data are for W(110) from [2.32]. Dipole moment values in parentheses are for $c(n_0)$, others for $c = 1/72$ (see text for definition of c)

Alkali	E_I[eV]		μ[Debye]		$\hbar\omega$[MeV]	r_0[a.u.]
	Theory	Experiment	Theory	Experiment	Theory	Theory
Li	4.74	-	.94	-	56	2.56
Na	2.85	2.44 ± 0.3	2.4 (4.34)	3.24 ± .36	24.7	3.7
K	2.08	1.87 ± 0.3	3.96	5.04 ± 1.2	13.7	4.5
Rb	1.70	-	5.98	-	6.0	5.5
Cs	1.29	1.65 ± .05	7.63 (10.68)	6.12 ± 1.6	4.0	6.25

performed for a substrate characterized by r_s = 2.07 [2.29]. This electronic density corresponds to that of aluminum. The results do not differ significantly from those for r_s = 1.5.

At the equilibrium position of the alkali ions, the wave functions of the conduction electrons in the metal still have a considerable amplitude. Thus the self-consistent potential seen by an electron around the adion is quite different from that of the original pseudopotential. In many model Hamiltonian studies of chemisorption [2.33], the potential seen by the electron is taken to be the isolated adatom potential plus a constant shift due to image effects. This simple model potential is clearly a bad approximation at equilibrium separation. Another point worthy of note is the strong asymmetry of the potential in the direction normal to the surface. Thus if one describes the charge density around the alkali adions in terms of partial occupancy of local orbitals, one must include hybridization of higher angular momentum states with s states as for example done in the calculation by MUSCAT and NEWNS [2.34].

At finite coverages, the binding energy and the induced dipole per adatom differ from the results for single adatom because of adatom-adatom interaction. Within the linear response formalism, the total energy is calculated to second order in the coverage, thus one can study the effect of this interaction among adatoms in the low coverage limit. For example in [2.29] it was calculated that the binding energy per Cs adatom on a jellium substrate of r_s = 2.07 decreases from 1.45 eV in the zero coverage limit to a value of 1.05 eV at a coverage of 2.6×10^{14} atoms/cm^2. This indicates a repulsive interaction among the alkali adatoms. Because of the dipole-dipole interaction, the induced dipole per adatom also decreases from 7.9 Debye to 5.8 Debye.

KAHN and RASOLT [2.35] have recently investigated the effect of including the gradient correction of the local exchange correlation energy on the chemisorption properties. They found a relatively small change in the binding energy but a significant change in the induced dipole. This is in contrast to the corresponding effect on the clean surface properties [2.36], where the corrections to surface energy are found to be larger than that for the work function. At present, there are some doubts whether the inclusion of the gradient correction is appropriate for surface calculations [2.37].

2.2.3 Effect of Field Emission Fields on Chemisorption

In the interpretation of experimental results obtained from field emission experiments and the extraction of various values for the chemisorption properties as quoted in Table 2.1, the electric field is usually assumed to be unscreened outside the metal and to terminate abruptly on the metal surface. It is usually also assumed that the equilibrium position and other characteristics of adion are unperturbed by the field. In this section we shall examine these questions more closely.

The uniform field can be viewed as resulting from an infinite sheet of charge parallel to the surface placed at infinity. In the linear approximation, the resultant change in the electrostatic potential is related to the response function as

$$V_1(Q = 0,x) \propto L(0;,x,\infty) \tag{2.38}$$

Thus, in the presence of the electric field, a screening charge centered at the position of the image plane (~ 2 a.u. for a substrate of r_s = 1.5) is built up outside the jellium edge. Consider now the influence of the applied field on the adion itself. The assumption that the field does not alter the dipole moment in the process of measuring the current is unjustified, since fields of a strength sufficient to extract electrons will force the adion closer to the metal surface. It is the dipole moment associated with the new equilibrium position which is in fact measured. The closer the adion sits to the metal surface, the smaller is the induced dipole moment as can be seen from Fig.2.2. This could account for the generally smaller dipole moments that TODD and RHODIN [2.32] obtained compared with corresponding results of other experiments in the absence of applied fields. To ascertain the shift in position of the adion and the resultant change in the dipole moment, KAHN and YING [2.29] calculated the response of the metal substrate to a perturbation consisting of a linear superposition of an alkali adion, represented by the pseudocharge distribution previously described, and a uniform sheet of charge at infinity to effect the electric field. Through this procedure the shift in the equilibrium position of cesium adsorbed on a substrate of r_s = 1.5 was calculated. When a charge sheet giving rise to a field of .42 V/Å was present, the equilibrium position was shifted from 6.25 to 5.5 a.u., measured from the jellium edge. This causes the dipole moment to be reduced from 10.6 to 8.2 Debye. A field of 0.8 V/Å, which is the upper limit of fields used in the experiments of TODD and RHODIN, shifts the equilibrium position to 4.3 a.u. from the surface, resulting in an induced dipole moment of approximately 6 Debye, bringing the theoretical value much closer to the experimental value as quoted in Table 2.1. This also explains the discrepancies of the values obtained in experiments with and without the presence of an electric field. So, we conclude that the electric field significantly alters the equilibrium position and the resulting dipole moment associated with the adsorbed alkalis.

The other quantity which TODD and RHODIN measured was the electric field required to desorb an alkali adion from the tungsten substrate. They determine the minimum strength of the electric field at which desorption first occurs, and identify this field as corresponding to the value at which the Schottky barrier vanishes as shown in Fig.2.5. To derive the value of the binding energy E_I as quoted in Table 2.1 from the knowledge of the desorption field, one still requires a detailed knowledge of the potential energy of the ion as a function of distance from the surface. In the absence of a detailed first-principle calculation, one usually calculates the height of the Schottky hump (relative to the energy of an ion at equilibrium) from the superposition of the linear potential due to the electric field and an image potential form. The position of the image plane is usually chosen arbitrarily as the boundary of the substrate surface. If one further assumes that the adion is situated at a distance x_0 equal to the ionic radius from this fictitious surface and that the ion is completely ionized, the following energy balance equation can be derived when the field F_d is just strong enough to cause the Schottky barrier to vanish [2.32,38]

$$E_I = F_d^{\frac{1}{2}} - F_d x_0 + O(F_d^2) \quad .$$

(2.39)

Here E_I is the ionic binding energy as defined in (2.30). From our earlier discussions, it is clear that the assumptions involved in the derivation of (2.39) are not well justified. The image potential $-[4(x - d)]^{-1}$ is only correct asymptotically when the adion is far away from the substrate. In addition, the position of the image plane is at a distance 2 a.u. from the jellium edge (see Fig.2.2) which is itself separated from the first lattice plane by half an interplanar spacing. There is no reason why the adion is just situated simply at an ionic radius distance away from this image plane. The value of the binding energy derived from (2.39) therefore involves large uncertainties even though the desorption field itself is rather precisely measured. KAHN and YING [2.29] have computed the self-consistent potential energy of the adion $W(x)$ in the presence of the field. By searching for the value of the field at which the Schottky barrier vanishes, the theoretical value for the desorption field was directly determined. This can be compared with the experimentally measured desorption field without going through (2.39) and avoids the various assumptions involved. In Table 2.2, we list the theoretical and experimental values for F_d. The agreement is good and the trend going from the lighter

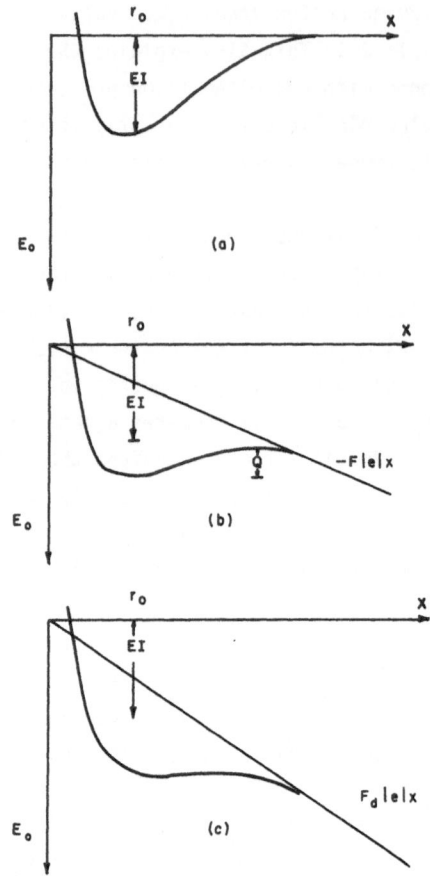

Fig.2.5a-c. Schematic diagram illustrating the field required to desorb adion from the surface. E_I is the binding energy in the absence of a field and Q the Schottky barrier. The figure shows the adion in (a) the absence of a field (b) a small field (c) under the influence of desorbing field, F_d. [2.29]

Table 2.2. Direct comparison of experimental and theoretical desorption fields. Experimental data are from [2.32] for the W(110) substrate, and theoretical values from [2.29]

	Desorption field [V/Å]	
Alkali	Theory	Experiment
Cs	0.26	0.318
K	0.425	0.361
Na	0.71	0.601

alkali to the heavier ones is correctly produced by the theory. For the systems studied by TODD and RHODIN, it turns out that the binding energies derived from (2.39) are also in fair agreement with the theoretical values as shown in Table 2.1. This is due to a fortuitous cancellation of errors involved in the two terms on the right-hand side of (2.39).

2.2.4 Perturbative Introduction of Single Crystal Lattice Structure for the Substrate: Application to Aluminum Substrate

So far we have discussed theoretical calculations in which the jellium model is used to represent the substrate. For this model, the properties of the various crystal faces cannot be distinguished. Also, no information can be obtained on the preferred adsorption sites over the surface plane. These properties can only be studied through the introduction of a discrete lattice structure for the substrate. For this purpose, use can again be made of the pseudopotential concept, this time to represent the ions of the substrate [2.14]. In this problem, the perturbation is the ionic pseudocharge minus the uniform jellium charge density. We denote this difference by $\Delta\rho_s(r)$ and the corresponding difference in the potential as $\Delta V_s(r)$. It is convenient to separate out from $\Delta V_s(r)$ the part which is the effective repulsion due to orthogonality requirements to the core electrons and denote the rest by $\Delta V_c(r)$. The energy functional before the introduction of $\Delta V_s(r)$ is already stationary with respect to the density obtained in the jellium calculation. Therefore, to first order in $\Delta V_s(r)$, the change in the binding energy can be expressed as

$$\Delta W(r_0) = \int \Delta V_s(r) n_{ad}(r,r_0) \, dr + Z\Delta V_c(r_0) \tag{2.40}$$

in which $n_{ad}(r)$ is the difference in the electronic density before and after the introduction of the adatom in the jellium model.

In the model for alkali chemisorption described in the previous sections, $n_{ad}(r)$ refer to the change in the valence electron density induced by the perturbing pseudocharge distribution $\rho_{ad}(r)$ representing the alkali ion. In the linear approximation, (2.40) can be reduced to the following expression involving the response function $L(r,r')$,

$$\Delta W(r_0) = \int \rho_{ad}(r - r_0) L(r,r') \Delta\rho_s(r') \, dr dr'$$

$$+ \{ Z\Delta V_c(r_0) - \int [\rho_{ad}(r - r_0) \Delta\rho_s(r')/|r - r'|] dr dr' \} \ . \tag{2.41}$$

The subtraction of an additional term in (2.41) corrects for the bare interaction between $\rho_{ad}(r)$ and $\Delta\rho_s(r)$ which is included in the first term through the definition of $L(r,r')$. For the systems studied in [2.29], there is little overlap between $\rho_{ad}(r)$ and $\Delta\rho_s(r)$ at equilibrium separations and the effects of the correction term are negligible. In the following discussions we shall concentrate on the effect of the first term in (2.41) alone. This can also be written in the form

$$\Delta W(r_0) = \int \rho_{ad}(r - r_0) \Delta V_0(r) dr \quad ,$$

with

$$\Delta V_0(\underline{r}) = \int L(\underline{r},\underline{r}')\Delta\rho_s(\underline{r}')d\underline{r}' \quad . \tag{2.42}$$

In (2.42), $\Delta V_0(r)$ represents the additional screened surface potential over that already present in the jellium model. This potential now depends on the coordinate parallel to the surface as well as the one perpendicular to it. The periodicity along the surface implies that it can be expressed in Fourier components as

$$\Delta V_0(\underline{u},x) = \frac{1}{A} \sum_{\underline{Q}} \Delta V(\underline{Q},\underline{x}) \exp(i \underline{Q} \cdot \underline{u}) \quad , \tag{2.43}$$

where A is the surface area and the $\underline{Q}$'s are the two-dimensional reciprocal lattice vectors. KAHN and YING [2.29] have applied (2.41-43) to study the effect of dis-crete lattice structure of the substrate for alkali chemisorption on aluminum. The choice of this substrate is based on the consideration that the pseudopotential picture has proved to provide a realistic description for bulk aluminum [2.30]. Also, there are some experimental results for chemisorption on this substrate avail-able for comparison. The jellium model for Al corresponds to r_s = 2.07. These authors then represented the substrate ions through the following pseudocharge density

$$\rho_I(\underline{r}) = (4\pi R_c^2)^{-1}\delta(r - R_c) \quad . \tag{2.44}$$

This generates a pseudopotential similar to the empty core potential but without the discontinuous jump at the core radius. The value for the core radius R_c was chosen to be 1.21 a.u. This is the value for Al as tabulated by ASHCROFT [2.31].

Aluminum, the substrate under consideration has a fcc structure and, therefore, has as its most densely packed surface the (111) plane. The result for this plane is therefore expected to show the smallest correction to the jellium model. The corrections for the (110) and (100) faces should be correspondingly larger. This is clearly demonstrated when one examines the change in the surface potential $\Delta V_0(r)$ as defined in (2.42). In Fig.2.6 we have plotted $\Delta V_0(r)$ obtained in [2.29] for sev-eral distances from the surface and different low index planes. For the (100) plane, it is seen that the minimum of ΔV_0 is located over the center site (C) which is directly above an ion in the second layer. For the (110) plane, the position of minimum depends on the distance x from the surface at which (x,u) is evaluated. For equilibrium position close to the metal surface (x < 3 a.u.), the minimum is located at the center site (C) over the ionic position in the second layer, whereas for x > 3 a.u., the minimum shifts to short bridge position. In the case of the (111) plane, two almost equivalent equilibrium positions are found, both of trigonal symmetry (sites C and C'). Here C refers to a site above an ion in the second layer and C' referes to that above a hole. The difference in the energy at these two

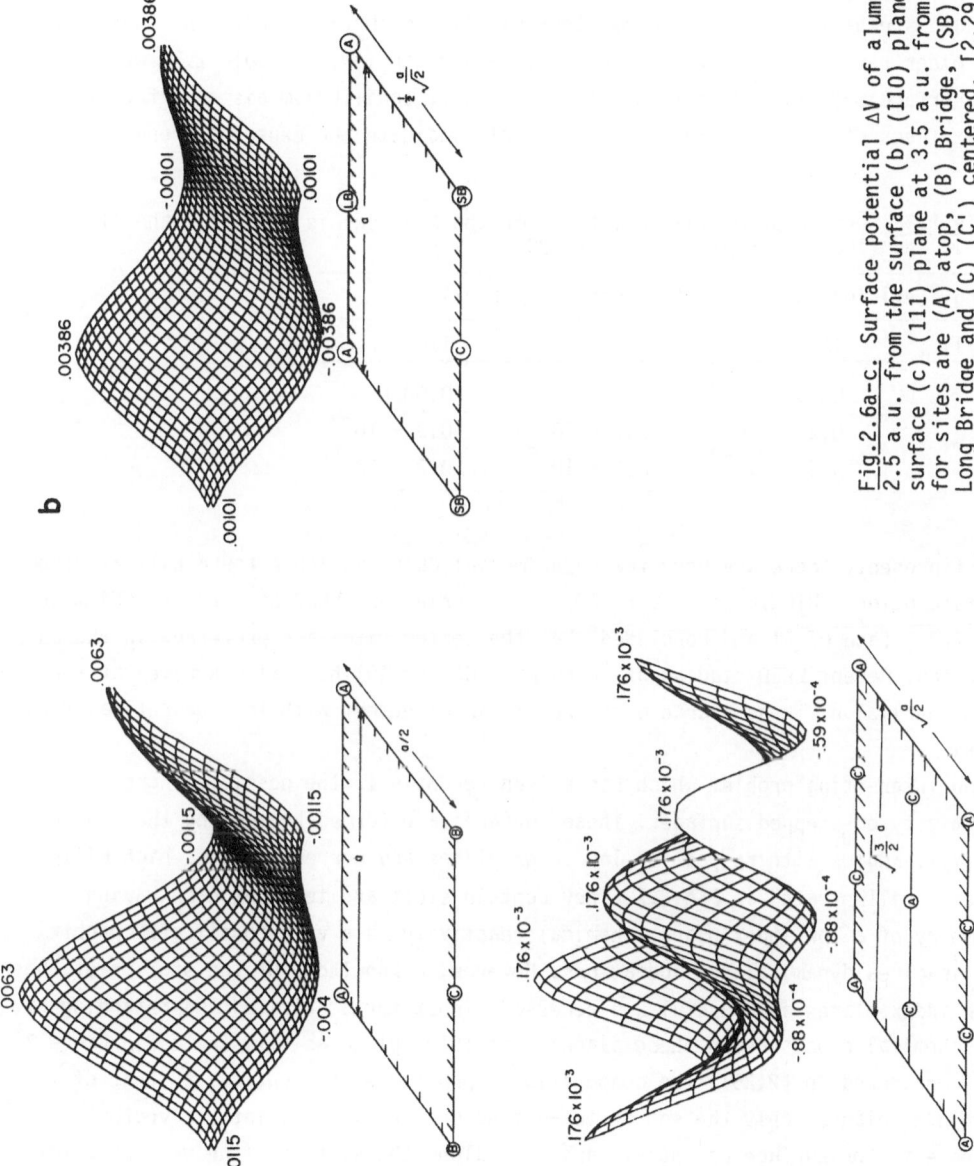

Fig.2.6a-c. Surface potential ΔV of aluminum (a) (100) plane at 2.5 a.u. from the surface (b) (110) plane at 2.5 a.u. from the surface (c) (111) plane at 3.5 a.u. from the surface. Notation for sites are (A) atop, (B) Bridge, (SB) Short Bridge, (LB) Long Bridge and (C) (C') centered. [2.29]

sites is small because it only arises from the influence of the second and third layers inside the substrate, whose effects are almost completely screened outside the substrate. The correction to the binding energy of the adatom is given by (2.42). For alkali chemisorption, after integrating over the pseudocharge distribution, the preferred adsorption sites are found to be the same as the minima of $\Delta V_0(x)$. The magnitude of lateral variation of potential energy, however, depends on the specific alkali under consideration. The peak to peak variation of $W(x,u)$ within the surface plane at equilibrium distance x is a rough measure of the activation energy for diffusion along the surface. Cesium, with the largest core radius is expected to have the smallest barrier to surface migration. Sodium, on the other hand, should have a relatively larger barrier. In Table 2.3, we summarize the surface potential variation for Na, K, Cs at equilibrium distance from the 3 low index faces of aluminum. The results clearly indicate the expected trend.

Table 2.3. Surface potential variation for the Alkalies is given for the (111), (110), and (100) faces of aluminum [2.29]

Energy [eV] for face (fcc structure, $r_s = 2.07$)		
Alkali	111	100	110
Na	0.13×10^{-3}	0.006	0.0076
K	0.2×10^{-5}	0.2×10^{-2}	0.3×10^{-3}
Cs	0.5×10^{-8}	0.1×10^{-5}	0.2×10^{-5}

At present, there are very few experimental data available for alkali-aluminum chemisorption. HUTCHINS et al. [2.39] have carried out LEED studies of sodium on the (100) face of Al and concluded that the center sites are preferred in the adsorption. Recent LEED studies of Na on Ni (100) [2.40] have also arrived at the same conclusion. These scarce data are all in agreement with the theoretical findings.

An interesting problem which has arisen recently is the possible increase of reactivity on stepped surfaces. These surfaces are formed by cutting the crystal at small angles with respect to low index planes and correspond to a high Miller index as illustrated in Fig.2.7. They contain steps and terraces each having a geometry of a low index plane. Chemical reactivity is a very complicated problem and involves dynamical considerations. However, a knowledge of the statics of these high index planes is a useful and necessary input for a detailed understanding of the chemical processes on these planes. The (910) plane as illustrated in Fig.2.7 was considered in [2.29]. The computational procedure is identical to that of the low index planes, only the set of two-dimensional reciprocal lattice vectors is different. The surface potential variation along the surface plane was found to be larger than that of the low index planes. For example for potassium, the peak to

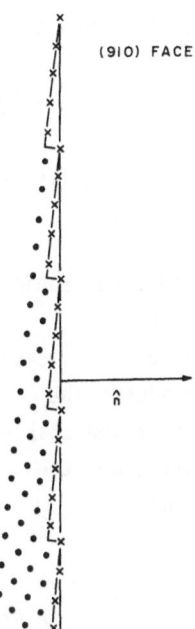

$\hat{n}$

Fig.2.7. The (910) surface for a fcc structure, $\underline{n}$ is the normal to the plane. The terraces, denoted by x, correspond to the (100) plane. [2.29]

peak variation for adsorption on (910) plane is 0.9 eV, compared with approximately 1.6×10^{-3} eV for low index planes. The preferred absorption sites are found to be in the region of the steps. These results indicate that chemisorption properties on these stepped surfaces can be significantly different from that of low index faces.

2.2.5 Nonlinear Response for Sodium Chemisorption

In the previous sections, we have described theoretical studies of chemisorption on simple metals based on the linear response formalism. For strong perturbations, the linear approximation is no longer sufficient and one has to include nonlinear effects. The solution of the nonlinear Euler's equation (2.4) is extremely difficult. However, one can solve the problem approximately by assuming a variational form for the electronic density and determining the parameters involved through direct minimization of the energy functional. This is the procedure adopted by HUNTINGTON et al. [2.41] to study the adsorption of a single sodium adatom on a jellium substrate with $r_s = 4$ appropriate for bulk sodium. We describe their work briefly in this section.

These authors adopted an energy functional similar to the one described in previous sections. Besides the local and first gradient term of the kinetic energy functional, they include the local exchange and correlational energy of the form

$$\varepsilon_{xc}(n) = -(3/4)(3/\pi)^{1/3} n^{1/3}(\underline{r}) - 0.056\, n^{1/3}(\underline{r})(n^{1/3} + 0.079)^{-1} \quad . \tag{2.45}$$

The last term in (2.45) is the standard Wigner interpolation expression. They took
the empty core potential of the form

$$
V_{Ash}(\underline{r}) = \begin{cases} Z_a/r & \text{for} \quad r > R_c \\ 0 & \text{for} \quad r < R_c \end{cases} \tag{2.46}
$$

to describe the adion. The initial valence electron density of the adsorbed sodium
atom was assumed to be of the form $(\sinh\beta r/\beta r)^2$ for $r < R_c$ and for $r > R_c$, the
Herman-Skillman atomic density was fitted by an exponential multiplied by a factor
$(1 - C/r)^2$. The two expressions were matched at R_c in slope and magnitude. The
value of R_c was determined by requiring the pseudoatom so constructed to be stable
against charge deformation and came out to be 0.9 Å. On the substrate side, the
electronic density for the clean surface is assumed to be of the form

$$
n(x) = \begin{cases} (n_+/2)\ exp(-\alpha x) & \text{for} \quad x > 0 \\ n_+[1 - exp(\alpha x/2)] & \text{for} \quad x < 0 \end{cases}. \tag{2.47}
$$

In the zeroth order, the total electronic density is just the superposition of the
valence electron density of the adatom and the clean surface electronic density.
The changes in the total electronic density due to the interaction were described
by dipole and quadrupole modulations. A dipolar term and a quadrupolar term with
the radial part proportional to the original valence electron density were centered
at the adatom while another dipolar term was centered on the atom-metal axis at a
small distance inside the metal. The parameters involved as well as the distance
of the adatom from the substrate are then determined through minimization of the
total energy.

The resultant binding energy is 0.37 eV at an equilibrium distance of 1.66 Å.
The effect of the charge readjustments on the total binding energy are relatively
small, amounting to about 10 percent. The two dipoles centered at the adatom and
inside the metal are of opposite sign, the net moment being approximately 0.32
Debye.

There are no experimental data on this system to compare with. For an estimate
of the accuracy of the theoretical value of the binding energy, HUNTINGTON et al.
compared it with twice the surface energy per atom of sodium which is about 0.38 eV.
This close agreement with theory is probably fortuitous since the two quantities
are comparable only if the formation of surfaces just involves the breaking of
localized interatomic bonds. This pairwise bonding picture is not quite valid for
a highly conducting metal like sodium. The theoretical value of the equilibrium
distance compared favorably with half the interplanar spacing for the closest

packed (110) plane in sodium of the value 1.52 Å. However, as discussed in the
previous section, the equilibrium distance of the adatom from the substrate is
actually a function of the coverage as a result of the adatom-adatom interaction.
Thus, there is an unknown error involved in comparing the single adatom result
with interplanar spacing with corresponds to the full coverage limit.

Besides the approximations involved in the choice of the energy functional,
the main uncertainties in the calculation arise from the somewhat arbitrary
choice of the form of modulation of the charge density. Whether a more complex
form of modulations is required for an accurate description can only be decided
when more applications of the formalism are made. It is also not clear how the
discrete nature of the lattice can be easily included in this formalism in a non-
perturbative manner.

2.3 Approach Based on the Solution of Effective One-Electron Schrödinger Equation

In this section, we shall describe density functional theory of chemisorption in
which the kinetic energy functional is treated exactly rather than through an
approximate gradient expansion. Recent works on the ground state properties of
various chemisorption systems as well as core excitation spectrum are discussed.

2.3.1 Methods of Solving the Kohn-Sham Equation for Chemisorption Systems

In Sect.2.1 we have discussed how the ground state properties of an interacting
inhomogeneous electron gas can be obtained through the solution of a set of
single-particle equations described by (2.9-11). In this approach, the kinetic
energy functional is treated exactly and there is no need of linearization. Thus
if the set of Kohn-Sham equations are solved self-consistently, the only error
will come through the approximations in the choice of the exchange correlation
functional, such as the local approximation discussed earlier.

In the absence of full three-dimensional periodicity, the solution of the self-
consistent equation is extremely difficult. For monolayer chemisorption, the
system still possesses periodicity in the dimensions parallel to the surface,
and recent self-consistent studies have been made for chemisorption on semiconductor surfaces and then for transition metal films as described in other chapters.
(See also a study by LANG [2.42] in which a layer of adsorbed Cs is represented
by a jellium slab.) For single-atom chemisorption on a semi-infinite metal, the
two-dimensional periodicity is also lost and the situation is more complicated. .
To date, the only self-consistent studies based on (2.9) for this geometry have
been for jellium model substrates. In this subsection we briefly outline the methods
developed by LANG and WILLIAMS [2.43], and by GUNNARSON et al. [2.44,45] for solving

the Kohn-Sham equation in this geometry. LANG and WILLIAMS noted that for the continum states (2.9) is equivalent to the Lippman-Schwinger equation

$$\psi^{MA}(\underline{r}) = \psi^M(\underline{r}) + \int d\underline{r}'G^M(\underline{r},\underline{r}')\delta V_{eff}(\underline{r}')\psi^{MA}(\underline{r}') \quad , \tag{2.48}$$

where the index M denotes a clean metal quantity, and the index MA stands for a quantity belonging to the metal-adsorbate system, and

$$\delta V_{eff}(\underline{r}) = V_{eff}[n^{MA}(\underline{r})] - V_{eff}[n^M(\underline{r})] \quad . \tag{2.49}$$

Because of the cylindrical symmetry of the entire system, the eigenfunctions of the bare metal surface are written in cylindrical coordinates as

$$\psi^M_{E,m,k}(\underline{r}) = e^{im\phi}J_m(k\rho)u^M_{Ek}(x) \quad . \tag{2.50}$$

The quantum numbers E, m and k are conserved by the symmetry of the bare metal surface. The functions $\psi^M(\underline{r})$ can be calculated with the method of LANG and KOHN [2.14] and are normalized to $\delta(E - E')\delta(k - k')\delta_{mm'}$. The Green's function in (2.48) is the solution of

$$\{\nabla^2/2 + E - V_{eff}[n^M(\underline{r})]\}G^M(\underline{r},\underline{r}') = \delta(\underline{r} - \underline{r}') \quad . \tag{2.51}$$

It should be noted that for the total system, k only serves to label the eigenfunctions and does not refer to a conserved quantity.

The central idea is to exploit the localization of δV_{eff} around the chemisorbed atom. LANG and WILLIAMS [2.46] showed that if S denotes a surface enclosing the adatom such that δV_{eff} is negligible outside S, then (2.48) leads to the following equation

$$\int_S dS'[\delta\psi(\underline{r}')\underline{\nabla}_{\underline{r}'}G^M(\underline{r},\underline{r}') - G^M(\underline{r},\underline{r}')\underline{\nabla}_{\underline{r}'}\delta\psi(\underline{r}')] = 0 \quad ,$$

and

$$\delta\psi(\underline{r}) = \psi^{MA}(\underline{r}) - \psi^{(M)}(\underline{r}) \quad , \tag{2.52}$$

for $\underline{r}$ within S. In the region around the adatom, both ψ^{MA} and ψ^M are expanded as

$$\psi^{M(A)}_{E,m,k}(\underline{r}) = \sum_{\ell'=|m|}^{\infty} \alpha^{M(A)}_{E,m,k,\ell'}\psi^{M(A)}_{E,m,\ell'}(\underline{r}) \quad , \tag{2.53}$$

where the functions $\psi^{M(A)}_{E,m,}(\underline{r})$ are obtained by direct numerical integration of (2.9) outwards from the origin chosen at the position of the adatom [2.47]. The only differences between the k- and ℓ-indexed functions are the boundary conditions satisfied by the latter

$$\lim_{r \to 0} r^{-\ell'} \int d\Omega Y^*_{\ell'm}(\Omega) \psi^{(M)}_{E,m,\ell}(\underline{r}) = \delta_{\ell\ell'} \quad . \tag{2.54}$$

Since $\psi^M(\underline{r})$ are obtained from the self-consistent calculations for the bare surface described by LANG and KOHN [2.14], the only quantities that need to be determined are the difference coefficients δa. This is achieved by substitution of
(2.53) into (2.52). After taking the limit $r \to 0$ and using the spherical harmonic
decomposition, it results in a set of simultaneous linear equations which determines
the coefficients.

This completes the solution of the continuum eigenstates for the entire system.
The bound state solutions are constructed by taking appropriate combinations of
the numerical solutions that lead to an exponentially decaying behaviour at large
distances. For a more detailed discussion of the method as well as a discussion of
the corrections due to the disturbances outside the surface S, the reader is referred to [2.46].

Given the eigenfunctions, one can calculate (within the surface S)

$$\delta n(\underline{r},E) \equiv 2 \sum_{m=-\infty}^{\infty} \theta(E - E_0) \int_0^{(E-E_0)^{\frac{1}{2}}} dk \left[|\psi^{MA}_{E,m,k}(\underline{r})|^2 - |\psi^M_{E,m,k}(\underline{r})|^2 \right]$$

$$+ \sum_c |\psi^{MA}_{E_c,m}(\underline{r})|^2 \delta(E - E_c) \quad . \tag{2.55}$$

Here $E_0 \equiv v_{eff}(-\infty)$ is the bottom of the metal band, and $\theta(E - E_0)$ is the Heaviside
function. The factor 2 is due to spin degeneracy for a spin unpolarized system.
The difference in electron density between the metal-adatom system and the bare
metal is

$$\delta n(\underline{r}) = \int_{-\infty}^{E_f} dE \delta n(\underline{r},E) \quad . \tag{2.56}$$

It is also convenient to define a difference of eigenstate density as

$$\delta n(E) = \int d\underline{r} \delta n(\underline{r},E) \quad . \tag{2.57}$$

GUNNARSON et al. [2.44,45] have developed an alternate method of solving the
self-consistent equation. They also utilized the fact that the change in the density and effective potential should be localized around the adatom. If one writes
the equation obeyed by the Green's function as an operator equation, one has

$$(z - H)G^{MA}(z) = 1 \tag{2.58}$$

where $H = H^M + V$, with V defined by (2.49), z stands for a complex variable, and the indices M and MA have the same meaning as before. Equation (2.58) leads to the equation

$$G^{MA} = G^M(1 - V G^M)^{-1} \quad .$$

(2.59)

A finite set of localized functions $\{\phi_n\}_{n=1}^N$ is then introduced, which spans a Hilbert space S. These authors showed that within the space S, the $N \times N$ matrix of the Green's function (with matrix elements $G_{nn'}$) can be written as

$$G_S^{MA} = [z - H_S - q(z)]^{-1}$$

(2.60)

with

$$q(z) = H_{s\underline{s}}(z I - H_{\underline{s}})^{-1} H_{\underline{s}s} \quad .$$

(2.61)

The semi-infinite matrices $H_{s\underline{s}}$ and $H_{\underline{s}s}$ have matrix elements coupling the space S and its complementary space $\underline{S}$ for the entire system. So far, however, calculations have been performed only with the assumption that for the region around the adatom, the Hilbert space S is almost complete, then (2.59) reduces to

$$G_S^{MA} = G_S^M(I - V_S G_S^M)^{-1} \quad .$$

(2.62)

The subscript s now indicates that the quantities appearing in (2.62) are $N \times N$ matrices in the space S. The problem is then reduced to the inversion of an $N \times N$ matrix. The knowledge of the Green's function immediately generates the density of states given as

$$n(\underline{r},E) = -\frac{1}{\pi} \sum_{nn'} \text{Im}\{G_{nn'}(E + i\delta)\}\phi_n^*(\underline{r})\phi_{n'}(\underline{r}) \quad .$$

(2.63)

The density follows from the integration of this equation over energy as in (2.56). This is then used to generate the new potential until self-consistency is reached.

In the calculation performed so far, the localized region around the adatom is prescribed as a sphere of radius $R \sim 5$ Å. The localized functions are chosen of the form $F(r)Y_{\ell m}(\Omega)$. The set is kept finite by restricting to 2-3 angular functions and ~ 10 radial functions for each. In principle, this method can be applied beyond the jellium model. A major obstacle here is the prerequisite knowledge of the Green's function G^M for the clean metal surface.

2.3.2 Single-Atom Chemisorption on Jellium; H, Li, O, Na, Si and Cl

LANG and WILLIAMS [2.43,46] have applied their formalism described in the previous sections to the study of single-atom chemisorption on jellium substrate. So far,

the list of adatoms studied includes H, Li, O, Na, Si and Cl. GUNNARSON et al.
[2.45] have also applied their method to study hydrogen chemisorption on jellium.
Since the results they obtained for this system are very similar to those obtained
by LANG and WILLIAMS, we shall concentrate in this section on summarizing and discussing the various results obtained in the series of works by the latter authors.

Table 2.4 lists the results for the two main quantities, the equilibrium distance d, and the atomic binding energy E_a [2.46]. For the atomic binding energy,
the energetics of the isolated atom is computed separately by the spin density
functional formalism [2.6]. While it is again difficult to compare these results
directly with experimental data on transition metals, the values in Table 2.4 are
physically realistic. For example, the theoretical predictions for the adatom-
metal separations are close to the values deduced from LEED studies [2.48]. More
important, however, are the trends exhibited by these detailed calculations on the
wide range of different adsorbates.

Table 2.4. Equilibrium distance d (adatom nucleus to positive-background edge)
and atomic binding energy ΔE_a for adsorption on a high-density (r_s = 2) substrate
[2.46]

Adatom	d_{eq} [bohr]	E_a [eV]
H	1.1	1.5
Ii	2.5	1.3
O	1.1	5.4
Na	3.1	0.9
Si	2.3	3.0
Cl	2.6	3.6

These trends are best illustrated by comparing the results for the three adsorbates Li, Si, and Cl. First, it is useful to compare the adsorbate induced
change in the density of states $\delta n(E)$. This is illustrated in Fig.2.8. The 2s resonance of Li lies primarily above E_F and the 3p resonance of Cl lies below E_F,
providing clear examples of positive and negative ionic chemisorption. The direction
of charge transfer is consistent with the electronegativities of Li, Cl, and an
Al surface (these calculations are for a jellium of r_s = 2 corresponding to the
density of Al). The resonance of Si, however, embraces E_F and is only partially
filled. In this case, the adsorption is covalent in character. These different
behaviors are also reflected in the spatial electron density. Figure 2.9 shows
contours of constant density for both the total electron density for the chemisorption system and the total minus the superposition of the metal and atom electron densities. It is seen that the metal contours deflect toward the Li and away
from the Cl, indicating the ionic attraction/repulsion, whereas for Si, there is

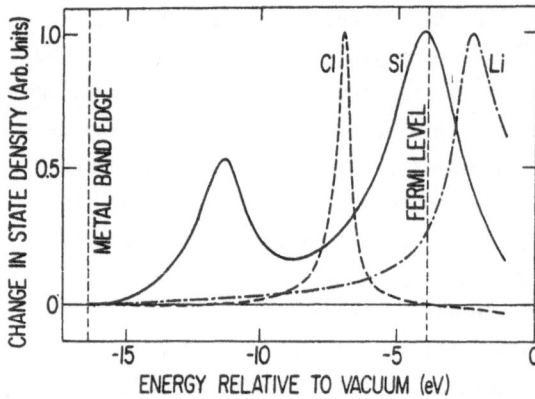

Fig.2.8. Change in state density $\delta n(E)$ from the chemisorption of an adatom at equilibrium separation from $r_s = 2$ jellium substrate [2.46]

more prominent projection of charge into the bond region. It is also interesting to note that these contours bear some similarities to those of corresponding molecules. For example, the kidney-shaped depletion contour on the vacuum side of the Li-metal system is similar to that of LiH and LiF molecules [2.49], and for Si, the depletion of charge near the nucleus and accumulation on both the bond and vacuum sides is typical of diatomics which exhibit p-orbital covalent binding [2.49]. In all cases, the sequence of contours continues into the metal in the form of Friedel oscillations. These features illustrate the dual solid state and chemical nature of chemisorption systems. This is what makes the problem so interesting and challenging.

Let us now consider how $\delta n(E)$ changes as a function of metal-atom separation. At the largest distance, the metal-adatom interaction is weak and the resonance below the Fermi level is relatively narrow. At closer separation, the interaction increases and the resonance broadens. When the atom is moved still closer, the resonance is generally shifted downwards and begins to narrow due to the decreasing density of metal states as the energy moves downward. It is easily demonstrated in first-order perturbation theory that the center of the adatom valence resonance tends to follow the bare-metal surface potential [2.12,46] provided that the total electronic density in the relevant region is much greater than the density distribution in the free atom.

HJELMBERG et al. [2.50] have recently studied the sensitivity of the results of hydrogen chemisorption on the parameter r_s characterizing the substrate density. They found that inside the metal surface, a localized level on the hydrogen is "locked" to the band bottom, lying just below for r_s between 2 and 4 as shown by bulk calculations [2.51]. Also, just as in the case of $r_s = 2$, for all these r_s values, the peak of the resonance level tends to follow the total surface potential. In going from $r_s = 2$ to $r_s = 4$, there is a factor of 8 change in the electronic

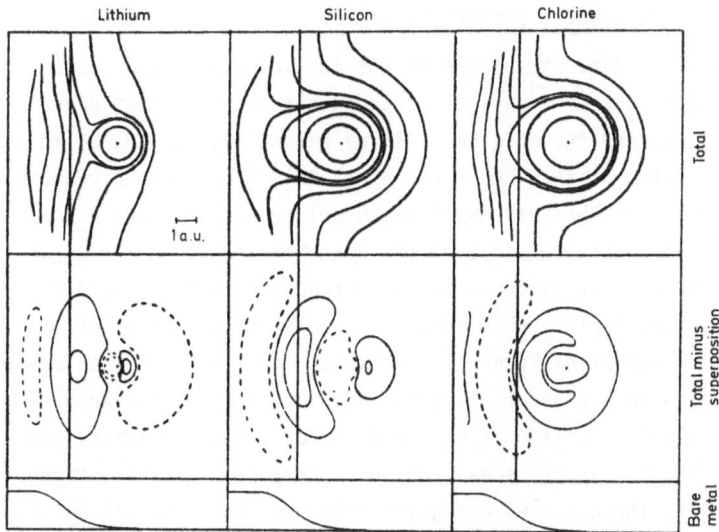

Lithium   Silicon   Chlorine

1 a.u.

Total

Total minus superposition

Bare metal

Fig.2.9. Electron-density contours for chemisorption on a high-density (r_s = 2) substrate, at equilibrium separation. *Upper row:* Contours of constant electron density in (any) plane normal to the metal surface containing the adatom nucleus (indicated by "+"). Metal is to the left-hand side; solid vertical line indicates positive-background edge. Contour values were selected to be visually informative. *Center row:* Total electron density minus the superposition of atomic and bare-metal electron densities (electrons/bohr^3). The polarization of the core region, shown for Li, has been deleted for Si and Cl. *Bottom row:* Bare-metal electron-density profile (shown to establish physical distance scale). [2.46]

density, yet these same general features persist. Thus they are very likely a general phenomena for adsorbate resonances.

The total binding energy curve for H on Na (r_s = 4) has a very shallow minimum situated just at the jellium edge. When first-order pseudopotential corrections are included (see section below), no minimum is found when the H atom is moved into the metallic surface. Thus it seems that the H atom would be absorbed rather than adsorbed. The theoretical model is, however, less reliable for this case. The local approximation is less justified for large values of r_s. Moreover, for the less densely packed metals like Na, higher order corrections of the effects of discrete lattice may be important. In fact, the calculated energy gain for H absorbed by sodium is twice the cohesive energy of the sodium metal, indicating a possibility of surface reconstruction.

LANG and WILLIAMS [2.46] have also investigated the effect of changing the substrate density from r_s = 2 to r_s = 4 for Si chemisorption. They found that there is a narrowing of the 3p resonance due to smaller metallic density of states at this energy and the 3s state splits off as a discrete state below the metal band edge. In this work, the adatom was placed at a fixed distance (d ~0.75 Å) from the jellium edge.

2.3.3 Reintroduction of Single-Crystal Lattice Structure; Si, O, and H on Al

As discussed in Sect.2.2.4, the discrete nature of the lattice can be introduced by representing the substrate ions as arrays of pseudopotential. To lowest order in the difference between this and the original jellium potential, the change in the binding energy of the adatom is then described by (2.40). Since one only needs to know the electronic density in the adatom-jellium model to evaluate the expression in (2.40), the difficult problem of solving the Kohn-Sham equation for an adatom interacting with a substrate of discrete lattice structure is thus avoided.

HJELMBERG et al. [2.50] have applied this method to study the variation of hydrogen chemisorption on different Al faces. The Ashcroft model potential as described by (2.46) was adopted to represent the substrate ions and the difference between this representation and a jellium of $r_s = 2$ ($\sim$ Al density) was treated as perturbation. For each of the low index faces, they have computed the corrections to the binding energy curves. These curves are quite different when the adatom approaches the surface along different normals. From symmetry considerations, one would expect that the largest binding energy would be obtained for a path going through a high symmetry site on the surface plane. Figure 2.10 summarizes the main results obtained by these authors. According to this work, the bridge positions correspond to stable configurations on the (100) and (110) surfaces while atop positions are predicted to be stable on (111) surfaces. The Al-H bond lengths in the stable configurations [1.9 Å on (100), 1.8 Å on (110), and 1.6 Å on (111)] correlate well with the distance 1.65 Å for the AlH molecule [2.52]. On the (100) surface, diffusion is predicted to occur from one bridge position to another along a

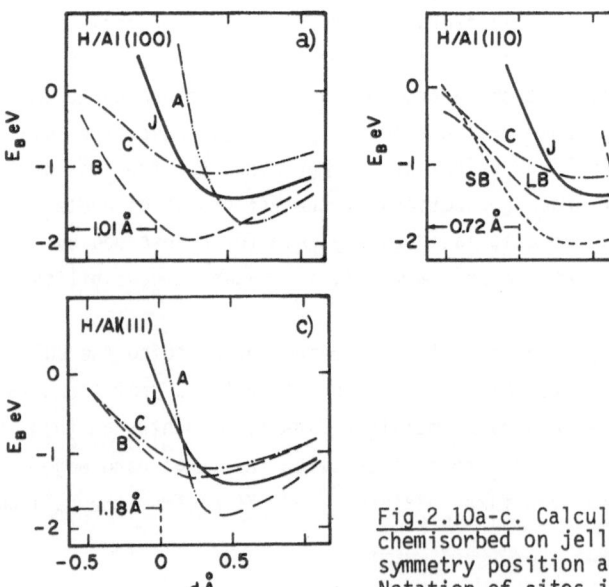

Fig.2.10a-c. Calculated binding-energy curves for H chemisorbed on jellium ($r_s = 2$) (J) and in various symmetry position above different faces of Al. Notation of sites is the same as in Fig.2.6. [2.1]

path which is 1.6-1.9 Å away from the nearest Al atom, with an activation energy of 0.1-0.2 eV. These authors have also improved their calculations by including part of the pseudopotential corrections nonperturbatively. This is based on a method first introduced by PERDEW and MONNIER [2.53] for clean surface calculations. The idea in this method is to add to the surface potential for the jellium model a step potential of the form c (-x). Since the system then still posesses translational invariance parallel to the surface, the self-consistent equations can be solved in the usual manner. This step potential allows one to include the major effects of the pseudopotential averaged parallel to the surface (Q = 0 component). The remainder can then be treated perturbatively. The parameter c is determined by minimizing the total energy (pseudopotential corrections included). In their improved calculation, HJELMBERG et al. [2.50] first did a calculation for the chemisorption system with the clean surface input as determined by PERDEW and MONNIER (for r_s = 2). They then included the remainder of the pseudopotential corrections to the binding energy perturbatively as described by (2.40). The qualitative conclusions are similar to the results they obtained earlier. There are significant increases in the binding energy over the bridge and center sites for adsorption on the (111) face.

The predictions of preferred adsorption sites from these works are somewhat surprising since experimentally, it is found that the center sites are preferred for most chemisorption systems. It is also quite different from the picture obtained by YING and KAHN (see Sect.2.4) in their linear response study of alkalies chemisorbed on Al. This could be due to the fact that the chemisorbed hydrogen has a much smaller size than other adsorbed atoms and as a consequence, feels a stronger lateral variation in the surface potential. With this in mind, one expects that higher order terms of $Q \neq 0$ contributions of the pseudopotentials may play an important role in this case. LANG and WILLIAMS [2.46] have recently applied the perturbative method to study the effect of lattice structure for Si chemisorbed on the (111) face of Al. In contrast to the behavior found for H on Al, they determined that the most favorable adsorption site is the center site C' (over a hole in the second layer of atoms). There is no crossover between the binding energy vs distance along different normals. The distance d from the jellium edge is decreased from 1.15 Å to 0.7 Å and the binding energy increases from about 3 eV to 3.8 eV relative to jellium results. Noting that d is measured from a point half an interplanar spacing in front of the outermost lattice plane, the Al-Si bond length for adsorption at this site was determined to be 2.6 Å, which is just about the sum of the covalent radius of the adatom and the metallic radius (half the nearest neighbor distance) of the substrate.

These different studies on the preferred adsorption sites indicate the lack of universality of this quantity for different chemisorption systems. The preferred adsorption site depends on a subtle balance between relative increases and decreases in the repulsive contribution such as the increase in kinetic energy and the attractive contributions such as the tail of the screened pseudopotential. Thus it may indeed vary from one particular combination of adsorbate-substrate to another. While the simple models for chemisorption described in the previous sections bring out many features which are not model sensitive, it may still be too simple for this particular aspect.

Another interesting study belong to this category is that by YU et al. [2.54] for oxygen chemisorption on Al. Experimentally, they found that the induced dipole of the oxygen atom on Al is practically zero (as reflected by the lack of change in the work function in the low coverage limit) and the 2p resonance of oxygen was located at 7.2 eV below the Fermi energy. This is quite contrary to the theoretical result for an oxygen-jellium system [2.43] which predicts a substantial induced dipole and a 2p resonance at 2.2 eV below E_F. The absence of work function changes indicated that the oxygen atom might have penetrated the metal surface (refer to Fig.2.2). Also, previous experimental studies of Sr and Cs indicated that oxygen is absorbed rather than adsorbed on these materials. Now, in the jellium model, the penetration of the adatom is prevented by the sharp increase in the electrostatic energy when the adatom moves into the positive background. When first-order pseudo-potential corrections are included, the barrier for penetration into a tetrahedral hole in the bulk is reduced to only 3 eV as opposed to ~16 eV for the jellium case. Thus, a stable position of the adatom could presumably result from inclusion of higher order terms. In the jellium calculation, when the adatom is placed inside the metal, the dipole moment rapidly disappears and the resonance drops quickly from 2 eV below E_F to an interior limit of about 10 eV below E_F. If one assumes that the effect of the pseudopotential on the individual electronic states is small, then the experimental result is consistent with the interpretation that the oxygen atom has penetrated the substrate. It is worthwhile to note here also that HARRIS and PAINTER [2.55] have studied the adsorption of an oxygen atom on a five atom Al-cluster with the X_α method. Their spectral results are in good agreement with those of LANG and WILLIAMS. They also concluded that the adsorbed oxygen will be closer to the Al substrate than indicated by the results of the jellium model.

2.3.4 Core Hole and Relaxation Effects; O, Na, Si, and Cl

One of the useful experimental probes in the study of chemisorption is X-ray photoemission. By measuring the energy spectrum of electrons ejected from the core levels of the chemisorbed atom, information on the adatom-metal binding and the symmetry environment of the bonding sites, etc., can be obtained. In the absence of electronic interaction, the threshold for exciting a core electron is simply related to its eigenvalue relative to the vacuum. In the interacting system, however, following the removal of a core electron, the remaining core hole perturbs the system and results in extra-atomic screening. If one assumes that during the process of excitation, the metal-adatom distance is held constant (Franck-Condon principle), then the extra-atomic reduction in the binding energy can be expressed as the difference of two total energy differences:

$$-\Delta = [E \text{ (chemisorbed atom with core hole)} + \phi - E \text{ (chemisorbed atom)}]$$

$$- [E \text{ (free atom with core hole)} - E \text{ (free atom)}] \quad . \tag{2.64}$$

Here E (chemisorbed atom) means the total energy of the metal-adatom system, and the sum E (chemisorbed atom with core hole) + ϕ is the total energy of the metal-adatom system with an adatom core electron removed to the vacuum level. The quantity ϕ is the metal work function, so that E (chemisorbed atom with core hole) is the energy of the metal-adatom system with a core electron removed and an additional electron at the Fermi level, which corresponds to the usual experimental situation. Δ is then the difference between the core binding energy of a free atom and that of the chemisorbed atom.

LANG and WILLIAMS [2.56] have studied the quantity Δ by the "Δ - SCF" approach. They applied the density functional theory outlined in the previous sections to study all the configurations (four in total) involved in the definition of Δ. When a core hole is present in the state i, the set of self-consistent equations is solved with the occupation number of the core-orbital n_i set equal to zero. Although the density functional theory is only a rigorous formulation for the ground state, there is some empirical evidence that it should also provide a reasonable description for the system in the presence of a core hole [2.57].

It is customary to separate the extra-atomic relaxation into a chemical shift Δ_c and a relaxation shift Δ_r. LANG and WILLIAMS introduced the following definition for Δ_c:

$$\Delta_c = \varepsilon_i \text{ (chemisorbed atom)} - \varepsilon_i \text{ (free atom)} \quad , \tag{2.65}$$

where the index i refers to a deep core level and the ε_i are energy eigenvalues. Equation (2.64) then defines the remainder as the relaxation shift. As defined by these equations, Δ_c is a ground state property whereas the relaxation shift depends on the contraction of the core valence orbitals in response to the reduced screening of the nuclear charge and the transfer of additional screening charge to the excited atom or its immediate vicinity.

In Fig.2.11, the results obtained by LANG and WILLIAMS for the chemical shifts and the relaxation shifts for a representative set of adatoms on high density ($r_s = 2$) substrate are displayed.

These results are best interpreted by inspecting the spatial forms of the extra-atomic screening charge density. This is defined in [2.56] as

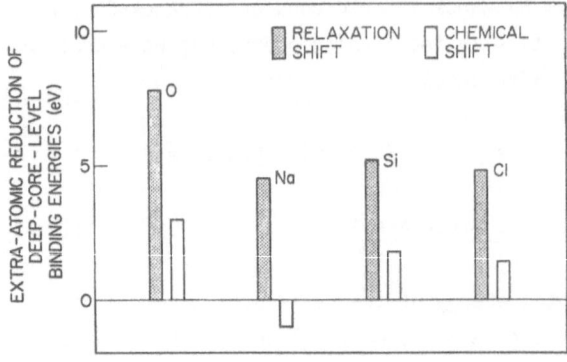

Fig.2.11. Decomposition of extra-atomic reduction of deep-core-level binding energy into its two components. Single atoms of O, Na, Si, and Cl chemisorbed on a high-density ($r_s = 2$) simple-metal surface. Oxygen was placed inside the metal. Hole state is 1s for O, 2s for Na, Si and Cl. [2.56]

$$n(\underline{r}; \text{ extra-atomic}) = [n(\underline{r}; \text{ chemisorbed atom with core hole})$$
$$- n(\underline{r}; \text{ chemisorbed atom})]$$
$$- [n(\underline{r}; \text{ free atom with core hole}) - n(\underline{r}; \text{ free atom})]. \quad (2.66)$$

This density for Na and Cl is displayed in Fig.2.12. The results indicate that the screening charge fills up mainly the lowest otherwise unoccupied atomic orbital. Thus for Na, the 3s resonance is being filled, leading to the nearly spherical character of the screening charge. For Cl, the situation is quite different, because its 3p resonance lies well below the Fermi level and thus is unable to accept any final state screening charge and the additional final state charge lies in the metal. The situation for Si is similar to that for Na. Accordingly, it is con-cluded in [2.56] that the "excited-atom approximation" provides an adequate de-scription only when there is sufficient "room" in the valence shell for the screen-ing charge. In this approximation, the screening charge is obtained by subtracting the electron density of the ion (i.e., the free atom with a core hole) from that of the excited atom, that is, an atom in which a core electron has been transferred (also self-consistency) to the lowest unoccupied valence orbital. The densities for Na and Si thus computed are indeed quite close to the full "Δ - SCF" method while for Cl there are considerable discrepancies. It seems that to obtain a better estimate for the screening charge within the excited atom scheme in a case like Cl, one has to fill the 3p level first and transfer the core electron to the 4p level.

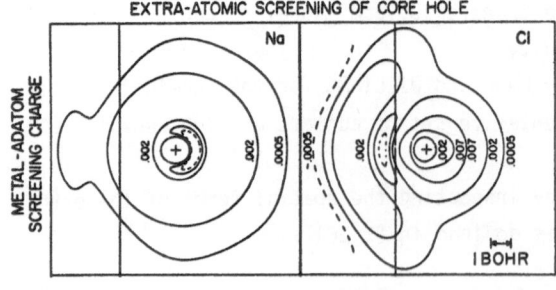

Fig.2.12. Contours of constant electron density showing the spatial distribution of extra-atomic screening charge for Na and Cl chemisorbed on a high-den-sity ($r_s = 2$) simple-metal surface. Metal on the left; vacuum on the right. Adatom nucleus indicated by +. Solid vertical line indi-cates positive-background edge. Contours in core regions have been deleted for clarity. Hole state is 2s. [2.56]

There exists a useful approximate formula for the relaxation shift [2.56-58]. This is given as

$$\Delta_r = (1/2) \int n(\underline{r}; \text{ extra-atomic}) \, |\underline{r}|^{-1} d\underline{r} \quad . \tag{2.67}$$

That is, the extra-atomic relaxation shift can be approximated by half the electrostatic potential at the position of the nucleus due to the extra-atomic screening charge. This quantity is not sensitive to the internal structure of the screening charge. This suggests that an approximate calculation of the screening charge, such as that obtained in a linear response study regarding the adatom with core hole as a point charge, can provide a useful estimate for the magnitude of the relaxation shift.

The sensitivity of these results to the parameters characterizing the substrate was also tested. For the case of Si, changing the substrate r_s from 2 to 4 while holding the adatom-metal separation fixed, changed Δ_r by only 0.3 eV. This insensitivity seems to suggest that in determining the relaxation shift, the metal substrate just acts as a reservoir of electrons. If this holds true beyond the particular theoretical models considered so far, then the usual interpretation of differences in core level binding energies of a given atom as chemical or initial state effects is justified.

The major conclusion in this study is that the relaxation shift depends mainly on the atomic number of the excited atom. This dependence of course is not a simple one, but comes in indirectly through charge transfer between the adatom and the metal and the nature of the final screening charge. This conclusion is quite different from those reached in other works in the literature which are basically perturbative in character [2.59].

2.4 Concluding Remarks

In the above sections, we have presented some theoretical approaches to the problem of chemisorption based on the density functional theory. Most of the applications so far are limited to very simple models for the substrate. The strength of these models lies in the fact that, because of their simplicities, a detailed description of various chemisorption quantities, including the spatial distribution of electronic density, can be obtained from the theory. Also, self-consistency between the charge density and the effective potential can be strictly maintained. This is of the utmost importance, sometimes even just to get a qualitative picture. Although one cannot expect a high accuracy in comparing the theoretical predictions with experimental results for any specific quantity, we believe that many qualitative features and trends exhibited by these model calculations should persist in future studies in which a more realistic representation of substrates is employed.

References

2.1 O. Gunnarson, H. Hjelmberg, B. Lundquist: "Adsorbate-Induced Electronic Structure", in *Photoemission from Surfaces*, ed. by B. Feuerbacher, B. Fitton, R.F. Willis (Wiley, New York 1977)
2.2 J.P. Muscat, D.M. Newns: "Chemisorption on Metals" (to be published)
2.3 D.G. Dampsey, L. Kleinman: J. Phys. F7, 113 (1977)
2.4 P. Hohenberg, W. Kohn: Phys. Rev. 136B, 864 (1964)
2.5 W. Kohn, L.J. Sham: Phys. Rev. 140A, 1133 (1965)
2.6 O. Gunnarson, B. Lundquist: Phys. Rev. B13, 4274 (1976)
2.7 M. Rasolt, D.J.W. Geldart: Phys. Rev. Lett. 35, 1234 (1975)
     A.K. Gupta, K.S. Singwi: Phys. Rev. B15, 1801 (1977)
2.8 O. Gunnarson, M. Jonson, B. Lundquist: Phys. Lett. 59A, 117 (1976)
2.9 J. Slater: *The Self-Consistent Field for Molecules and Solids: Quantum Theory of Molecules and Solids*, Vol.4 (McGraw-Hill, New York 1974)
2.10 S.C. Ying, J.R. Smith, W. Kohn: J. Vac. Sci. Technol. 9, 575 (1971)
2.11 J.R. Smith, S.C. Ying, W. Kohn: Phys. Rev. Lett. 30, 610 (1975)
2.12 S.C. Ying, J.R. Smith, W. Kohn: Phys. Rev. Lett. B11, 1483 (1975)
2.13 E.P. Wigner: Trans. Faraday Soc. 34, 678 (1938)
2.14 N.D. Lang, W. Kohn: Phys. Rev. B1, 4555 (1970)
2.15 J.R. Smith: Phys. Rev. 181, 522 (1969)
2.16 N.D. Lang, W. Kohn: Phys. Rev. B3, 1215 (1971)
2.17 C.F. Von Weizäcker: Z. Phys. 96, 431 (1935)
2.18 K. Kahn, S.C. Ying: Surf. Sci. 59, 333 (1976)
2.19 T.E. Madey, T.T. Yates, Jr.: *Structure et Properties des Surface des Solides* (Editions du Centre National de la Recherche Scientifique, Paris 1970) No.187, p.155
     T.W. Hickmott: J. Chem. Phys. 32, 810 (1960)
2.20 E.W. Plummer, A.E. Bell: J. Vac. Sci. Technol. 9, 583 (1972)
2.21 E.M. Propst, T.C. Piper: J. Vac. Sci. Technol. 4, 53 (1917)
2.22 H. Froitzheim, H. Ibach, S. Lehward: Phys. Rev. Lett. 36, 1549 (1976)
2.23 N.D. Lang, A.R. Williams: Phys. Rev. Lett. 34, 531 (1975)
2.24 J.R. Smith, S.C. Ying, W. Kohn: Solid State Commun. 15, 1491 (1974)
2.25 P.W. Anderson: Phys. Rev. 124, 41 (1961)
2.26 E.W. Plummer, B.J. Waclawski: Proc. Physical Electronics Conf., 1973 (unpublished)
2.27 T.E. Madey: Surf. Sci. 36, 281 (1973)
2.28 L.M. Kahn, S.C. Ying: Solid State Commun. 16, 799 (1975)
2.29 L.M. Kahn, S.C. Ying: Surf. Sci. 59, 333 (1976)
2.30 M.L. Cohen, D. Weaire: In *Solid State Physics*, Vol.24, ed. by F. Seitz, D. Turnbull, H. Ehrenreich (Academic Press, New York 1970)
2.31 N.W. Ashcroft: Phys. Lett. 23, 48 (1966); J. Phys. C , 232 (1968)
2.32 C.J. Todd, T.N. Rhodin: Surf. Sci. 42, 109 (1974)
2.33 D.M. Mewns: J. Chem. Phys. 50, 4512 (1969)
2.34 J.P. Muscat, D.M. Newns: Solid State Commun. 11, 737 (1972); J. Phys. C7, 2630 (1974)
2.35 L.M. Kahn, M. Rasolt: Solid State Commun. 20, 1073 (1976)
2.36 J.H. Rose, Jr., H.B. Shore, D.J.W. Geldart, M. Rasolt: Solid State Commun. 19, 619 (1976)
2.37 J.P. Perdew, D.C. Langreth, V. Sahni: Phys. Rev. Lett. 38, 1030 (1977)
2.38 R. Gomer, L.W. Swanson: J. Chem. Phys. 38, 1613 (1963)
2.39 B.A. Hutchins, T.N. Rhodin, J.E. Demuth: Surf. Sci. 54, 419 (1976)
2.40 J.E. Demuth, D.W. Jepsen, P.M. Marcus: J. Phys. C8, L25 (1975)
     S. Andersson, T.B. Pendry: J. Phys. C5, 41 (1972)
2.41 H.B. Huntington, L.A. Turk, W.W. White III: Surf. Sci. 48, 187 (1975)
2.42 N.D. Lang: Solid State Commun. 9, 1015 (1971); Phys. Rev. B4, 4234 (1971)
2.43 N.D. Lang, A.R. Williams: Phys. Rev. Lett. 34, 531 (1975); 37, 212 (1976)
2.44 O. Gunnarson, H. Hjelmberg: Phys. Scripta 11, 97 (1975)
2.45 O. Gunnarson, H. Hjelmberg, B. Lundquist: Phys. Rev. Lett. 37, 292 (1976); Surf. Sci. 63, 348 (1977)

2.46 N.D. Lang, A.R. Williams: Phys. Rev. B *18*, 616 (1978)
2.47 A.R. Williams, J. Van W. Morgan: J. Phys. C*7*, 37 (1974)
2.48 T.E. Demuth, D.W. Jepsen, P.M. Marcus: Phys. Rev. Lett. *31*, 540 (1973)
2.49 R.F.W. Bader, W.H. Nenneker, P.E. Cade: J. Chem. Phys. *46*, 3741 (1976);
     R.F.W. Bader, I. Keaverry, P.E. Cade: J. Chem. Phys. *47*, 3381 (1967)
2.50 H. Hjelmberg, O. Gunnarson, B. Lindsquist: Surf. Sci. *68*, 158 (1977)
2.51 C.O. Almbladh, U. Von Barth, Z.D. Popovic, M.J. Scott: Phys. Rev. B*14*, 2250
     (1976)
2.52 P.E. Cade, W.M. Huo: J. Chem. Phys. *47*, 649 (1967)
2.53 J.P. Perdew, R. Monnier: Phys. Rev. Lett. *37*, 1286 (1976)
2.54 K.Y. Yu, J.N. Miller, P. Chye, W.E. Spicer, N.D. Lang, A.R. Williams: Phys.
     Rev. B*14*, 1446 (1976)
2.55 J. Harris, G.S. Painter: Phys. Rev. Lett. *36*, 151 (1976)
2.56 N.D. Lang, A.R. Williams: Phys. Rev. B*16*, 2408 (1977)
2.57 C.O. Almbladh, U. Von Barth: Phys. Rev. B*13*, 3307 (1976)
2.58 L. Hedin, S. Lundquist: In *Solid State Physics*, Vol.23, ed. by F. Seitz,
     D. Turnbull, H. Ehrenreich (Academic Press, New York 1969) p.1
2.59 J.W. Gadzuk: J. Vac. Sci. Technol. *12*, 289 (1975)
     G.E. Laramore, W.J. Camp: Phys. Rev. B*9*, 3270 (1974)
     B. Gumhalter, D.M. Newns: Phys. Lett. *57*A, 423 (1976)

3. Chemisorption on Semiconductor Surfaces

J. A. Appelbaum and D. R. Hamann

With 24 Figures

The close contact between theory and experiment in semiconductor surface studies
has shaped the form of recent theoretical studies to a large extent. As a result,
these studies have concentrated on specific substrate surfaces and adsorbates, with
little attention to "generalized" models. The systems were typically chosen on the
bases of simplicity and experimental accessibility. These criteria led to a concentration on low index surfaces, and ordered adsorbed layers with simple unit
cells. Since independent-electron models, with correlation treated in an averaged
way [3.1,2] have proven adequate to interpret experiment, there has been little
work on specifically many-body effects [3.3].

3.1 Background Information

Semiconductor surfaces display a bewilderingly complex set of interactions with
foreign atoms and molecules. This fact became known in the early 1960s, largely
through research stimulated by the introducing of fluorescent screen displays in
apparatus for low energy electron diffraction (LEED) [3.4-6]. Such apparatus en-
abled the experimenter to make a quick visual identification of the formation of
ordered surface phases by an adsorbate, and provided an immediate characterization
of their two-dimensional periodicity. Unfortunately, it was much easier to find
new surface phases than to explain them. For example, LANDER reported eight surface phases formed by interacting In with the Si(111) surface under varying conditions [3.5b]. Such surface phases typically form two-dimensional lattices in registry with the substrate. The unit cell is some multiple of the surface unit
cell of the terminated ideal bulk lattice, and is conventionally specified by giving the factors by which the ideal unit cell is multiplied in each of two directions, such as 2×1, 7×7, etc. While it is in principle possible to determine the
exact atomic geometry of ordered surface phases by analysis of the voltage variation
of LEED intensities, in practice it is very difficult. A small number of such
analyses have been carried through for ordered overlayers on transition metal surfaces [3.7], but only very recently have a few comparable analyses appeared for

semiconductors [3.6]. Lacking such information, it was only possible to speculate
on possible geometries through chemical reasoning, a process which seldom produced
unique results. Compounding the difficulties was the fact that most clean semiconductor surfaces "reconstruct", that is, display surface phases with superlattice
structures. Determining the detailed nature of these phases is a problem comparable
to that for chemisorbed phases. Adsorption on a reconstructed clean surface might
or might not change its periodicity, depending on adsorbate, exposure, and thermal
treatment.

A new thrust toward understanding chemisorption on semiconductors was provided
by the introduction of ultraviolet photoemission spectroscopy (UPS) in the photon
energy range of 15 to 40 eV in the early 1970s [3.5,8,9]. Electrons ejected from
states associated with chemical bonds by photons in this energy range typically
have a very short mean-free path with respect to collisions which will cause them
to lose a substantial fraction of their energy. Thus the energy distribution curves
of the photoemitted electrons are primarily related to the electronic structure
of the first few atomic layers, and show marked changes in their features upon
chemisorption. While the total photoemission process is quite complex, it was
readily apparent that features in the spectra which did not change with photon
energy (when referred to constant initial state energy) could be associated with
the density of states in the surface and near-surface region.

The existence of such spectroscopic data provided a strong motivation for theoretical studies of surface electronic structure based on sufficiently detailed
models that useful information could be gained by comparison between theory and
experiment [3.1,7]. The kind of information initially sought, of course, was veri-
fication of the accuracy of the physical model and the adequacy of the computational
techniques applied to it. Subsequently (although to some extent simultaneously)
the questions of bonding geometry raised by LEED studies and chemical arguments
were addressed. These efforts have been extremely successful, and have produced a
wealth of detailed information.

3.2 Bulk Properties

We shall be primarily concerned in this review with the chemisorptive properties
of Si, Ge and GaAs surfaces. Since some understanding of the bulk properties of
these semiconductors is essential for our later surface discussions we briefly review in this section their electronic structure. A modern review relating electronic
structure and many other properties is given by PHILLIPS [3.11]. The semiconductors
of interest to us here all crystallize in the diamond lattice which is an open
structure, or the closely related zincblende and wurtzite structures. The bonding,

which proceeds primarily via s-p hybridization, gives rise to energy band structures for Si, Ge and GaAs which are shown in Figs.3.1-3. The four valence bands have a total width of 12.0-13.0 eV, and are separated by an average band gap (for fixed crystal momentum) of 3.0-4.0 eV. The minimum gap, by contrast, ranges from 0.7 to 1.5 eV.

The valence bands of Si and Ge are quite similar, while GaAs has a well-developed band gap within the valence bands separating the essentially As-like s states from the remaining bonding orbitals. In spite of this heteropolar gap, bonding in GaAs in predominantly covalent, with effective charge transfer between anion and cation of ≈ 0.2 electrons [3.11]. Unlike the valence bands, the conduction bands of Si and Ge differ significantly in the ordering of their energy levels.

By far the most accurate description of the energy bands of these semiconductors has been achieved using nonlocal pseudopotential methods [3.12,13], although rather good descriptions of the energy bands are obtainable with local pseudopotential theory [3.14]. Empirical tight-binding calculations, on the other hand, involving s and p states, and including near and next nearest neighbor contributions to the Hamiltonian, describe the valence bands and the lowest lying conduction bands satisfactorily, but fail badly for the higher lying states [3.15,16]. This limitation stems from the artificial cutoff of overlap and the neglect of d character in the wave function expansion, which plays an important role in the conduction bands of all three semiconductors. Both pseudopotential and tight-binding methods have been extended to treat the problem of surface band structure.

3.3 General Surface Properties

The clean surfaces of Ge, Si and GaAs occur in a variety of structural forms. Si(111), as an example, exhibits a stable 7×7 superlattice pattern below 840° C, at which temperature it converts to a 1×1 structure. In addition to these stable phases, a metastable 2×1 superlattice is formed by cleavage, which converts ir-reversibly to 7×7 near 350° C [3.17]. Similar geometric complexity is exhibited by the other semiconductors [3.4b]. The electronic properties of these surfaces are in general sensitive to the details of the surface's geometric structure. An exception to this is the ionization potential, which varies little for these semi-conductors from one crystal face to another [3.18]. A schematic of the surface barrier, of which the ionization potential is a measure, is shown in Fig.3.4.

The surface barrier, ≈ 14.5 eV, is similar to that in Al, which has the same average electron density and is Si's neighbor in the periodic table. It is made up of two contributions [3.2]. The exchange and correlation contribution accounts for two-thirds of the surface barrier, the remainder is electrostatic in origin

Fig.3.1. Energy band structure of Si, showing the valence bands and low-lying conduction bands on principal symmetry lines in the Brillouin zone calculated using a self-consistent pseudopotential [3.23]

Fig.3.2. Energy band structure of Ge, showing the valence bands and low-lying conduction bands on principal symmetry lines in the Brillouin zone. The bands are calculated from a self-consistent pseudopotential similar to that in [3.23]

Fig.3.3. Energy band structure of GaAs, showing the valence bands and low-lying conduction bands on principal symmetry lines. The bands are calculated from a self-consistent pseudopotential similar to that in [3.23]

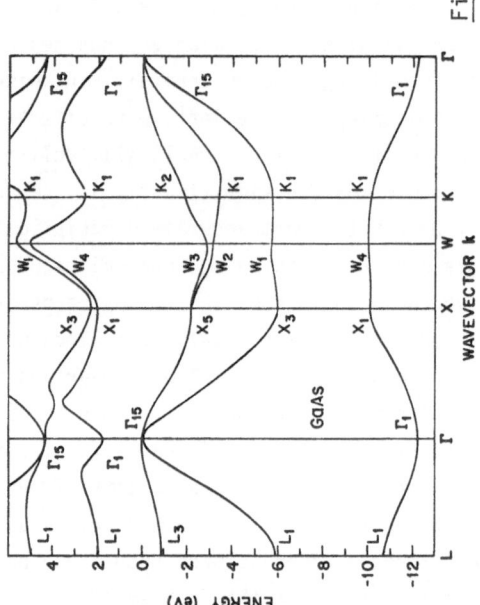

Fig.3.3

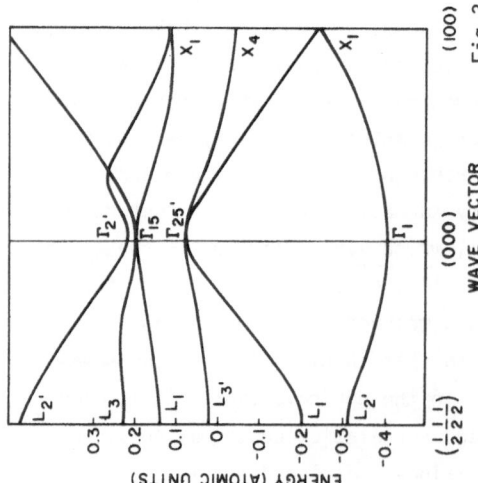

Fig.3.1

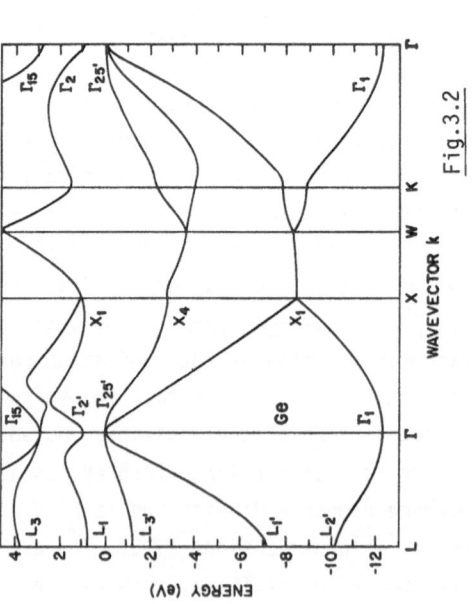

Fig.3.2

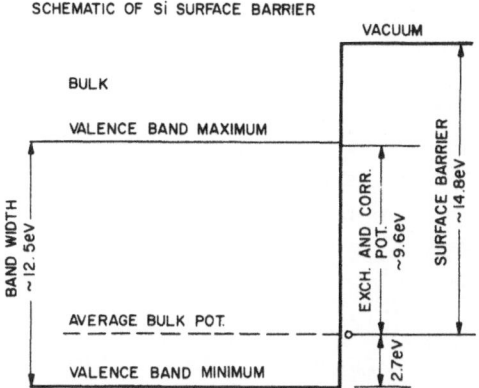

SCHEMATIC OF Si SURFACE BARRIER

Fig.3.4. A schematic energy level diagram of the surface barrier appropriate to the Si(111) surface, constructed from experimental work function measurements and experimental bulk band structure data [3.20]

[3.19,20]. It is only this latter contribution that can vary from one surface to the next and that changes with chemisorption. The surface barrier or potential develops primarily within 1 or 2 Å of the surface atom plane [3.19]. This remarkable localization comes from the highly polarizable electron density characteristic of bulk semiconductors, which makes them in this regard similar to metals of comparable electron density.

While the surface barrier of Si and Al may be quite similar, their spectral properties are entirely different. To understand the general spectral properties of a surface and its relationship to bulk band structure as well as the surface barrier, we must examine how the energy bands are modified in going from the three-dimensional periodicity of the bulk to the two-dimensional periodicity of the surface.

The energy bands of the surface $\epsilon_s^n(k_{\shortparallel})$ depend on a two-dimensional wave vector $k_{\shortparallel}$ and band index n. These bands are formally divided into continuum bands derivable from the bulk band structure and surface states or localized bands. The former are obtained by expressing the bulk band structure $\epsilon_B^n(k)$, where k is the bulk Bloch wave vector, as $\epsilon_B^n(k_{\shortparallel},k_z)$ where $k_{\shortparallel}$ is the corresponding surface Bloch vector, and allowing k_z, no longer a good quantum number, to run over all its allowed values in the bulk Brillouin zone. Such a projection for the Si(111) surface is shown in Fig.3.5. The two-dimensional continuum spectrum will in general have band gaps in which localized surface states can exist. The position of these states or bands depends on the details of the surface potential, i.e., broken bonds, new chemisorptive bonds, etc. It quite often occurs that the formation of new chemisorptive bonds produces essentially localized states within an energy band continuum. This results in a continuum state resonance which can be remarkably sharp if the spatial character of the continuum and "localized" states are sufficiently distinct. Of course, along symmetry lines in the surface Brillouin zone (SBZ) the localized states may be forbidden to mix with the continuum states, in which case we have bona fide surface states in the continuum.

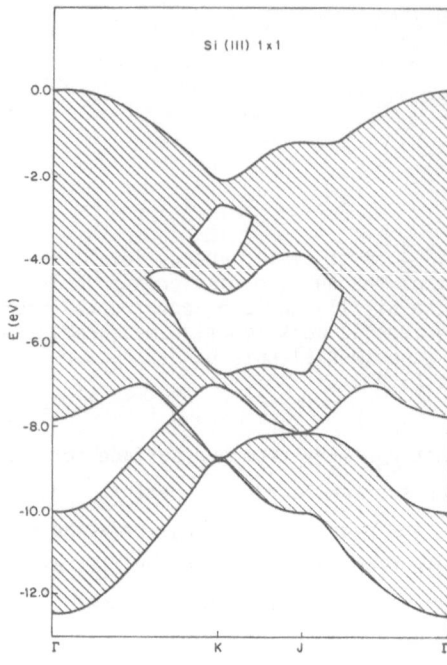

<u>Fig.3.5.</u> Projected bulk valence bands of Si shown as a function of the two-dimensional Bloch vector $k_{\shortparallel}$ along symmetry lines in the surface Brillouin zone. Shaded portions represent allowed energies. [3.35]

Surface states and resonances are the spectral signature of new chemisorptive bonds. These chemisorptive surface state or resonant bands may be comparatively broad, but they reveal themselves through their two-dimensional critical point structure [3.21]. In contrast to three-dimensional systems, two-dimensionally periodic systems have densities of states with logarithmic singularities and band edge discontinuities which makes their experimental identification simpler than would otherwise be the case. With improved angular photoemission techniques [3.8b], it has become possible to map out the dispersion relation of specific surface states and identify their orbital character. The tools that are needed to calculate these surface bands are described in the next section.

3.4 Computational Approaches

We review here the two primary computational methods that have been employed for calculating the electronic structure of semiconductor surfaces — —either clean or with chemisorbed overlayers. The methods are the self-consistent pseudopotential approach (SCPP) [3.22-25] and the semi-empirical tight-binding approach (ETB) [3.16,26-29]. We shall not cover the work on semiconductor cluster calculations [3.30].

The self-consistent calculations assume that the potential can be written as

$$V(\underline{x}) = V_{es}(\underline{x}) + V_{xc}(\underline{x}) + V_{ion}(\underline{x}) \tag{3.1}$$

where V_{es} is the electrostatic potential of the conduction electrons and smoothed ion charges, and V_{ion} represents the nonelectrostatic electron-ion interaction treated within a model or pseudopotential framework. The exchange and correlation potential, V_{xc}, is treated within the local density or statistical exchange and correlation approximation. While this approximation has been justified on a first-principles basis only in the limit of slowly varying electron density [3.31], its success in a wide variety of calculations on bulk solids and molecules provides strong empirical justification for its use in the surface context [3.32].

For a fixed surface potential, two approaches to solving the Schroedinger equation have been employed. In the first the surface is studied as the boundary of a semi-infinite solid [3.22]. The potential $V(\underline{x})$ and wave function $\Psi_{\underline{k}_{\shortparallel}}(x)$ are expanded in a Laue representation [3.33]

$$V(\underline{x}) = \sum_{\underline{G}_{\shortparallel}} \exp(i\underline{G}_{\shortparallel} \cdot \underline{x}_{\shortparallel}) v_{\underline{G}_{\shortparallel}}(z) \tag{3.2}$$

$$\Psi_{\underline{k}_{\shortparallel}}(\underline{x}) = \sum_{\underline{G}_{\shortparallel}} \exp(i\underline{G}_{\shortparallel} \cdot \underline{x}_{\shortparallel} + i\underline{k}_{\shortparallel} \cdot \underline{x}_{\shortparallel}) u_{\underline{G}_{\shortparallel}}(z) \quad . \tag{3.3}$$

The coordinate system is oriented so that $\hat{z}$ is normal to the surface, and $\underline{x}_{\shortparallel}$ parallel to it. $\{\underline{G}_{\shortparallel}\}$ is the set of reciprocal lattice vectors that characterizes the two-dimensional periodicity of the surface. Each wave function is labeled by a Bloch wave vector $\underline{k}_{\shortparallel}$, which, together with the energy E, constitute the basic quantum numbers of the problem. The Laue representation implies relatively rapid convergence of V in a plane wave basis. To achieve this one must use for V_{ion} a model or pseudopotential in which the strong atomic core potential has been effectively removed [3.14].

Substituting (3.2 and 3) into Schroedinger's equation results in a set of coupled differential equations for $u_{\underline{G}_{\shortparallel}}(z)$, viz,

$$\left[-\frac{1}{2}\frac{d^2}{dz^2} + \frac{1}{2}(\underline{k}_{\shortparallel} + \underline{G}_{\shortparallel})^2 - E \right] u_{\underline{G}_{\shortparallel}}(z) + \sum_{\underline{G}_{\shortparallel}'} v_{\underline{G}_{\shortparallel} - \underline{G}_{\shortparallel}'}(z) u_{\underline{G}_{\shortparallel}'}(z) = 0 \tag{3.4}$$

where the number of $\underline{G}_{\shortparallel}$'s retained is determined by the strength of V.

For a given E and $\underline{k}_{\shortparallel}$, (3.4) can be integrated numerically assuming $u_{\underline{G}_{\shortparallel}}(z)$ and $(d/dz)u_{\underline{G}_{\shortparallel}}(z)$ are specified on a plane, say $z = z_1$. Integration to another plane $z = z_2$ is represented concisely by defining a transfer matrix $\tilde{T}_{2,1}$,

$$\underline{u}_2 = \tilde{T}_{2,1}\underline{u}_1 \tag{3.5}$$

where $\underline{u}_i$ is a vector constructed from $u_{\underline{G}_n}(z_i)$ and $du_{\underline{G}_n}(z_i)/dz$,

$$
\underline{u}_i =
\begin{bmatrix}
u_{\underline{G}_1}(z_i) \\
\vdots \\
u_{\underline{G}_N}(z_i) \\
(d/dz)u_{\underline{G}_1}(z_i) \\
\vdots \\
(d/dz)u_{\underline{G}_N}(z_i)
\end{bmatrix} .
\tag{3.6}
$$

Knowledge of $\tilde{T}_{2,1}$ constitutes only a partial solution to the problem; satisfying the appropriate boundary conditions on $\underline{u}(z)$ for z in the bulk and in the vacuum represents the remainder. The boundary value problem is illustrated in Fig.3.6, where space is divided into three regions by two planes parallel to the surface. To the left of the bulk plane V is assumed equal to its infinite bulk solid value, and to the right of the vacuum plane V is assumed to have no spatial variation parallel to the surface. In both these regions the solutions $\underline{u}_i$ to the Schroedinger equation are known. In region I they are Bloch waves that either propagate or decay to the left, and in the region III they have the form $\exp[i(\underline{k}_n + \underline{G}_n) \cdot \underline{x}_n]\phi_{\underline{G}_n}(z)$, where $\phi_{\underline{G}_n}(z)$ decays exponentially with increasing z. Connecting the solutions of known form in region I to region III is the job of the transfer matrix. The boundary conditions can be written as

$$
\underline{u}_b = \tilde{T}_{b,v}\underline{u}_v
\tag{3.7}
$$

where $\underline{u}_v$ and $\underline{u}_b$ are sums of allowed vacuum and bulk solutions, respectively, for the given E and $\underline{k}_n$.

For E within a bulk band, $\underline{u}_b$ is assumed to contain an incident propagating wave of unit amplitude and (3.7) then uniquely specifies the remainder of $\underline{u}_b$ and $\underline{u}_v$, and through (3.5) $\underline{u}$ everywhere. For E in a bandgap no propagating Bloch waves exist, only evanescent waves, and (3.7) can have solutions only at special energies $E_n(\underline{k}_n)$, which define the spectral location of the surface states. These states, free to propagate along the surface direction, are bound to the surface region by the vacuum barrier on one side and the bandgap of solid on the other. Their energies are found by varying E in a bandgap and searching for possible zeros of the multidimensional vector $\underline{u}_b - \tilde{T}_{b,v}\underline{u}_v$. It must be borne in mind that $\underline{k}_n$ is always fixed, and that a bandgap for fixed $\underline{k}_n$ can and often does coincide with regions of allowed electron state density at other $\underline{k}_n$'s.

The second approach to solving the wave equation studies the surface as the boundary of a slab which is periodically embedded in a hypothetical empty tetragonal

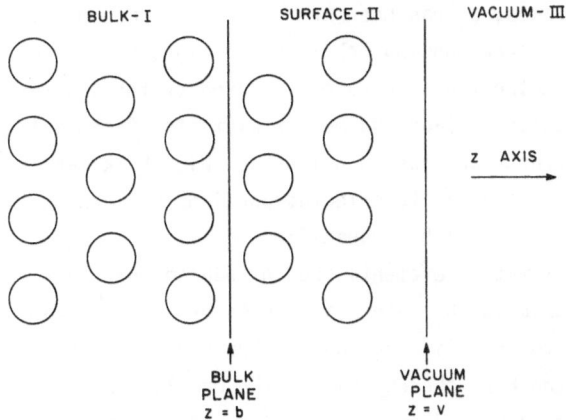

BULK-I   SURFACE-II   VACUUM-III

z AXIS

BULK   VACUUM
PLANE   PLANE
z = b   z = v

Fig.3.6. Schematic representation of the three regions into which space is divided in the method developed by the authors. In region I the disturbance in the potential produced by the surface is assumed to have been screened to negligible values. In region II, only the nuclear coordinates are specified and the potential is allowed to adjust itself self-consistently. In region III, the variation of the potential parallel to the surface is assumed negligible. [3.10]

lattice [3.24,25,34]. This lattice has the same periodicity parallel to the slab surface as that of the slab; normal to the surface the lattice has a repeat distance equal to the sum of the slab thickness and an "allowed" vacuum region, typically 8 A, which separates adjacent slabs.

With periodic boundary conditions the natural basis for expanding the eigenvectors becomes

$$u_{\underline{G}_{\shortparallel},n}^{\underline{k}_{\shortparallel}}(\underline{x}) = \exp[i(\underline{G}_{\shortparallel} \cdot \underline{X}_{\shortparallel} + \underline{k}_{\shortparallel} \cdot \underline{X}_{\shortparallel} + k_n z)] \qquad (3.8)$$

where

$$k_n = \frac{2\pi n}{L}, \qquad n = 0, \ \pm 1, \ \dots \qquad (3.9)$$

and L is the length (normal to the surface studied) of the tetragonal unit cell. A sufficiently large number of $u_{\underline{G}_{\shortparallel},n}^{\underline{k}_{\shortparallel}}(\underline{x})$ are retained for convergence, with Löwdin perturbation techniques used to simplify the matrix diagonalization.

Distinct from these methods is the empirical tight binding, or in chemical terms, the linear combination of atomic orbital approach [3.27-29]. In these methods the wave functions of a thin slab are expanded as

$$\psi_{\underline{k}_{\shortparallel}}^{j\ell}(\underline{x}) = \sum_m \exp(i\underline{k} \cdot \underline{R}_m) a_{\underline{k}_{\shortparallel}}^{i\ell} \phi_{j\ell}(\underline{x} - \underline{R}_m - \underline{A}_\ell)$$

where $\phi_{j\ell}(\underline{x} - \underline{R}_m - \underline{A}_\ell)$ are atomiclike orbitals centered about atom sites located at $\underline{R}_m + \underline{A}_\ell$ and the sum is over all lattice vectors $\underline{R}_m$ of the slab. In the application of these methods to semiconductor surfaces, ϕ_j has been restricted to s and p orbitals, assumed to be orthonormal to corresponding orbitals on neighboring sites. The Hamiltonian matrix elements H_{ij} are treated as parameters fixed by the assumption that for an ideal clean semiconductor surface the matrix elements of the bulk crystal and the slab are identical, except for missing neighbors. With bulk matrix elements fixed empirically by constructing a fit to selected portions of a known bulk band structure, the Hamiltonian for the slab is specified.

For chemisorbed surfaces, additional matrix elements are introduced that must be specified by other means. One of these is the molecular analog procedure [3.35] where a molecule with bonds analogous to those in the chemisorbed bond is found. Its spectrum is then calculated with the ETB, thereby fixing the Hamiltonian parameters for the molecule, which are in turn used for the chemisorbed system.

As we will see, both approaches have been successfully used to study chemisorption on semiconductor surfaces. The ETB method is primarily a spectroscopic tool with its strength lying in its relative simplicity. Its assumptions are severe, and rest (in the case of chemisorption studies) on the requirement of finding appropriate molecular analogs. The SCPP methods are more general. In addition to spectroscopic information they can yield work function changes, can study chemical bond strengths and equilibrium positions, and deal with real potentials and wave functions. On the other hand, they are considerably more complex to apply than the ETB method and depend on a careful choice of model or pseudopotential. For completely satisfactory results, the latter may have to be nonlocal in character, again complicating the methodology.

3.5 Clean Semiconductor Surfaces

We briefly review the electronic structure of clean semiconductor surfaces, beginning with the ideal Si(111) 1×1 surface. A ball-and-stick model of this surface, drawn in Fig.3.7, clearly shows the single broken bond per surface atom characteristic of this surface. This broken bond is the origin of a narrow band of surface states lying in the energy gap separating the valence and conduction bands. This band, whose dispersion is plotted in Fig.3.8, is half full.

It is extremely unlikely that the surface Si plane on an actual Si(111) 1×1 surface would maintain its nominal separation from second layer Si atoms. Theoretical studies [3.19,36] indicate a strong dependence of the surface spectrum on this separation coordinate. If the surface atoms relax inward, not only is the dangling bond surface state changed (it becomes more p_z-like), but the surface Si atom's backbonds are strenghtened, resulting in additional surface state bands below and within the valence bands (~ 2.4 eV below the valence band maximum).

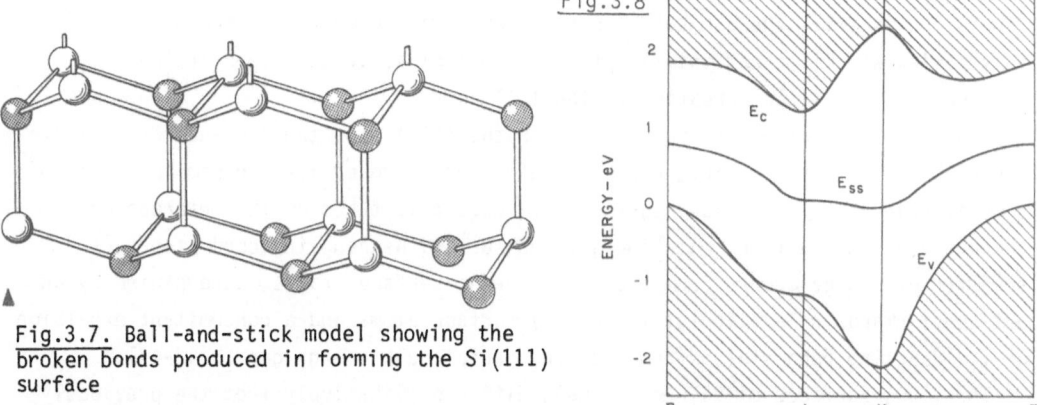

Fig.3.8

ENERGY- eV

E_c

E_{ss}

E_v

Γ   J   K   Γ

Fig.3.7. Ball-and-stick model showing the broken bonds produced in forming the Si(111) surface

Fig.3.8. Energy dispersion relation for the dangling bond surface state on relaxed Si(111), E_{ss}, along symmetry lines in the surface Brillouin zone. Projected valence and conduction band edges, E_v and E_c, are also shown. (From calculations by the authors reported in [3.19,40])

Chemisorption can be expected to directly effect the dangling bond bands, which should participate in bonding, and indirectly, the backbond surface states, through a modification of surface relaxation.

The ideal Ge(111) 1×1 surface has the same structure as that of ideal Si(111), but the dangling bond surface state band disperses quite differently from its counterpart on Si because of the very different ordering of the conduction band states in these two materials [3.37,38]. Deep lying backbonding states, by contrast, are nearly identical in Si and Ge. The difference between the dangling bond bands in these two materials is believed to be responsible for the fact that their room temperature equilibrium surface phases differ, with Si occurring in a 7×7 super-lattice structure, and Ge in a 2×8 superlattice. These differences can carry over to chemisorbed Si and Ge surfaces, as we will see in the case of chemisorbed Cl, which chemisorbs at different sites on these materials. When chemisorption of a foreign atom occurs at structurally identical locations on Si and Ge, and is strong, so that the energy of the bonds that form lie well within the valence band, the near identity of the bulk valence bands of these materials imposes similar con-ditions on the spectra of the chemisorbed systems.

Our discussions have focused primarily on the ideal or high temperature phase of Si(111) and Ge(111). The metastable 2×1 forms of these surfaces are character-ized by a structural buckling in which alternate rows of atoms move upward and downward — the so-called Haneman model [3.39]. The surface is ionic, with charge transfer between the raised and lowered dangling bond orbitals [3.40], and tends to revert to a 1×1 structure with chemisorption. The stable 7×7 and 2×8 forms of Si(111) and Ge(111) are not well understood, but spectroscopic and LEED evi-

dence suggests that they do not differ greatly from the 1 × 1 form [3.41]. This
conclusion is still the subject of considerable controversy. For the purposes of
this review we will treat chemisorption on 7 × 7 Si as if the superlattice was a
comparatively small perturbation on the 1 × 1 form.

The (110) face of Si and Ge shares with the (111) face the presence of a single
broken bond per surface atom, and their electronic properties are similar. The
same is true of the GaAs(110) surface, if account is taken of the heteropolar
character of the surface. With inequivalent Ga and As dangling bond states, a
transfer of charge occurs from the Ga to the As surface states, accompanied by an
outward (inward) relaxation of the As(Ga) surface atoms and a concomitant expulsion
of the dangling bond states from the fundamental bulk energy gap [3.42-44].

The (100) surfaces of Si, Ge and GaAs differ qualitatively from the previously
discussed surfaces in that there are two broken bonds per surface atom present on
the ideally terminated (100) crystal face. This surface has two bands of surface
states within the bandgap, whose spatial character is shown in Fig.3.9. The
"dangling bond" surface band is narrow, with dispersion comparable to the single
dangling bond surface band on Si(111), while the "bridge" state or lone orbital
surface state band, nearly empty of charge, disperses by nearly 3.0 eV, extending
into the conduction band [3.45]. The surface is unstable, always occurring in a
(2 × 1) or more complex superlattice structure. It is believed that these surfaces
are dimerized, so that the resulting surface has only one broken bond per surface
atom [3.46,47]. The surface charge density of the dimerized Si(100) surface is
contour plotted in Fig.3.10. In its dimerized form, the Si surface atoms have
(once again) a single broken bond per surface atom. Unlike the other faces, the
broken bond states are comparatively close together on Si(100) 2 × 1, so that they
interact relatively strongly, and chemisorbed atoms at these sites can be expected
to behave similarly.

3.6 H Chemisorption on Si Surfaces

Hydrogen chemisorbed on Si surfaces has been the subject in recent years of a large
number of theoretical and experimental studies [3.35,48-52]. The reasons for this
are best illustrated with respect to an ideal Si(111) 1 × 1 surface. The single
broken bond per Si surface atom is a likely candidate to participate with the singly
occupied H 1s level in the formation of a strong covalent bond. The new chemisorbed
H-Si bond, considerably stronger than that of Si-Si, is easily observed spectro-
scopically, as is the disappearance of broken bond derived spectral features characteristic of the clean surface.

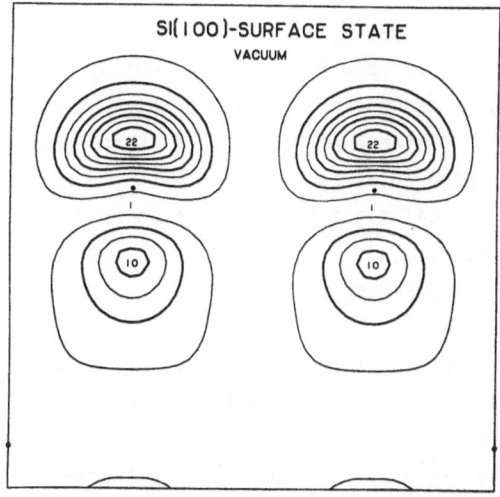

SI(100)-SURFACE STATE
VACUUM

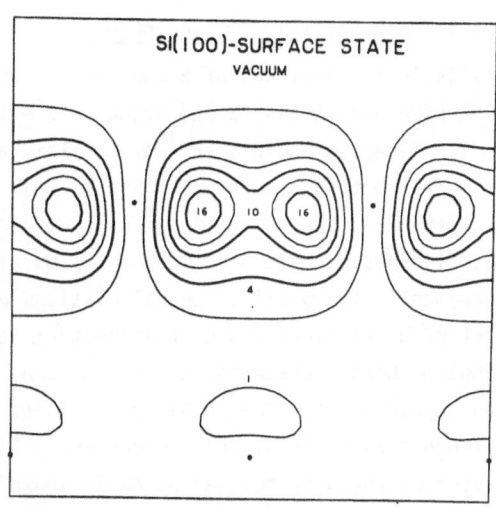

SI(100)-SURFACE STATE
VACUUM

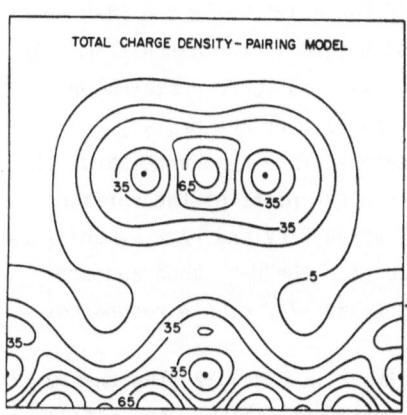

TOTAL CHARGE DENSITY—PAIRING MODEL

Fig.3.9a,b. Contour plots of the charge densities of the "dangling bond" (a) and "bridge" (b) surface states of the Si(100) surface at the $\underline{k}_\shortparallel = K$ (corner) point of the surface Brillouin zone. The plotting plane is normal to the surface and passes through a row of broken bonds. Dots indicate Si atoms in the first and fourth layers, and charge densities are in a.u. $\times 10^3$. [3.45]

Fig.3.10. Contour plot of the total valence charge density of the 2×1 reconstructed Si(100) surface for the dimer model. The plotting plane and units are as in Fig.3.9. [3.61]

Consider briefly the energies of H chemisorption. The Si-H bond strength is 3.2 eV, as inferred from molecular data (SiH$_4$ and Si$_2$H$_6$) [3.53]. This makes the reaction

$$H_2 + Si^* \rightarrow 2HSi^*$$

where Si* is a surface Si atom exothermic (the H$_2$ dissociation energy is 4.74 eV). Experimentally [3.51,52], H$_2$ does not chemisorb on Si with any appreciable probability, presumably due to an activation barrier whose origin is in the large separation between Si* atoms and the very compact nearly rare gas configuration of H$_2$. To experimentally study H chemisorption on Si, predissociation of H$_2$ is necessary. In its atomic form H rapidly chemisorbs on all Si surfaces.

3.6.1 H on Si(111) — Simple Phase

Realistic calculations of H chemisorbed on Si(111) were first carried out by
APPELBAUM and HAMANN [3.48] using SCPP methods described in Sect.3.4. H was assumed
to lie directly above a surface Si atom, at a distance of 2.8 a.u., a choice based
on the Si-H bond length in silane [3.53]. Subsequently calculations with the same
geometry have been carried out using the empirical tight-binding methods by PANDEY
[3.35] and supercell SCPP methods by HO et al. [3.50]. Broadly speaking, there is
agreement between all three calculations and experiment [3.52] concerning the effect of H chemisorption on the spectral features of the surface. H removes the
dangling bond surface state from the gap, suppresses the top of the valence band,
and introduces a strong H-Si spectral feature near -4.5 eV (referenced to the
valence band maximum) and a secondary H-Si structure near -7.0 eV. The density of
states on the H calculated by AH is shown in Fig.3.11; its comparison with experiments by SAKURAI and HAGSTRUM [3.52] is shown in Fig.3.12. While the Si-H spectral features are rather sharp, the charge in the Si-H bond is contained primarily
in a relatively broad (3.0 eV) band of states which exist as surface states only
over a portion of the SBZ. The features at -4.5 eV and -7.0 eV are primary and
secondary Van Hove singularities in this surface-state surface-resonance band. The
spatial character of this band is best seen in the contour plots of the surface
state charge densities obtained by HO et al. [3.50] at -4.5 eV and -7.2 eV and shown
in Fig.3.13. The states at -4.5 are highly localized near the H-Si bond and are
bona fide surface states near the SBZ perimeter, those at -7.2 eV are resonances
and are considerably more delocalized.

The excellent agreement between theory and experiment in Fig.3.12 (the theory
has no adjustable surface parameters and preceded experiment) allays the initial
fear that strong relaxation and correlation effects at a surface that are outside
an effective one-electron theory would make it impossible to compare directly
one-electron calculations with experiment. To the contrary, what we find here is
that when the chemisorbed bonds are strong, as is the case for H on Si(111), the
associated surface energy bands are broad, and reasonable success can be achieved
in explaining surface spectral features. In addition to yielding important spectral
information, the SCCP calculations yield information about work function changes
with chemisorption, tell us what the dynamic effective charge [3.54] on the H is,
and allow the calculation of an equilibrium Si-H separation and bond force constant.
The calculation of AH shows almost no change in ϕ_E (< 0.2 eV) with H chemisorption,
consistent with experiment, and a very small dynamic effective charge (as measured
theoretically by the change in work function with Si-H distance), consistent with
molecular data indicating that Si-H bonds have only a few percent ionic content.

Using the HELLMANN-FEYNMAN theorem [3.55], it was determined that the equilibrium
Si-H separation was 2.7 Bohr and the Si-H force constant was 0.17 a.u. The latter
is within 10 % of experiment [3.56,57] and the former consistent with molecular

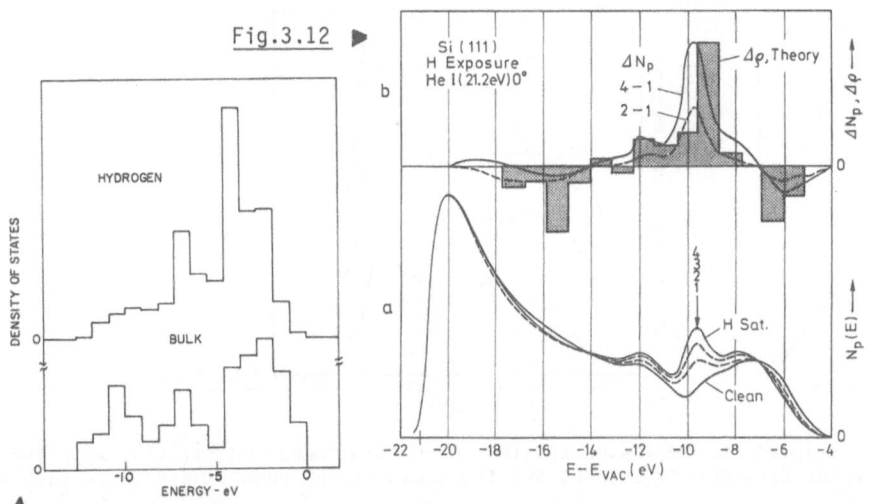

Fig.3.12 ▶

Fig.3.11. Local density of states at H site for H chemisorbed on Si(111), compared with bulk Si density of states. Energy origin is valence band maximum. [3.48]

Fig.3.12. Ultraviolet photoemission spectrum for Si(111) with increasing H coverage from curve 1 to curve 4 and the substrate at ~150° C. The energy origin is vacuum. Difference curves are compared to theoretical results from Fig.3.11 in inset. [3.52]

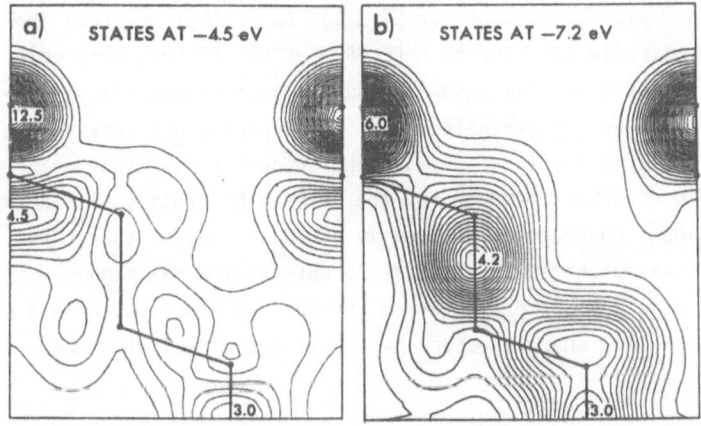

Fig.3.13a,b. Charge density contour plots for zone edge surface state (a), and zone center resonance (b) for H chemisorbed on Si(111). Plotting plane is normal to surface and includes H atoms (uppermost dots) and Si atoms in layers 1 to 4. [3.50]

data [3.53], which indicates a Si-H separation of 2.78 for almost all Si-H molecular prototypes.

3.6.2 H on Si(111) — Disordered

The previous studies were concerned with H chemisorbed as an ordered overlayer. At room temperature, H chemisorbs randomly on the Si(111) surface, achieving an ordered structure at saturation. During presaturation, the spectral signature of the H bond

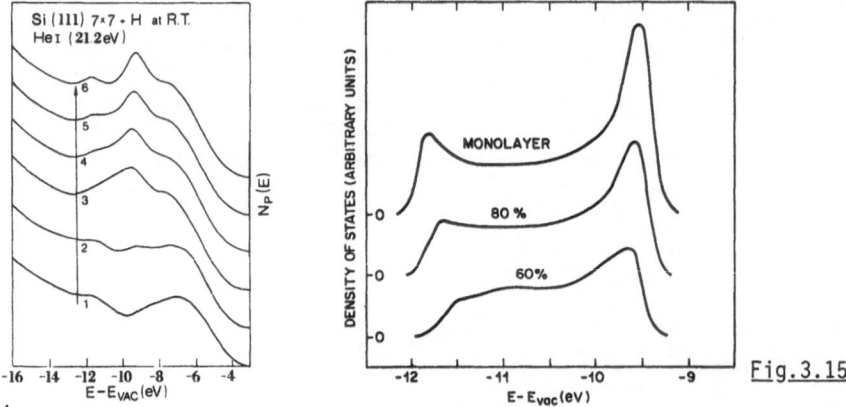

Fig.3.14. Ultraviolet photoemission spectra for H chemisorbed on Si(111), with increasing coverage from curves 1 to 6, and the substrate at room temperature. [3.57]

Fig.3.15. The theoretically calculated two-dimensional density of states of a model
H overlayer on Si(111) is plotted vs energy, referenced to the experimental vacuum
energy E_{VAC}. The three curves, uniformly shifted for clarity, were calculated
assuming random 60%, 80%, and 100% coverage. [3.57]

undergoes considerable evolution, as seen in Fig.3.14. It is not simply related to
the behavior of an isolated H atom chemisorbed on Si(111). An attempt to model the
effect of disordering theoretically was made by APPELBAUM et al. [3.58]. They calculated the density of states of a two-dimensional randomly distributed H layer described by an up to third neighbor tight-binding model whose parameters were chosen
so that in the saturated monolayer limit the primary and secondary spectral
features at -4.5 eV and -7.0 eV were reproduced by the model. The density of states
calculated at 60, 80, and 100 % coverage is plotted in Fig.3.15. Notice that the
spectral features (Van Hove singularities) evolve differently, with the high
energy feature, associated with the SBZ perimeter in the ordered phase, much less
sensitive to disorder than the low energy feature, which is associated with the
SBZ center.

3.6.3 H on Si(111) —Trihydride Phase

H chemisorption proceeds to saturation on a Si(111) 7 × 7 thermodynamically stable
surface. If the Si(111) surface is quenched from ~1000° C a metastable 1 × 1 structure is formed. This surface's behavior with H exposure is similar to the stable
phase at low exposure but with massive exposures to H a radical change in the
nature of the H-Si surface takes place [3.49]. PANDEY et al. have identified this
new structure as the formation of a H_3-Si surface phase made possible by the cor-
rosive removal of the first layer of Si, exposing a Si surface with three broken
bonds capable of forming a silane-like surface structure. Using an empirical tight-
bonding model, with silane and disilane as molecular input, PANDEY obtained excellent

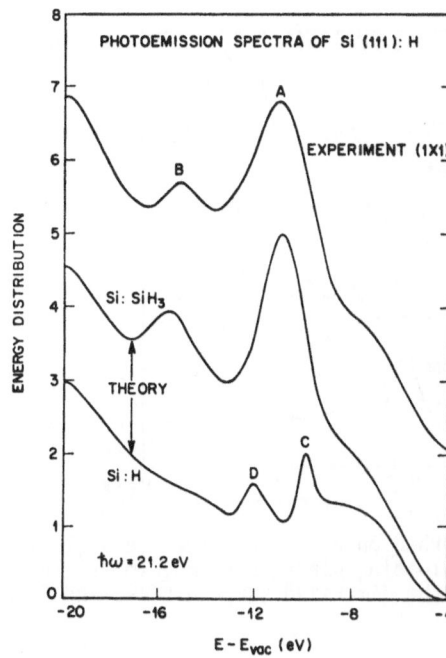

Fig.3.16. Ultraviolet photoemission spectrum for the Si(111)-H₃ "trihydride" (top curve) compared with the theoretical spectrum (center). The top curve is shifted up by 2 units for clarity. The lowest curve shows the theoretical spectrum for the Si(111)-H "monohydride". The theoretical spectra are broadened by a Lorentzian of half width 0.3 eV, and have a smooth background added. [3.58]

agreement between theory based on the trihydride phase and experiment. The comparison of theory and experiment is shown in Fig.3.16.

This model has also been studied by HO et al. using SCPP methods [3.50]. They found qualitatively the same behavior as PANDEY et al. [3.49] but differ with regard to secondary spectral features sensitive to H-H interactions between H atoms on neighboring H_3Si^* complexes.

3.6.4 H Chemisorbed on 2 × 1 Si(100)

A number of theoretical [3.59,60] and experimental [3.51,61,62] studies have been carried out of H interacting with a Si(100) surface. Experimentally Si(100) occurs in a 2 × 1 superlattice, and with exposure to H, either a 2 × 1 or 1 × 1 H saturated surface can form. It is believed [3.61] that the former consists of an underlying dimerized Si substrate with H atoms attached to the dangling bonds of the dimerized surface atoms and the latter consists of two H atoms bonded to each Si^* atom on an ideal Si(100) surface.

Theoretical studies of the 2 × 1 H covered Si surface were carried out by APPELBAUM et al. [3.60] using the SCPP method. In its dimerized form Si(100) surface atoms each have a single broken bond —with their backbonds forming nearly perfect tetrahedral angles with each other. The H atoms were assumed to bind so as to complete the tetrahedral environment of the Si^* atoms at a bond length of 2.8 a.u. . Hellmann-Feynman force studies on the H atoms confirmed this assumption. The total charge density calculated for this surface is shown in Fig.3.17. The Si-H bond

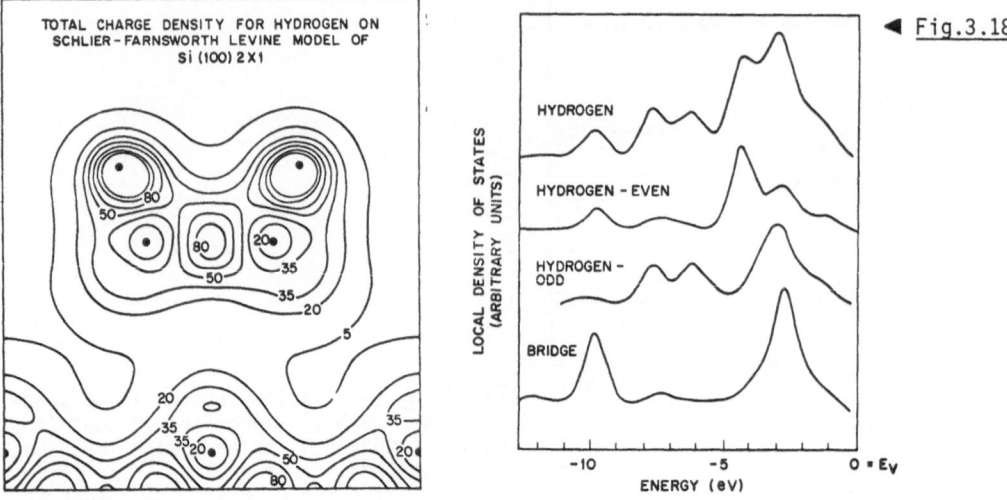

◀ Fig.3.18

▲Fig.3.17. Valence charge density for H chemisorbed on 2 × 1 reconstructed Si(100) assuming the dimer reconstruction model. The plotting plane is normal to the surface, passing through the two H's (top dots), the dimerized surface Si's, and fourth layer Si's. [3.61]

Fig.3.18. Local density of states at the H atom and at the middle of the dimer bond for H chemisorbed on 2 × 2 reconstructed Si(100). Even and odd refer to wave function symmetry with respect to a plane perpendicular to the surface and to the dimer bond. [3.61]

formed here is very similar to that on Si(111), seen in Fig.3.9, although a small polarization of the Si-H bond charges toward the Si[*] dimer can be detected. The dimerization bond appears uneffected by H chemisorption.

The spectral density on the H is plotted in Fig.3.18, as well as its decomposition into even and odd parts. The more complex nature of the H spectrum here comes about from H-H interactions on the same Si dimer doubling or splitting the two features seen for a Si-H bond on Si(111) surfaces at -7.0 and -4.5 eV, and the resonant interaction of the H s and p levels with those of dimer.

Experimentally, hydrogenated Si surfaces do not show this doublet structure. On the contrary, they exhibit strong enhancement near -5.5 eV, uncharacteristic of Si-H bonds. By contrast, hydrogenated Ge surfaces show the doubling expected for them. It appears likely to us that the failure of theory and experiment to agree in the case of Si may be indicative of a more complex H chemisorptive mode on Si than had been appreciated rather than a defect in the theory.

3.6.5 H Interacting with 1 × 1 Si(100)

The bonding of H to an unreconstructed Si(100) surface was studied by APPELBAUM and HAMANN [3.59]. Using SCPP methods they calculated the interaction of a monolayer of H positioned directly above and at varying distances from the Si surface atoms. The clean surface, which has two broken bond derived surface states, one dangling

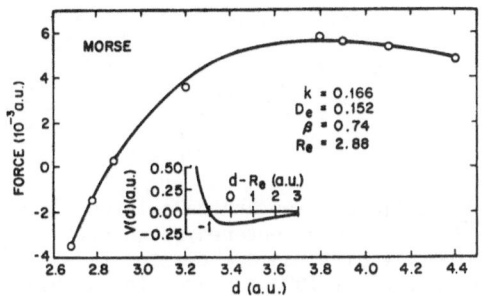

Fig.3.19. Open circles are calculated forces on the H layer for H centered above the surface atoms at distance d on a 1×1 Si(100) surface. The solid curve is a fit to the force using a sum of Morse potentials, and the inset shows the fitted potential $V = D_e\{\exp[-\beta(d - R_e)] -2\}\exp[-\beta(d - R_e)]$. K is the equilibrium force constant. [3.60]

bond-like, containing almost two electrons and one bridge-like and nearly empty [3.45], bonds the H via the dangling bond orbital. This requires the promotion of an electron into the bridge surface state band.

The bond of the H to Si has a similar bond length and force constant to that on Si(111), in spite of the fact that H bonding does not saturate the broken bonds of the Si* atom. The interaction potential of the H with the Si substrate over a wide range of Si-H separations was found to obey a simple diatom force law, as can be seen in Fig.3.19. This allows one to infer a dissociation limit for the effective Si-H surface bond which is within 0.4 eV of experiment.

The relatively simple interaction potential found where the H monolayer was positioned directly above the Si surface atoms should be contrasted with that recently found by APPELBAUM et al. [3.63] for the case in which H was above second layer Si atoms. The H atoms, relatively distant from the dangling bond arbitals and in a plane at right angles to the bridge bonds, interact weakly and via polarization forces at moderate distances from the surface. Similar behavior was found for H binding to the Al(100) [3.64] surface over the second layer.

At close approach (d < 2.2 a.u.) the H-Si interaction force reaches a plateau, then increases dramatically with decreasing Si-H separation. The force peaks near 1.0 a.u. and decreases rapidly to zero near d = 0.25 a.u. This true equilibrium places the H 2.76 a.u. above the second layer Si atoms, close to the covalent bond length of a Si-H bond. A plot of the force law experienced by the H between d = 0.2 and d = 4.0 is given in Fig.3.20. Superimposed on this figure is the Morse potential interaction found in the previous study of H interacting directly with the surface Si atoms, shifted only so its origin coincides with the second layer Si atoms. Near equilibrium, the slope of the dashed curve and a line through the calculated points are nearly the same, but at further distance the two depart significantly and imply a bond strength for H bond to second layer Si atoms 1.2 eV weaker than to surface Si atoms.

The nature of this unorthodox bond (the second layer Si atoms are saturated and should not be expected to bond additional H atoms) can be seen in Fig.3.21, where the total charge density is contour plotted at three Si-H separations. For large distance the charge density looks like a superposition of free H atoms and clean Si(100) 1×1 surface charge densities. By contrast, for d = 0.2, a multicentered

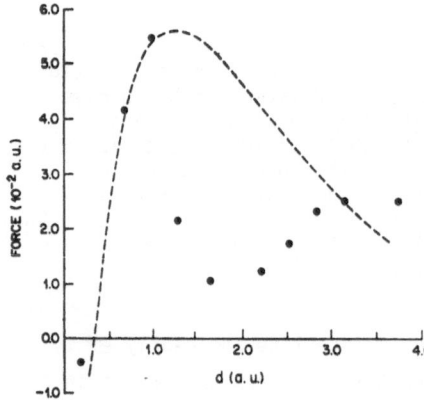

Fig.3.20. Solid circles are calculated forces on the H layer for H centered above second layer atoms at distance d above the first layer on a 1×1 Si(100) surface. The dashed curve is the Morse potential fit previously obtained for H interacting with surface Si atoms, Fig.3.19, but with its origin referenced to the second layer. [3.61]

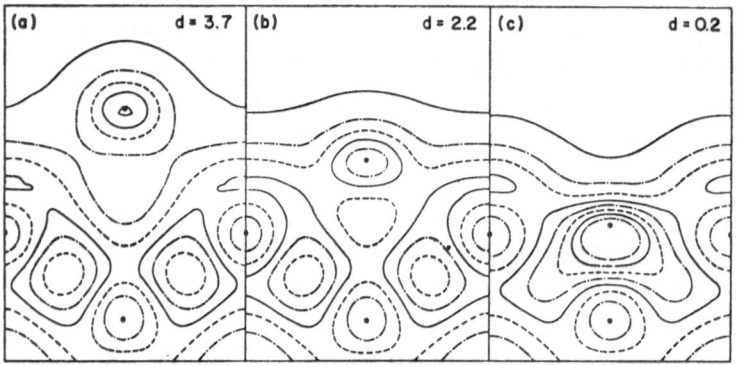

Fig.3.21a-c. Charge density contour plots for H centered above second layer Si atom on Si($\bar{1}$00) surface. Top center dot is H, edge dots are first layer Si, and bottom center dots are second layer Si. d is the H distance above the first Si layer, and the contours are spaced 1.5×10^{-3} a.u. apart, starting at 5×10^{-3} a.u. in the vacuum and alternating solid to dot-dashed to dashed. The "three-center" bond formed at equilibrium is seen in (c). [3.61]

bond is clearly seen which is analogous to those found in the boron hydrides. The formation of this bond weakens the Si* backbonds (there is 20% less charge in the backbonds), which should lead to their expansion. With additional exposure to H this weak bond could rupture and H induced corrosion of the surface take place.

3.7 Cl Chemisorption on Si(111) and Ge(111)

Chemisorption of Cl to saturation on either Si(111) or Ge(111) results in an ordered 1×1 Cl monolayer. This system has been studied extensively by a combination of theoretical and experimental tools involving angular and polarization dependent photoemission, self-consistent pseudopotential and empirical tight-binding calculations [3.65-67].

One of the primary objectives of these studies was to determine the local co-ordination of the Cl on the Si or Ge substrate from a study of its electronic structure. Two kinds of chemisorbed sites are possible, one directly above the sur-face atom, as was the case for H on Si(111), and two in threefold coordinated sites. One of these threefold sites would place the Cl above the second layer Si or Ge atoms, the other in a void. Self-consistent calculations for the onefold site and the second threefold void site were carried out for Si(111). The density of states found for these two cases is shown in Fig.3.22. The most prominent feature for both the onefold or threefold site is due to lone pair orbitals; p_x, p_y in the case of the onefold site, p_x, p_y, p_z for the case of the threefold site. Absent from the spectrum of the threefold site are the features marked A and B in Fig.3.22 and associated with σ bonding. This bonding proceeds via the Cl p_z and Si s-p band, with peak A having more Si-s admixture and peak B more Si-p_z admixture. Similar results are expected for theoretical calculations in-volving Ge(111) as a substrate, considering how similar the Si and Ge valence bands are. Experimentally, Si and Ge behave differently, with the photoemission spectrum of Cl on Si(111) indicative of a onefold bond site, and that of Cl on Ge(111) indicative of a threefold site. This is consistent with the angular de-pendence of the spectral features associated with the threefold and onefold sites.

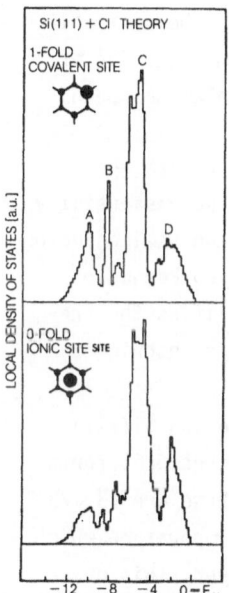

Fig.3.22. Calculated local density of states histograms for a Si(111) surface region containing one Si substrate layer and one Cl adsorbate layer. The upper curve corresponds to the onefold covalent site chemisorption bond geometry, where-as the lower curve shows the results for a threefold ionic site geometry. [3.65]

3.8 Al Chemisorption on Si(111)

Interfaces of semiconductors and metals have received a great deal of attention
in experimental studies because of their technological importance in electronic
devices. The formation of rectifying Schottky barriers at these interfaces is re-
gulated by one key parameter—the position at which the Fermi level is pinned relative to the semiconductor bands. This is in general different from the position
at which the Fermi level is pinned at the clean semiconductor surface (if it is in
fact pinned), and depends in a poorly understood way on properties of the metal
and the semiconductor.

The earliest attempt to explain Schottky barrier effects, by BARDEEN in 1947,
proposed Fermi level pinning by intrinsic surface states [3.68]. It was later
pointed out by HEINE [3.69] that this mechanism could not be literally correct,
since semiconductor surface states could couple into the continuum of states on
the metal side of the interface. Therefore at best these would become broadened
resonances. The first detailed calculation of such interfaces was carried out by
LOUIE et al. using the self-consistent pseudopotential method in a repeated slab
geometry, and replacing the metal ion cores by a uniform positive background
(jellium) [3.70]. Pinning trends were reasonably reproduced, and a qualitative
confirmation of Heine's proposal given, since tails of the metal wave functions
decaying as evanescent waves into the semiconductor gap play a major role in determining the interface potential. While the jellium model avoided the difficult
theoretical problem of lattice mismatch, it failed to deal with the question of
the interface chemical bonds, whose importance to the barrier problem has been
pointed out by PHILLIPS [3.71].

A new avenue of approach to the semiconductor-metal interface problem opened
with the experimental observation that Schottky barrier formation was essentially
complete with the formation of 1 to 2 monolayers of metal on a clean semiconductor
substrate [3.72-74]. This implies that a chemisorbed monolayer is an adequate
model for the essential physics of the interface, which both simplifies the theoretical treatment of this problem and allows surface spectroscopic techniques
to probe details of electronic structure.

The prototype semiconductor-chemisorbed metal system has been Al on Si(111).
This system was found by LANDER [3.4b] to exist in a variety of structural forms
depending on coverage and thermal treatment. The simplest among these are $\sqrt{3} \times \sqrt{3}$
structure at 1/3 monolayer coverage, and a phase formed at high temperature which
is interpreted as Al substitutionally replacing the surface Si layer. This phase
is reconstructed in a 5×5 pattern at room temperature, but this is probably only
a weak distortion of the basic substitutional structure.

The first theoretical study of Al chemisorption on Si(111) was reported by
APPELBAUM and HAMANN in 1974 [3.37]. The Lander substitutional model was chosen

for investigation, the 5 × 5 structure was ignored, and the Si-Al bond length was fixed by summing the Si covalent radius and the Al metallic radius of 1.25 Å. Since the Al is threefold coordinated and each Si fourfold coordinated, it was anticipated that a filled band (or saturated bonding) situation, and hence a semi-conducting surface might result. The calculations, carried out using the semi-infinite self-consistent pseudopotential method, gave quite a different result. Three bands of surface states were found, ranging from 10.3 eV below the valence band maximum (E_v) to 0.85 eV above E_v. Two bands are partially occupied, and the surface is thus metallic. One of these bands is narrow, lies within the absolute gap, and pins the Fermi level near the conduction band minimum (E_c). The others are quite broad and cut through bulk bands in some parts of the surface Brillouin zone.

The valence charge density for this surface is shown in Fig.3.23. The charge around the Al is significantly more spread out than around the Si, and the Si-Al bond charge is polarized toward the Si, consistent with the electronegativity difference. Neighboring Al can apparently communicate through their common Si neighbor to produce a two-dimensional metal, despite the fact that they are 3.8 Å apart, compared to 2.5 Å in Al metal.

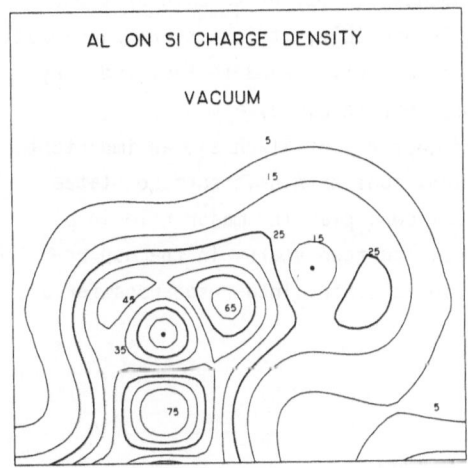

Fig.3.23. Charge density contour plot for Al substitutionally chemisorbed on Si(111). The top dot is Al in the first layer, the lower dot is Si in the second layer, and the third layer Si is directly below but just off the plotting area. Charge densities are a.u. × 10^3. [3.37]

Calculations on this model were repeated by CHELIKOWSKY, who used the periodic slab SCCP method, and shortened the Al bond radius to 1.17 Å, equal to that of Si, on the basis that nearly this much bond shortening would be predicted from ionicity corrections [3.75]. The results are in substantial agreement with the AH results, and considerable detail is given on the surface states and spectrum.

Predictions of the above calculations have not been directly tested on an Si-Al surface prepared according to Lander's specifications for the substitutional phase. Detailed recent experimental work was carried out using Al layers chemisorbed on

room temperature 7 × 7 Si(111) surfaces [3.72]. The detailed structure of this reconstructed surface is unknown, and, even if it were, a full calculation would be
prohibitive. To cover a variety of local bonding geometries other than the substitutional one which might exist on this surface, ZHANG and SCHLUTER considered
three additional geometries [3.76]. In the first or "covalent" geometry, the Al
was assumed to sit above the surface Si at the Si-Si bond length. In the second
and third "ionic" geometries, the Al was placed above the center of the six-membered
ring formed by the first and second layer Si, out of direct line of the broken
bonds. The Al-Si bond length is set equal to the Si-Si bond length in the second
case, and to 1.25 times this value in the third.

The electronic structure was studied using the periodic slab SCPP method, and
the same pseudopotentials as in [3.75], making the calculations on all four geometries exactly parallel. The three geometries treated here also all show several
bands of surface states in the gap, and a metallic surface in cases I and II, as
shown in Fig.3.24. The Fermi level is pinned near E_C in cases II and III, but near
E_V in case I, in contradiction with experiment. Comparison of experimentally observed spectral features with computed surface densities of states for all four
geometries indicates that the substitutional and shorter bond length "ionic" sites
give the best agreement. If one assumes that the vacancy model [3.5] may be correct
for the 7 × 7 surface, sites with local environments equivalent to both of these
would be available. While this interpretation is possible, it is rather roundabout,
and spectra taken on a simpler surface phase are clearly needed to test the very
thorough theoretical treatment that this surface has received.

Independent of the final details of bonding geometry of Al on Si, an important
fact emerges pertaining to Schottky barrier formation: extrinsic surface states
associated with the semiconductor-metal chemical bond play the major role in pinning the Fermi level. No vestige of the intrinsic surface states in the gap remains, and the ultimate broadening of the extrinsic states into resonances for a
thick metal is apparently inconsequential.

3.9 Al Chemisorption on GaAs(110)

The (110) surface of the compound semiconductor GaAs presents a distinctly different type of substrate from the elemental semiconductor surfaces considered
previously. There are two broken bonds per surface unit cell, but they are chemically distinct, one being associated with the Ga cation and the other with the As
anion. The clean surface does not reconstruct, simplifying the interpretation of
chemisorption geometries.

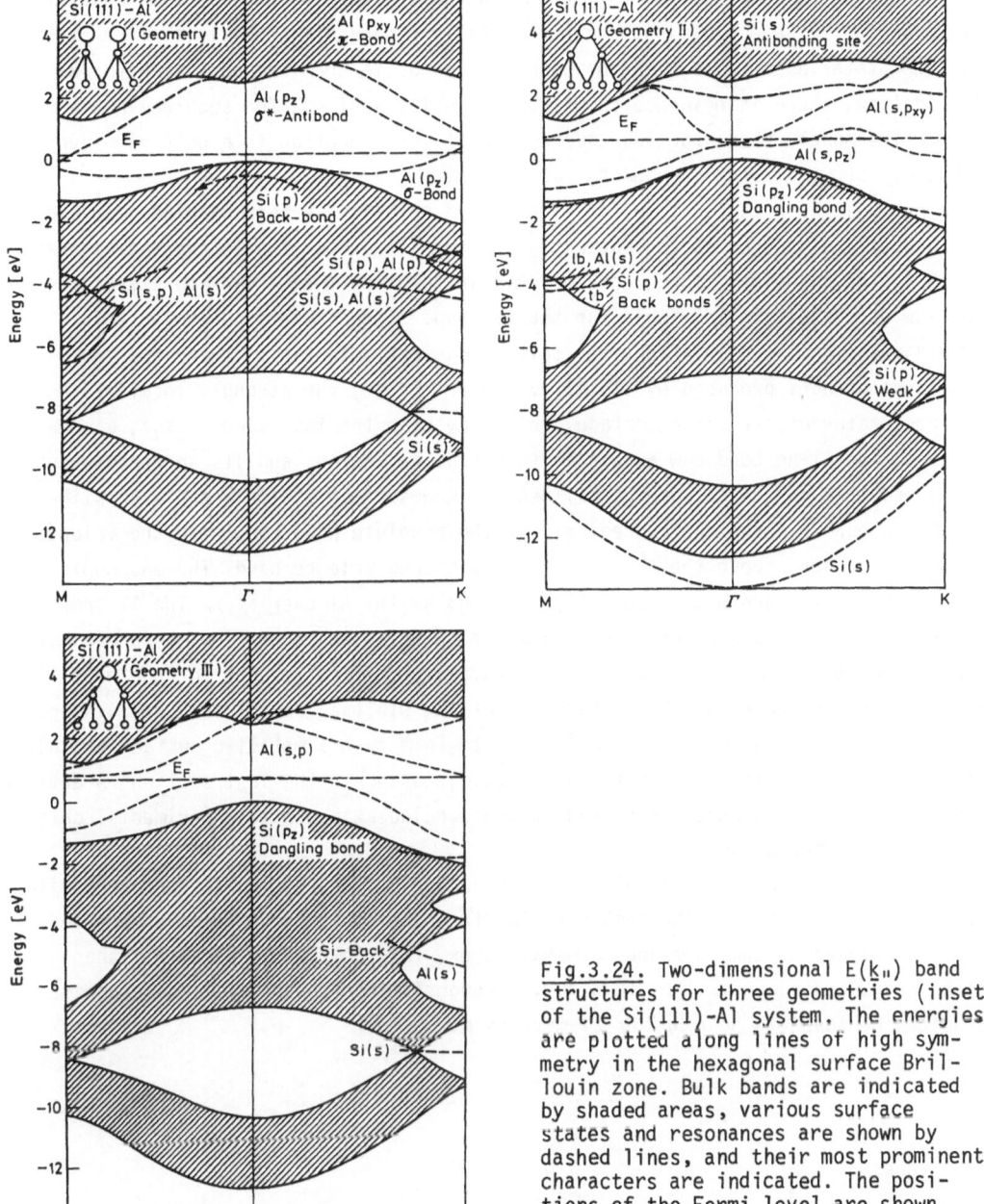

Fig.3.24. Two-dimensional $E(\underline{k}_{\shortparallel})$ band structures for three geometries (insets) of the Si(111)-Al system. The energies are plotted along lines of high symmetry in the hexagonal surface Brillouin zone. Bulk bands are indicated by shaded areas, various surface states and resonances are shown by dashed lines, and their most prominent characters are indicated. The positions of the Fermi level are shown. [3.76]

The study of metal chemisorption on this surface is of particular interest because the Bardeen mechanism for Schottky barriers [3.68] is ruled out, it having recently been shown that there are no surface states in the thermal gap for the clean surface [3.77]. Periodic SCPP calculations for the clean surface indicate that the gap is free of surface states only if the surface relaxes, with the As atoms moving above the ideal surface plane and the Ga atoms below [3.43]. Such relaxation was proposed to help explain the experimentally observed insensitivity of empty surface states associated with the Ga broken bonds to chemisorbed layers [3.78]. In addition, there is evidence from LEED intensity analysis for such relaxation [3.6]. The model, then, for low-coverage metal chemisorption is a half-monolayer structure of metal atoms bonded to surface As.

A periodic SCPP calculation for Al adsorbed on GaAs has been carried out using the geometry discussed above, but with the simplification of omitting the relaxation [3.79]. (The principal consequence of this is to place the Ga surface state band too low so that it falls within the thermal gap. Since it remains empty, this should not significantly affect the self-consistent potential.) There are two prominent effects produced by the Al overlayer. First, the strongly localized As surface states of the clean surface, which lay near the top valence band, are removed. The Al-As bond now exists in this spatial region, and its spectral weight is spread through the valence bands. Second, new surface states associated with the Al appear in three energy regions: in the absolute band gap, near the valence band maximum, and above the threshold of the second valence band. The unoccupied Ga surface states are not appreciably affected by the Al overlayer. The Al induced gap surface states are surprisingly metallic in their character, being spread along the Al chains with a density which is essentially uniform (except for the node at the atoms necessitated by the p-character of the states). The Al are over 4 Å apart on this surface, and the ability of the Al to form such a metallic surface at this distance parallels the results for Al on Si discussed in the last section. A detailed comparison of the spectrum predicted in these calculations with experimental data has not been carried out.

The surface Fermi level is pinned 0.5 eV above E_v, in reasonable agreement with Schottky barrier values. The mechanism for the pinning is the extrinsic surface states produced by the chemisorbed metal layer. This completely parallels the Si-Al case, and suggests that this may, through further theoretical and experimental studies, be found to be a universal mechanism.

References

3.1   J.A. Appelbaum, D.R. Hamann: Rev. Mod. Phys. *48*, 479 (1976)
3.2   N.D. Lang: "Local Density Formalism and the Electronic Structure of Metal Surfaces", in *Solid State Physics*, Vol.28, ed. by F. Seitz, D. Turnbull, H. Ehrenreich (Academic Press, New York 1973) p.225
3.3   D.C. Langreth, J.P. Perdew: Phys. Rev. B*15*, 2884 (1977)
3.4   J.J. Lander, J. Morrison: Surf. Sci. *2*, 553 (1964)
      J.J. Lander: Prog. Solid State Chem. *2*, 26 (1965)
3.5   H. Ibach (ed.): *Electron Spectroscopy for Surface Analysis*, Topics in Current Physics, Vol.4 (Springer, Berlin, Heidelberg, New York 1977)

3.6   C.B. Duke: "Atomic Geometry of Semiconductor Surfaces", to appear in *Crit. Rev. Solid State Sci.*, Proc. 3rd Intern. Summer Institute in Surface Science, Milwaukee, Wisconsin, August 22-26, 1977

3.7   J.A. Strozier, Jr., D.W. Jepsen, F. Jona: "Surface Crystallography", in *Surface Physics of Materials VI*, ed. by J.M. Blakely (Academic Press, New York 1975) pp.2-78

3.8   D.E. Eastman, M.I. Nathan: Phys. Today *28*, 4 (1975)

3.9   L. Ley, M. Cardona (eds.): *Photoemission in Solids II*, Topics in Applied Physics, Vol.27 (Springer, Berlin, Heidelberg, New York 1979) Chap.2

3.10 J.A. Appelbaum, D.R. Hamann: CRC Crit. Rev. Solid State Phys. *6*, 357 (1976)

3.11 J.C. Phillips: *Bonds and Bands in Semiconductors* (Academic Press, New York 1973)

3.12 J.R. Chelikowsky, M.L. Cohen: Phys. Rev. Lett. *33*, 1339 (1974); Phys. Rev. B*10*, 5095 (1974)

3.13 K.C. Pandey, J.C. Phillips: Phys. Rev. B*9*, 1552 (1974)

3.14 M.L. Cohen, V. Heine: "The Fitting of Pseudopotentials to Experimental Data and Their Subsequent Application", in *Solid State Physics*, Vol.24, ed. by H. Ehrenreich, F. Seitz, D. Turnbull (Academic Press, New York 1970) p.38

3.15 G. Dresselhaus, M.S. Dresselhaus: Phys. Rev. *160*, 649 (1967)

3.16 K.C. Pandey, J.C. Phillips: Phys. Rev. B*13*, 750 (1976)

3.17 W. Monch: "On the Physics of Clean Semiconductor Surfaces", in *Advances in Solid State Physics*, Vol.13, ed. by H.J. Queisser (Pergamon Press, Oxford 1973) p.241

3.18 J.C. Riviere: In *Solid State Surface Science*, ed. by M. Green (Dekker, New York 1970)

3.19 J.A. Appelbaum, D.R. Hamann: Phys. Rev. Lett. *32*, 225 (1974)

3.20 J.A. Appelbaum: "Electronic Structure of Solid Surfaces", in *Surface Physics of Materials*, Vol.1, ed. by J.M. Blakely (Academic Press, New York 1975) pp.79-119

3.21 A.A. Maradudin, I.P. Ipatova, E.W. Montroll, G.H. Weiss: *Theory of Lattic Dynamics in the Harmonic Approximation*, 2nd ed. (Academic Press, New York 1971)

3.22 J.A. Appelbaum, D.R. Hamann: Phys. Rev. B*6*, 2166 (1972)

3.23 J.A. Appelbaum, D.R. Hamann: Phys. Rev. B*8*, 1773 (1973)

3.24 J.P. Alldredge, L. Kleinman: Phys. Rev. B*10*, 559 (1974)

3.25 M. Schluter, J.R. Chelikowsky, S.G. Louie, M.L. Cohen: Phys. Rev. B*12*, 4200 (1975)

3.26 J.C. Slater, G.F. Koster: Phys. Rev. *14*, 1408 (1954)

3.27 K. Hirabayashii: J. Phys. Soc. Jpn. *27*, 1475 (1969)

3.28 K.C. Pandey, J.C. Phillips: Phys. Rev. Lett. *32*, 1433 (1974)

3.29 D.J. Chadi, M.L. Cohen: Phys. Status Solidi (b) *68*, 405 (1975)

3.30 I.P. Batra, S. Ciraci: Phys. Rev. Lett. *34*, 1337 (1975)

3.31 P.C. Hohenberg, W. Kohn: Phys. Rev. *136*, B864 (1964)
      W. Kohn, L.J. Sham: Phys. Rev. *140*, A1133 (1965)

3.32 J.C. Slater: *The Self-Consistent Field for Molecules and Solids* (McGraw-Hill, New York 1974)

3.33 M. von Laue: Phys. Rev. *37*, 53 (1931)

3.34 M. Schluter, J.R. Chelikowsky, S.G. Louie, M.L. Cohen: Phys. Rev. Lett. *34*, 1385 (1975)

3.35 K.C. Pandey: Phys. Rev. B*14*, 1557 (1976)

3.36 J.A. Appelbaum, D.R. Hamann: Phys. Rev. Lett. *31*, 106 (1973)

3.37 J.A. Appelbaum, D.R. Hamann: *Proc. 12th Intern. Conf. on the Physics of Semiconductors* (Teubner, Stuttgart 1974) p.675

3.38 J.R. Chelikowsky: Phys. Rev. B*15*, 3236 (1977)

3.39 D. Haneman: Phys. Rev. *121*, 1093 (1961)

3.40 J.A. Appelbaum, D.R. Hamann: Phys. Rev. B*12*, 1410 (1975)
      K.C. Pandey, J.C. Phillips: Phys. Rev. Lett. *34*, 1450 (1975)
      M. Schluter, J.R. Chelikowsky, S.G. Louie, M.L. Cohen: Phys. Rev. Lett. *34*, 1385 (1975)

3.41 J.E. Florio, W.D. Robertson: Surf. Sci. *24*, 17 (1971)
      P. Mark, J.D. Levine, S.M. McFarlane: Phys. Rev. Lett. *38*, 1408 (1977)

3.42 J.D. Joannopoulos, M.L. Cohen: Phys. Rev. B10, 5075 (1974)
      K.C. Pandey, J.L. Freeouf, D.E. Eastman: J. Vac. Sci. Technol. 14, 904 (1977)
3.43 J.R. Chelikowsky, M.L. Cohen: Phys. Rev. B13, 826 (1976); Phys. Rev. B14,
      4724 (1976)
3.44 D.E. Eastman, J.L. Freeouf: Phys. Rev. Lett. 33, 1601 (1974)
3.45 J.A. Appelbaum, G.A. Baraff, D.R. Hamann: Phys. Rev. B11, 3822 (1975); 12,
      5749 (1975)
3.46 R.E. Schlier, H.F. Farnsworth: J. Chem. Phys. 30, 917 (1959)
3.47 J.A. Appelbaum, G.A. Baraff, D.R. Hamann: Phys. Rev. Lett. 35, 729 (1975);
      Phys. Rev. B, 588 (1976)
3.48 J.A. Appelbaum, D.R. Hamann: Phys. Rev. Lett. 34, 806 (1975)
3.49 K.C. Pandey, T. Sakurai, H.D. Hagstrum: Phys. Rev. Lett. 35, 1728 (1975)
3.50 K.M. Ho, M. Schluter, M.L. Cohen: Phys. Rev. B15, 3888 (1977)
3.51 H. Ibach, J.E. Rowe: Surf. Sci. 43, 481 (1974)
3.52 T. Sakurai, H.D. Hagstrum: Phys. Rev. B12, 5349 (1975)
3.53 R.T. Sanderson: *Chemical Bonds and Bond Energy* (Academic Press, New York 1971)
3.54 S.E. Trullinger, S.L. Cunningham: Phys. Rev. B8, 2622 (1973)
      E. Evans, D.L. Mills: Phys. Rev. B5, 4126 (1972)
      S.L. Cunningham, A.A. Maradudin: Phys. Rev. B7, 3870 (1973)
3.55 H. Hellmann: *Einführung in die Quanten Theorie* (Franz Deuticke, Leipzig 1937)
      p.285
      R.P. Feynman: Phys. Rev. 56, 340 (1939)
3.56 G.E. Becker, G.W. Goebeli: J. Chem. Phys. 38, 2942 (1963)
3.57 H. Froitzheim, H. Ibach, S. Lehwald: Phys. Lett. A55, 247 (1975)
3.58 J.A. Appelbaum, H.D. Hagstrum, D.R. Hamann, T. Sakurai: Surf. Sci. 58, 479
      (1976)
3.59 J.A. Appelbaum, D.R. Hamann: Phys. Rev. B15, 2006 (1977)
3.60 J.A. Appelbaum, G.A. Baraff, D.R. Hamann, H.D. Hagstrum, T. Sakurai: Surf.
      Sci. 70, 654 (1978)
3.61 T. Sakurai, H.D. Hagstrum: Phys. Rev. B14, 1593 (1976)
3.62 S.J. White, D.P. Woodruff: Surf. Sci. 63, 254 (1977)
3.63 J.A. Appelbaum, D.R. Hamann, K. Tasso: Phys. Rev. Lett. 34, 1487 (1977)
3.64 O. Gunnarson, H. Hjelmberg, B.I. Lundqvist: Phys. Rev. Lett. 37, 292 (1976)
3.65 M. Schluter, J.E. Rowe, G. Margaritondo, K.M. Ho, M.L. Cohen: Phys. Rev. Lett.
      37, 1632 (1976)
3.66 K.C. Pandey, T. Sakurai, H.D. Hagstrum: Phys. Rev. B16, 3648 (1977)
3.67 P.K. Larsen, N.V. Smith, M. Schluter, H.H. Farrell, K.N. Ho, M.L. Cohen:
      Phys. Rev. B17, 2612 (1978)
3.68 J. Bardeen: Phys. Rev. 71, 717 (1947)
3.69 V. Heine: Phys. Rev. 138, A1689 (1965)
3.70 S.G. Louie, M.L. Cohen: Phys. Rev. Lett. 35, 866 (1975); Phys. Rev. B13,
      2461 (1975)
      S.G. Louie, J.R. Chelikowsky, M.L. Cohen: J. Vac. Sci. Technol. 13, 790 (1976)
3.71 J.C. Phillips: J. Vac. Sci. Technol. 11, 947 (1974)
3.72 J.E. Rowe, S.B. Christman, G. Margaritondo: Phys. Rev. Lett. 35, 1471 (1975);
      Phys. Rev. B14, 5396 (1976); Phys. Rev. B15, 2195 (1977); J. Vac. Sci. Technol.
      13, 329 (1976)
3.73 D.E. Eastmann, J.L. Freeouf: Phys. Rev. Lett. 34, 1624 (1975)
3.74 P.W. Chye, I.A. Bababola, T. Sukegawa, W.E. Spicer: Phys. Rev. Lett. 35, 1602
      (1975)
      W.E. Spicer, P.E. Gregory, P.W. Chye, I.A. Bababola, T. Sukegawa: Appl. Phys.
      Lett. 27, 617 (1975)
      P.E. Gregory, W.E. Spicer: Phys. Rev. B12, 2370 (1975)
3.75 J.R. Chelikowsky: Phys. Rev. B16, 3618 (1977)
3.76 H.I. Zhang, M. Schluter: To be published
3.77 A. Huijer, J. Van Laar: Surf. Sci. 202 (1975)
3.78 J.E. Rowe, S.B. Christman, G. Margaritondo: Phys. Rev. Lett. 35, 1471 (1975)
3.79 J.R. Chelikowsky, S.G. Louie, M.L. Cohen: Solid State Commun. 20, 641 (1976)

4. Chemisorption on d-Band Metals

F. J. Arlinghaus, J. G. Gay, and J. R. Smith

With 35 Figures

This chapter is devoted to a discussion of the surface properties of the noble metals and transition metals. There is far and away more experimental data on the surfaces of these metals than on any other class of solid surfaces. The reasons primarily are that their surfaces are relatively easy to clean and they are very important technologically. With the recent advent of angular photoemission, experiment is well ahead of theory in studies of the surface electronic structure of these metals.

These experimental advances have recently sparked considerable interest among theorists to deal with noble and transition metal surfaces. Commensurate with the experimental interest, the surfaces of these metals are receiving more theoretical attention than any other. There are now a growing number of instances where theoretical surface band structure is being used to explain angular photoemission data on clean and adsorbate covered surfaces. Very recently, a surface state band was predicted prior to being isolated experimentally.

It would seem then that theoretical studies of transition and noble metal surfaces are in an accelerated growth stage, and are on the verge of being able to explain a large body of experimental data. Ultimately a predictive capability is desirable, and early indications are that it may be within our grasp.

We have given some reasons for why experiment has forged ahead in this field. There are also reasons why the theory of these surfaces has lagged behind that of some other material classes.

The electronic structure of d-band metals is a field apart from that of the simple metals or elemental semiconductors. They are more complex and of course more interesting than the simple metals or elemental semiconductors; and if their electronic structure is complicated, that of their surfaces is more so, and their chemisorbtive properties far more so still.

These difficulties arise from the unique character of the d electrons themselves, whose intermediate nature is not shared by any other entity in the whole periodic table. In general, electrons in solids may be easily divided into core electrons and valence or bonding electrons. The core electrons are tightly bound to their nuclei and are little affected by the presence of the other atoms in the solid. They are so well localized near the nucleus that they essentially do not overlap the

electrons of other atoms and do not notice their presence. In most calculations
they are easily and routinely ignored, orthogonalized out of the problem or "frozen",
and of little interest.

The valence electrons, usually s and p electrons of the outermost shell, behave
just the opposite. They overlap so strongly from one atom to another as to lose
any real identification with any specific atom. As a consequence, it becomes a good
description to treat them as "itinerant", as free-running plane waves traversing
the whole solid. The "jellium" model, the "nearly-free-electron" approximation, and
all the array of pseudopotentials are all very successful but essentially simple
concepts depending on this delocalized character.

This desirable situation deteriorates in the case of d electrons. Their wave
functions have maxima well inside the atom and decay rapidly further out; what is
more, the changes in the central region of the atom on putting the atom into a
solid are perceptible but small. Clearly this is rather core-electron-like behavior.
However, the tails of the d electron wave functions extend a long way out; the
tails on neighboring atoms overlap considerably and are considerably modified on
going from isolated atom to solid. This is valence-electron behavior.

This intermediate behavior precludes treating a d electron as either core or
valence and thus complicates matters. The d electrons of course cannot be ignored,
but neither can they be treated by simple approximations. The "jellium" model
presumes a uniformity which does not obtain, since the d electrons have appreciable
localized character. The band structure no longer even remotely resembles "nearly-
free-electron" modified parabolas, and simple pseudopotentials must be augmented
by a considerable complication to account for d resonances. Thus full scale treatments like APW and KKR are necessary. And for the surface case, even more elaborate
development becomes necessary.

But even worse, most d-band metals are characterized by a partially filled
d shell; the extent of that filling changes between atom and solid, and requires
some kind of self-consistent calculation to get the correct occupancy. For the surface case, further charge rearrangement is possible, making self-consistency even
more important. And finally chemisorption, the subject of this article, introduces
the severest problem in allowing charge transfer between substrate and adsorbate.

From all this we may conclude that chemisorption on d-band metals will strain
our computational resources to the utmost, and that self-consistency will probably
be the most important requirement (and a severe burden) for an accurate treatment
of chemisorption at the surface of a d-band metal.

4.1 Moments Method

We begin by discussing some fairly simple computational approaches to the chemisorption problem. First of these is a method which avoids detailed calculation of the individual eigenstates of the system.

The so-called "moments method" [4.1] relies on a mathematical artifice for determining densities of states. The density of states can be expanded in terms of its moments. The mth moment of the density of states n(E) is defined as

$$\mu_m = \int n(E)E^m dE \quad , \tag{4.1}$$

μ_0 gives the total charge, μ_1 the shift, and μ_2 the width of the density distribution. Within the tight-binding approximation, it is possible to calculate these moments of n(E) without ever calculating the eigenstates themselves. The moments are a function of the overlap integrals and certain correlation functions which characterize the local order in the system. Disordered systems can be calculated as well as ordered systems. Various complicated mathematical techniques have been used to improve the convergence of the method, in particular a continued fraction technique [4.2].

The method has been used to study the surface density of states at low index surfaces of fcc, hcp, and bcc d-band metals by HAYDOCK and KELLY [4.3]. Figure 4.1 shows local density of states (LDOS) curves for the (111) surface of an fcc d-band metal. It has been applied to the calculation of the surface states of Mo and W by DESJONQUÈRES and CYROT-LACKMAN [4.4]. They found a surface state on Mo(100) and W(100), but not on the respective (110) surfaces. The results are shown to be independent of spin-orbit coupling, in contrast with the work of STURM and FEDER [4.5]. The results do agree with experiment.

The method has also been used to calculate surface densities of states for cleaved transition metals [4.6] and to calculate the electronic structure at the surface of ferromagnetic Ni and Fe [4.7]. In the latter case, no large variation of magnetic moment near the surface is found.

The method has been used to calculate the binding energy of a single 5d transition metal atom on various planes of bcc and fcc 5d metals [4.8]. This was a much simplified calculation in which only the first two moments of the density of states were used, due to the numerical complexity of this case. The curve of binding energy versus atomic number for 5d atoms on the (100) and (111) planes of bcc tungsten agrees well with the experimental results of PLUMMER and RHODIN [4.9].

The principal drawback to the method appears to be the necessity of computing very high order moments in order to obtain reasonable accuracy. In spite of its elegance, the method therefore seems unlikely to be able to compete with developing methods for surface calculations. In particular, it does not seem useful for

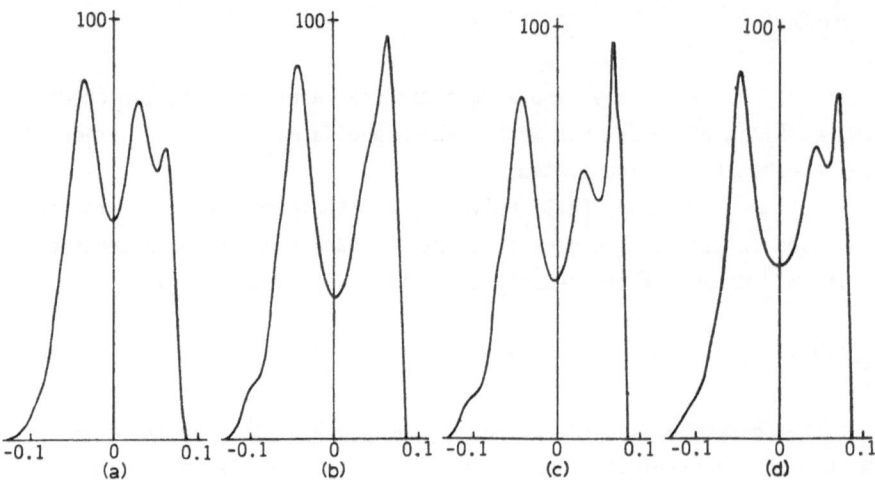

Fig.4.1a-d. The local density of states of an atom in (a) the (111) surface plane where Z = 9, (b) the first plane and (c) the second plane below the surface, and (d) in the bulk where Z = 12 for the fcc d band, by the moments method

chemisorption calculations on realistic systems as opposed to model ones. In addition there has been no attempt to achieve self-consistency with this method, although a procedure for doing so has been proposed [4.10].

4.2 Muffin-Tin Methods

A number of authors have adapted the muffin-tin potential approach used for bulk solids [4.11] to d-band metal surface calculations. The muffin-tin potential surrounds each atom with a sphere within which the potential is spherically symmetric, while the potential between spheres is a constant. For bulk d-band metals, which are close packed, the muffin-tin potential has been successfully used in a vast number of calculations [4.11].

The problem in the surface adaptations lies in devising a surface barrier which is realistic while being computationally convenient. The first such calculation [4.12] was for a monolayer film of Cu in which the surface barrier was approximated by requiring the wave functions to vanish in a plane on either side of the monolayer. KOHN [4.13], and KAR and SOVEN [4.14] have presented a formalism for multilayer slabs in which the barrier is an arbitrary function of the coordinate perpendicular to the slab, and KAR and SOVEN [4.14] have carried through a calculation for a single monolayer of Cu. Their non-self-consistent band structure is shown in Fig.4.2. The results are quite similar to those of bulk copper.

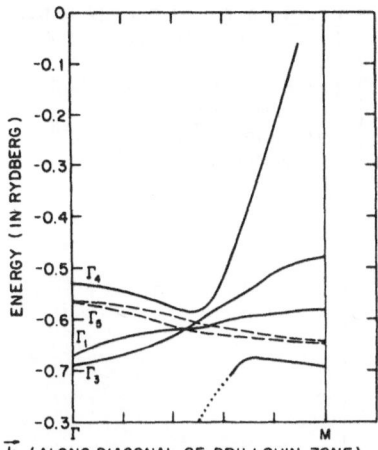

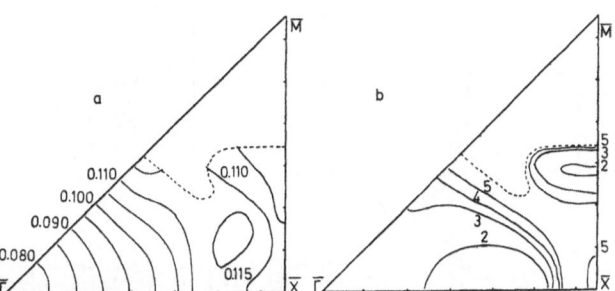

Fig.4.3a,b. Surface states on the copper (001) surface. (a) Energy contours, in hartrees. (b) Localization contours, in units of the layer spacing. There are no surface states around $\bar{M}$. The dotted line is at a localization length of 10 atomic layer spacings taken as the limit of the surface state region

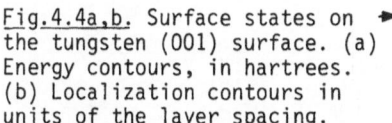

Fig.4.2. Band structure along diagonal of Brillouin zone for a single-layer film Cu(100) face. Dashed lines represent states odd under reflection on the plane of the film

Fig.4.4a,b. Surface states on the tungsten (001) surface. (a) Energy contours, in hartrees. (b) Localization contours in units of the layer spacing.

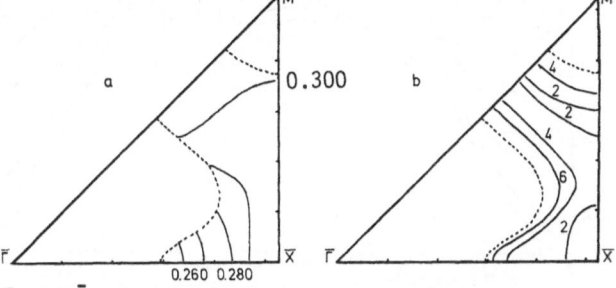

There are no surface states near $\bar{\Gamma}$ and $\bar{M}$: The dashed line is at a localization length of 10 atomic layer spacings, taken as the limit of the surface state region

GURMAN and PENDRY [4.15,16] have developed a related method in which they match bulk muffin-tin solutions with plane waves across an abrupt potential barrier. This method has been used to find surface states on Cu(100) and W(100) surfaces. They found many such states on Cu(100), existing over 75% of the surface Brillouin Zone (Fig.4.3). There are no surface states near $\bar{M}$, and the states are not sensitive to the form of the surface barrier. For W(100), they found surface states over 60% of the zone, but none near either $\bar{\Gamma}$ or $\bar{M}$ (see Fig.4.4).

A less restricted "warped" muffin-tin potential has been used by CARUTHERS et al. [4.17] to calculate the electronic structure of a 13-layer Fe(100) slab. The warped muffin-tin potential has the usual spherically symmetric form inside the muffin-tin spheres but is unrestricted outside. The calculated two-dimensional energy bands are shown in Fig.4.5. A number of surface states were found, as also shown in Fig.4.5.

All these methods suffer from the arbitrariness in the placing of the surface barrier, to a greater or lesser degree. Even where the results are not very sensitive to barrier shape or position, there is no guarantee that charge rearrangement upon going to self-consistency would not be important. None of these calculations are self-consistent, even within the muffin-tin constraint.

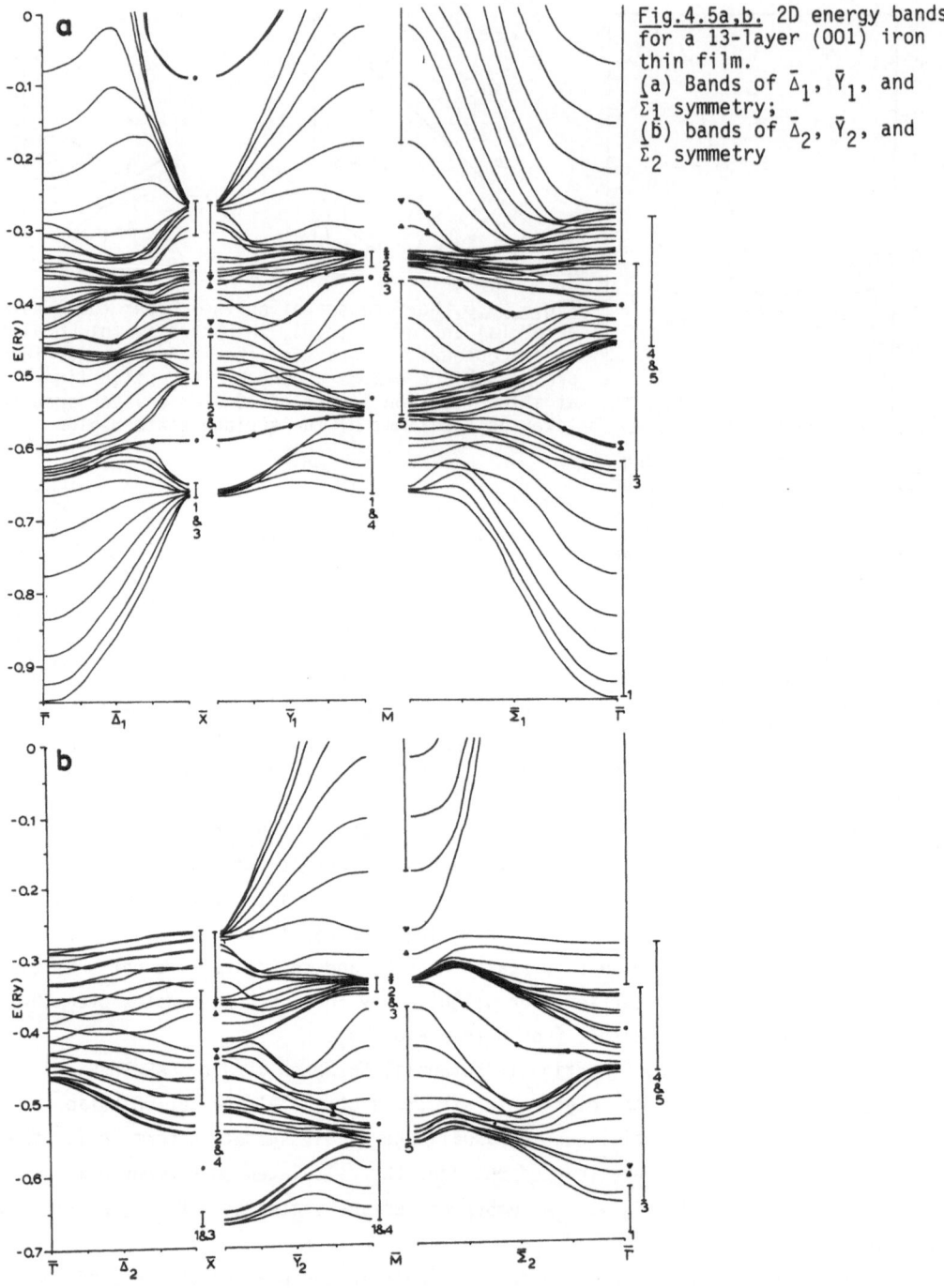

Fig.4.5a,b. 2D energy bands for a 13-layer (001) iron thin film.
(a) Bands of $\bar{\Delta}_1$, $\bar{Y}_1$, and Σ_1 symmetry;
(b) bands of $\bar{\Delta}_2$, $\bar{Y}_2$, and Σ_2 symmetry

4.3 Parametrized LCAO Methods

A large number of methods have been based on the linear combination of atomic orbitals (LCAO) approach. This method is attractive because its localized orbitals seem the obvious way to deal with a local feature like a surface. Several such methods are now discussed which have in common that the LCAO integrals are treated as parameters to be derived from auxiliary information of one kind or another, rather than being directly evaluated.

4.3.1 Slater-Koster Method for Clean Surfaces

A parameterized Slater-Koster [4.18] LCAO procedure has been developed and used extensively by KLEINMAN et al. in calculations on clean surfaces of ferromagnetic iron [4.19-21] and copper [4.22-24]. In this method matrix elements between 3d, 4s, and 4p orbitals on bulk lattice sites are found by fitting the computed bulk energy bands of the metal. These are used to construct the Hamiltonian matrix for a slab with specified surfaces. For all their calculations the slabs are approximately 50 Å thick.

The principal shortcoming of the method is that there is no recipe for determining the changes in matrix elements which accompany the charge rearrangement at the slab surface—i.e., there is no way to make the calculation self-consistent. This point has been exhaustively discussed by the authors [4.19-25]. While the method is inexpensive enough to permit the use of very thick multilayer films and thus to examine wave functions in great detail, the lack of self-consistency casts some doubt on any predictions about surface states.

4.3.2 Extended Huckel Theory

Turning to an actual chemisorption calculation, FASSAERT and VAN DER AVOIRD [4.26] developed a simple treatment for an adsorbed atom on a surface. They have investigated the case of H on Ni by means of an Extended Huckel Theory method. They use a layer geometry with periodicity in the two directions parallel to the surface. Nickel 3d, 4s, and 4p orbitals are included. For the hydrogen chemisorption a full monolayer of H atoms is added, with the same periodicity as the substrate, though in varying registries with the surface. The hydrogen 1s orbital is the only extra orbital required.

Overlap integrals are calculated using Slater orbitals, with certain assumptions as to orbital exponents. The diagonal Hamiltonian matrix elements for nickel orbitals are approximated by the ionization energies [4.27]; for hydrogen an adjusted value is used. The off-diagonal elements are taken to be proportional to the overlaps and to the average of the diagonal matrix elements is a way standard in such calculations:

$$H_{pq}(\underline{R}) = 1.75 \, S_{pq}(\underline{R})[H_{pp}(0) + H_{qq}(0)]/2 \quad . \tag{4.2}$$

Four shells of neighbors were included.

The band structure for clean nickel computed this way agrees moderately well with more elaborate band calculations, though the d band is somewhat too narrow and positioned too low relative to the sp band. Results are presented for chemisorption on (100) and (111) surfaces, and for several adsorption sites. Detailed results are given for changes in total and projected densities of states for all these cases. Figure 4.6 shows such results for the Ni(111) surface.

Bonding is observed between hydrogen orbitals and both Ni 4s and 3d orbitals, the hydrogen bonding most strongly to the d orbital directed toward it in all cases. The sp contribution cannot be neglected. Bonding is most stable when the hydrogen has the fewest nickel neighbors, but the energy differences are not large. The author's final point is that the strength and character of chemisorption on nickel is not determined by bulk metal properties or by clean surface properties, but by geometrical features of the surface-adsorbate interaction.

This model has the advantage of computational simplicity, but the arbitrariness of the assumptions used is disturbing. Moreover, there is no attempt at self-consistency except to adjust parameters when apparently unreasonable results arise. Any conclusions about relative strengths of chemisorption on different sites could easily be reversed by going to genuine self-consistency.

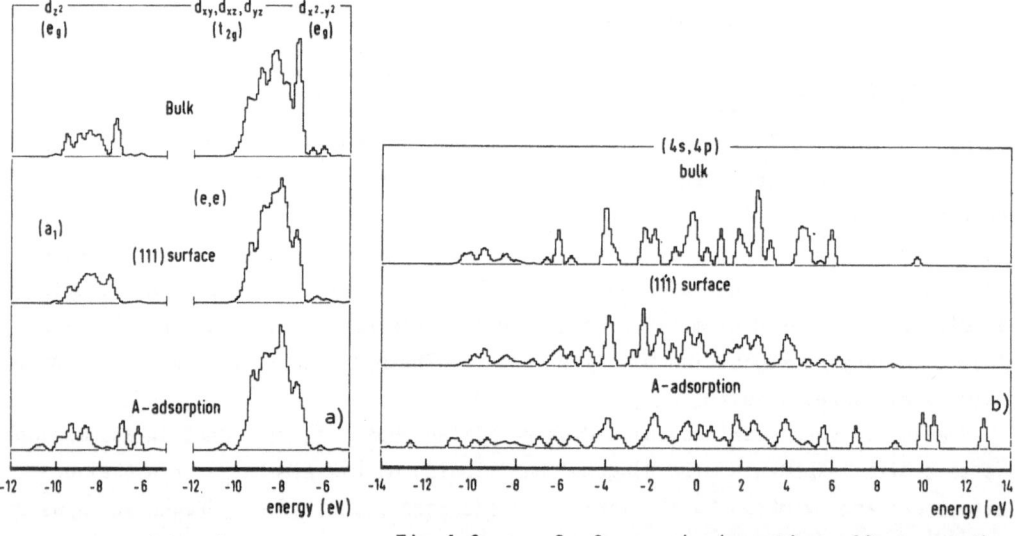

Fig.4.6a-c. Surface and adsorption effects on the partial local densities of states at the Ni(111) surface. (a) Surface layer density of states of different 3d orbitals denoted by the representation symbols of the point group C_{3v}. For comparison the bulk density of states is given. (b) Surface layer density of states of the 4s and 4p orbitals. (c) Local density of states in the adsorbed hydrogen layer, as compared to the density of states of an isolated hydrogen layer with the same periodicity

The preceding calculation makes fairly simple assumptions about the interaction between adsorbate and substrate. The next calculation uses more sophisticated assumptions and obtains rather more satisfying results.

4.3.3 Parametrized LCAO for Chemisorption

SMITH and MATTHEISS [4.28] used a parametrized LCAO model to study the adsorption of H_2 on W(100). They derived their parameters from first principles APW calculations in a novel and clever way. Their model for the (100) tungsten surface was a 16-layer slab of W atoms. The hydrogen adsorption is treated by adding an extra layer of H atoms (corresponding to a full monolayer coverage of hydrogen) on one side of the slab with the hydrogens located in bridge positions as per the ESTRUP-ANDERSON model [4.29].

A standard Slater-Koster parametrized LCAO model [4.18] was set up. This model used nine W orbitals corresponding to the 5d, 6s, and 6p levels and one hydrogen 1s orbital for each of the two H atoms in the unit cell. The W-W interaction parameters were given by the fit of CHAKRABORTY et al. [4.30] to the bulk tungsten band structure. That is usual enough. The novel part is the method used to obtain W-H and H-H interaction parameters. Gaseous hydrogen values are not very useful, and the self-consistency problem is severe for any arbitrary procedure.

SMITH and MATTHEISS circumvented this difficulty by performing a standard bulk APW calculation for a quite fictitious tungsten hydride crystal with an assumed CsCl structure and W-H bond length of 2.14 Å. An LCAO fit was then done to these results to get W-H and H-H parameters. They then used these parameters, all determined from first-principles calculation, in the slab calculation. One empirical adjustment is made: the center of gravity of the H 1s band was half-filled. This was done to prevent large transfer of charge from the H atoms to the W substrate, which the authors felt would be physically unreasonable.

Figures 4.7 and 8 show the two-dimensional band structures for the clean and hydrogen-covered W(001) surfaces, respectively. Only the most important surface states are shown, that is, those with large eigenvector components in the outermost layers. The states are labeled by the dominant eigenvector component. Figure 4.9 shows the dispersion of the hydrogen-induced resonances in the photoemission spectra of a W(100) surface [4.31]. This curve is also taken from SMITH and MATTHEISS [4.28]. The resonance C in Fig.4.9 they identify with the dz^2 state of Fig.4.8, which is quite obviously related to the dz^2 state of the clean surface in Fig.4.7. Similarly the states labeled A in Fig.4.9 are identified with the H band at the bottom of Fig.4.8.

SMITH and MATTHEISS discussed these and other correlations with experiment in some detail. Their conclusions must be regarded as tentative and preliminary. The details are sensitive to the choice of LCAO parameters. The inclusion of the H-H interactions they see as absolutely necessary, with single H atom results being completely inadequate.

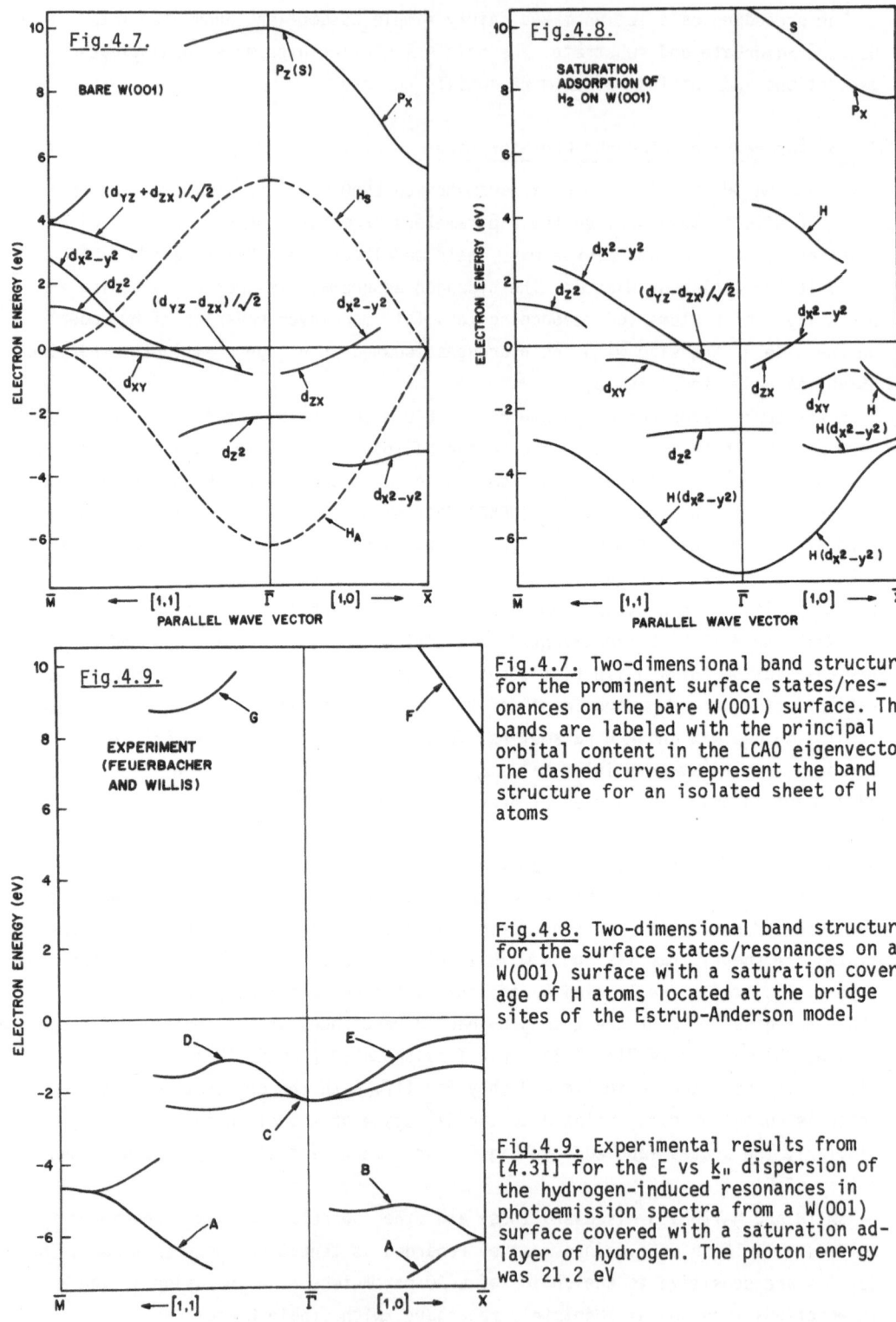

Fig.4.7. BARE W(001)

Fig.4.8. SATURATION ADSORPTION OF H₂ ON W(001)

Fig.4.9. EXPERIMENT (FEUERBACHER AND WILLIS)

Fig.4.7. Two-dimensional band structure for the prominent surface states/resonances on the bare W(001) surface. The bands are labeled with the principal orbital content in the LCAO eigenvector. The dashed curves represent the band structure for an isolated sheet of H atoms

Fig.4.8. Two-dimensional band structure for the surface states/resonances on a W(001) surface with a saturation coverage of H atoms located at the bridge sites of the Estrup-Anderson model

Fig.4.9. Experimental results from [4.31] for the E vs $k_\parallel$ dispersion of the hydrogen-induced resonances in photoemission spectra from a W(001) surface covered with a saturation adlayer of hydrogen. The photon energy was 21.2 eV

This calculation suggests the usefulness of the LCAO approach and the slab geometry. The authors' use of a tungsten hydride APW calculation to obtain reasonable interaction parameters is ingenious. But again the calculation does contain arbitrary adjustments and is not properly self-consistent. The authors' cleverness in making reasonable parameter choices does not completely substitute, particularly since they note the sensitivity of their results to the choice of parameters. And ultimately the values derive from bulk results.

4.3.4 Tight Binding

BULLETT and COHEN [4.32] employed some novel methods in a tight-binding calculation for carbon monoxide chemisorbed on nickel (100) and (111). The calculation is based on the ANDERSON localized orbital pseudopotential method [4.33] and employs potentials and orbitals obtained from Hartree-Fock-Slater atomic calculations [4.34]. The calculation is done for slabs of nickel sandwiched between fractional monolayers of carbon monoxide. There are two nickel layers for the (111) surface and three for the 100). The slab potential is constructed by surrounding each atom with a cell within which the potential is the atomic potential of that atom: the cell boundaries being defined as the points at which atomic potentials of neighboring atoms are equal. The potential is therefore continuous but with discontinuities of derivative at the cell boundaries.

No attempt is made to achieve point-by-point self consistency, but charge transfer among the nickel, carbon, and oxygen atoms is accounted for approximately by shifting atomic levels (i.e., diagonal elements of the Hamiltonian) linearly according to the charge on the atom. The proportionality constants are obtained by a comparison of atomic and ionic atomic energy levels. The theory contains no adjustable parameters since the Hamiltonian matrix is completely determined from Hartree-Fock-Slater atomic and ionic solutions.

Since the potential is periodic in the plane of the slab, the eigenstates are characterized by wave vectors in a two-dimensional Brillouin zone in the slab plane. In diagonalizing the Hamiltonian a special points scheme was used [4.35] whereby only eigenstates at certain points in the zone were found. This limited set of eigenstates [four for (100) and six for (111)] was used in determining the atomic charge for the approximate self-consistency procedure and in the calculations of density of states to be discussed below.

The main purpose of the calculation is to determine the equilibrium geometries of CO on the two nickel surfaces by selecting from several plausible geometries the one with minimum one electron energy sum. For Ni(100) the lowest energy geometry is with the CO located, carbon down, at a fourfold site above the center of a square of nickel atoms. For Ni(111) the lowest energy geometry is with the CO located, carbon down, at a thresfold site above the center of a triangle of nickel atoms.

In addition to giving these reasonable results for geometries, the calculations predict electron transfer to the CO in accord with work function measurements and, for Ni(100), give densities of states in reasonable agreement with ultraviolet photoemission spectra. The comparison is shown in Fig.4.10. Note that, while the CO levels are shifted upward by about 1 eV and acquire substantial width upon chemisorption, there is no tendency for the CO levels to smear into broad bands as is the case with nitrogen chemisorbed on Cu(100) (see Sect.4.6.2). This is presumably a consequence of the relatively weak interaction of CO with the nickel surface.

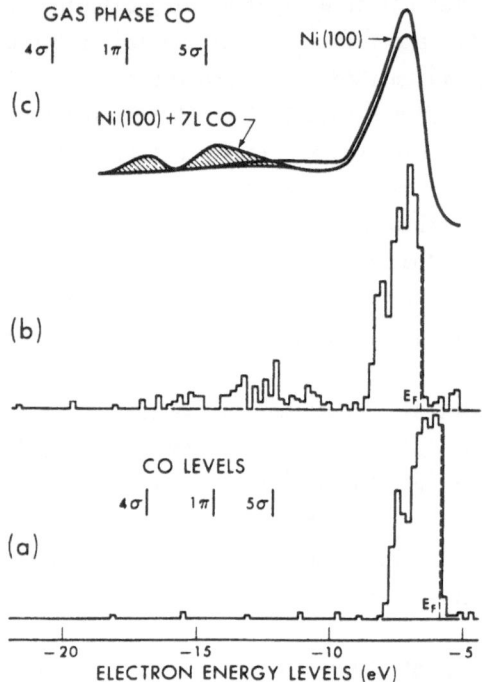

Fig.4.10a-c. Valence band electronic densities of states calculated by BULLETT and COHEN from 10 k points for the three-layer Ni(100) slab (a) before and (b) after chemisorption of CO. Calculated levels of the isolated CO molecule are included in (a). (c) UPS spectra at hv = 40.8 eV for (1) Ni and (2) Ni +7L of CO

4.4 Linear Combination of Muffin-Tin Orbitals (LCMTO) Method

The last non-self-consistent method we want to discuss uses basis orbitals defined for a muffin-tin, but a more general potential. It has several computational advantages and could indeed be made self-consistent, though as yet this has not been done.

KASOWSKI has used the linear combination of muffin-tin orbitals (LCMTO) method [4.36] to calculate the electronic structure of clean Cu(001) and Ni(001), as well as that of Ni(001) with various adsorbates.

The LCMTO method [4.36] is a generalization of the familiar Korringa-Kohn-Rostocker (KKR) method, based on a multiple scattering formalism. It reduces essentially to the KKR method for a bulk muffin-tin potential. The wave function is expanded in terms of muffin-tin orbitals (MTOs), each of which is derived from the solution of Schrödinger's Equation for a single, isolated muffin-tin potential. The MTO for site i is then

$$X_{\ell m}(\underline{r} - \underline{R}_i) = \phi_{\ell m}(\underline{r} - \underline{R}_i) + C_\ell J_{\ell m}(\kappa|\underline{r} - \underline{R}_i|) \quad \text{inside}$$

$$= d_\ell K_{\ell m}(\kappa|\underline{r} - \underline{R}_i|) \qquad \text{outside the sphere}$$

(4.3)

where $\phi_{\ell m}$ is the single muffin-tin solution, and J and K are wave equation solutions, J regular at the origin and K regular at infinity; each is multiplied by a spherical harmonic of angle. The coefficients C_ℓ and d_ℓ are determined by requiring continuity of logarithmic derivatives at the sphere radius, and κ is a variational parameter. LCMTOs are constructed as a Bloch sum of MTOs on different sites, and the standard multiple-scattering or KKR condition is applied.

While a single muffin-tin potential is used in defining the orbitals, the method is quite general and handles general potentials. The potential is derived from overlapping atomic charge densities and is expanded in spherical harmonics. The LCMTO method has a great deal in common with the standard LCAO (linear combination of atomic orbitals) method, but the orbitals used get around the problem of calculating the very difficult multicenter integrals of that method, by allowing use of summation theorems for spherical Bessel and Neumann functions.

KASOWSKI used a 20-layer slab in his study of W(100) and Mo(100) surfaces [4.37]. He found well-localized surface states in both cases. The surface state he found in W(100) at -0.22 Ry relative to the Fermi energy is sensitive to surface relaxation. If the outer layer relaxes such that the interlayer bond distance decreases by 3.22%, the surface state moves to -0.14 Ry. In Mo(100), a surface state appears at -0.05 Ry when the surface bond distance is decreased by 3.22%. These results contrast with those for Cu and Ni, where relaxation effects are small.

With regard now to chemisorption, KASOWSKI has calculated the effect of an oxygen overlayer on the surface states of Ni(100) [4.38]. He found surface states induced, localized in the oxygen overlayer, which correspond well with surface states observed in photoemission and ion-neutralization spectroscopy. And for CO adsorbed on Ni(100) [4.39], he again found good agreement with photoemission experiments.

KASOWSKI's results for the chemisorption of O and Na on Ni(100) are summarized in Fig.4.11. For clean nickel, he used a five-layer slab of Ni atoms with an outer layer to represent surface charge. For the chemisorption case, he again used five Ni layers, with an outer layer of adsorbed atoms —a full monolayer, with the atoms in fourfold sites above the hollows.

The states in Fig.4.11 are all for the Γ point in the two-dimensional Brillouin zone. The abscissa corresponds to the different layers. An oxygen 2p surface state

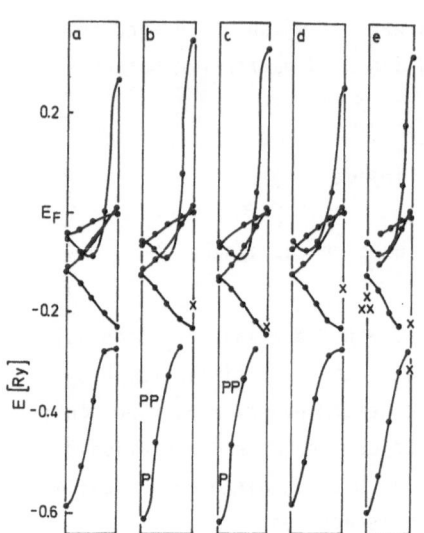

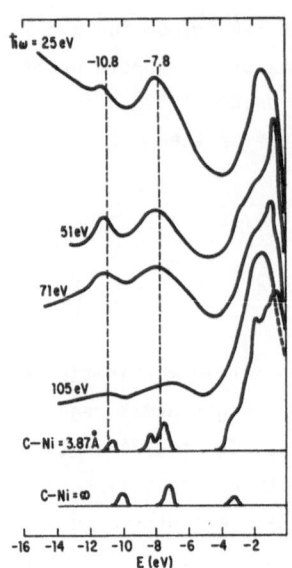

Fig.4.11a-e. The energy states of (a) five-layer Ni(001) thin film and the effect on these thin film states of (b) O overlayer at d = 0.90 Å, (c) O overlayer at d = 1.50 Å, (d) Na overlayer at d = 2.87 Å, and (e) Na overlayer at d = 1.76 Å. O 2p states are represented by a p while induced surface states are represented by an X

Fig.4.12. Comparison of experimental photoemission spectra of GUSTAFSSON et al. [4.42] with Kasowski's theoretical spectra for C-Ni distances at 3.87 a.u. and ∞

appears well below the Ni d states for both Ni-O separations (marked P in the figure), while another surface state is induced from a band state near the bottom of the d band. Experimentally, a surface state is found about 5 eV below the Fermi energy by both photoemission [4.40] and ion-neutralization spectroscopy [4.41]. For the Na case surface states are only induced near the bottom of the d-band; more of these states are induced at the closer Na-Ni separation.

The results for CO on Ni(001) are shown in Fig.4.12. A four-layer thick Ni slab was used, with a half-monolayer of CO on the surface in the c(2 × 2) structure. The CO is perpendicular to the surface, C atom nearer, above the fourfold hollow sites. Figure 4.12 shows the excellent agreement of the calculated density of states (DOS) with photoemission results [4.42]. In the same figure is shown the calculated DOS for a CO monolayer far removed from the Ni surface. The broadness of the levels for this last case indicates the importance of CO-CO interactions. The effect of the presence of the Ni surface is to shift the 5σ level far lower, where it essentially merges with the 1π level to form a single peak. The 5σ level is primarily a C state and closest to the surface, thus the larger change in it compared to the other, primarily oxygen levels. Examination of the eigenvectors shows a sizeable perturbation and delocalization of the molecular states.

Finally on comparing Fig.4.12 and Fig.4.10, we see that these results agree well with those of BULLETT and COHEN [4.32], who also found two peaks well below the d band. It is interesting to note that the case of CO on Ni(100) is one where the molecular character of the adsorbate dominates the results. The CO molecular levels are clearly distinguishable in the calculated DOS for the chemisorbed case, though these levels are shifted due to the influence of the substrate. This contrasts with the N on Cu(100), which we take up presently, where the N levels broaden out and overlap the d band such that well separated N levels are not seen at all.

Thus the LCMTO method is a useful tool for chemisorption calculations. Its principal advantage is that it is rapid and inexpensive and can treat large unit cells without excessive difficulty. This is a notable advantage for slab models with many layers. Its principal drawback is one it shares with many other methods, the fact that it is not a self-consistent procedure. We will see that self-consistency is probably the single most important requirement for a reliable theory of chemisorption because of the charge rearrangements involved. The LCMTO method could be made self-consistent, but has not been so used thus far.

We see then that interesting results have been obtained by several methods, none of which go far enough to achieve self-consistency. They are all only first steps toward the reliable computational methods free from arbitrary parameters which must be the ultimate goal.

4.5 Self-Consistent Methods

All the methods described so far in this review are not self-consistent. Because of the complex charge rearrangements at the surface, these methods will of necessity be limited in their ability to describe the details of the interactions involved in chemisorption. Useful as they may be for conceptual purposes, detailed understanding is impossible. For that we must be able to describe the fine details. We now discuss a number of methods which do make a greater or lesser attempt at achieving self-consistency.

4.5.1 Gaussian Expansion Method

The first such method is a variation of the LCAO technique. APPELBAUM and HAMANN [4.43] have recently reported results for the surface electronic structure of clean Cu(111), using this method.

This method is an adaptation of the molecular calculation technique of SAMBE and FELTON [4.44] to the solid surface problem. The basis set consists of core functions plus two radial s, p, and d functions located on the atoms in a three-

or five-layer slab. These are orthogonalized to the core states, and the core charge density is frozen. The radial part of each orbital is expanded in a series of Gaussians. The charge density and potential are also expanded in Gaussians centered both at atomic sites and at interstitial or extra-surface sites. Calculations are carried to self-consistency within the constraint imposed by this form on the charge density.

For the case of Cu(111), APPELBAUM and HAMANN expanded the potential and charge density in Gaussians on 47 sites, most of them outside the surface layer of the three-layer slab. Extra s and p basis functions outside the surface were also used for increased variational freedom. The 5-layer basis is similar but scaled up. The computed 5-layer work function was 5.1 eV and the surface energy 0.75 eV (experiment 0.6 eV) [4.45].

These results were used to construct matrix elements for an 11-layer slab, using bulk matrix elements for the inner layers. The electronic structure of this slab was calculated but not iterated further; presumably it was reasonably self-consistent as it stood. The density of states for the surface layer, second layer, and bulk (represented by the central three layers) are shown in Fig.4.13. The surface layer DOS is 9% narrower and shifted upward by 0.4 eV, while the second layer DOS is a tiny 2% wider than the bulk and very similar to it.

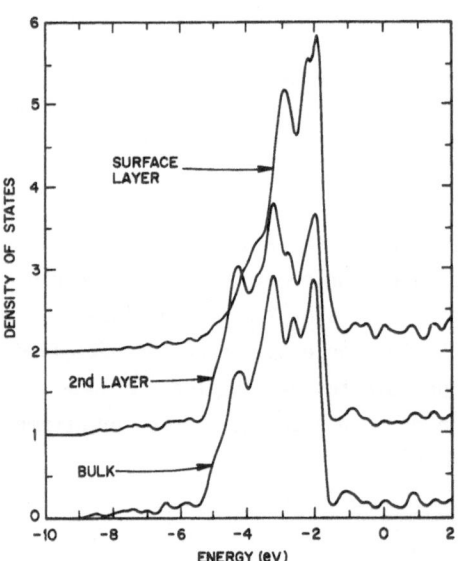

Fig.4.13. The density of states per atom calculated by APPELBAUM and HAMANN for an 11-layer Cu(111) film for the surface layer, second layer, and bulk (average of central three layers). The curves are displaced vertically from each other by one unit (1 unit = 40 states/Hartree = 1.47 states/eV

A d-like surface state is found 2.9 eV below the Fermi energy, and another, p-like, surface state at Γ 0.5 eV below the Fermi energy. This latter state has 50% of its charge density in the surface layer and attenuates rapidly. Its position agrees well with a state 0.4 eV below the Fermi energy found experimentally [4.46,47].

The method has not been used in a chemisorption calculation. While it is self-consistent, and it does find the experimentally observed surface state, the expansion of the charge density and potential in spherical Gaussians is worrisome on two counts. This is a restrictive form (a matter which can be obviated by using large number of Gaussians, as was in fact done) and it is unsystematic. Artful placement of the Gaussians is important, but no rules are laid down. Thus it is hard to generalize the method and hard to optimize a calculation or to know whether it is optimized.

4.5.2 Numerical Basis Set Method

Another method with similar features but which has been more widely applied, especially to chemisorption systems, is the numerical basis set method of WANG and FREEMAN.

WANG and FREEMAN have generalized the discrete-variational LCAO method of ELLIS and PAINTER [4.48] and of ZUNGER and FREEMAN [4.49], a bulk technique, to the surface case. The Coulomb potential used is a superposition of overlapping, spherical atomic potentials; the charge density is again a superposition of atomic charge densities, and a Kohn-Sham exchange potential is used [4.50].

A numerical basis set is used; that is, the LCAO-type basis orbitals are tabulated and not of any particular analytic form. These numerical Bloch functions are orthogonalized [4.51] to the atomic core states in order to reduce the size of the secular equation to be solved. The basis functions used were the (numerical) atomic valence orbitals generated by a Herman-Skillman atomic calculation, but with long-range tails suppressed. Matrix elements were calculated by direct numerical integration using the Diophantine integration method [4.52]. It was thus not necessary to break up the problem into the usual multicenter integrals. The reader is referred elsewhere for details.

The surface is here also modeled by a multiple atomic layer slab. The core charge density is frozen at its atomic value. The valence charge density is computed from the solutions of the secular equation. Energies and wave functions are calculated at a regular grid of points in the irreducible part of the two-dimensional Brillouin zone. Analytic interpolation is used to obtain accurate Fermi energy, densities of states, and charge density using a triangle scheme which is a generalization of the bulk analytic linear energy tetrahedron method [4.53].

Self-consistency is achieved by varying the atomic configuration, using the occupation numbers of the various orbitals as parameters to be adjusted iteratively, using a least-squares fitting procedure. The potential is constrained at all times to be due to a sum of overlapping spherical charge densities, although there appears to be no reason why a more general form could not be used (at some increase in complexity). The method should handle without difficulty the larger effects of going to self-consistency, the charge transfer and changes of orbital occupancy.

WANG and FREEMAN applied this method to the clean Ni(001) surface, using one-, three-, and five-layer slab models [4.54]. The basis set consisted of the numerical 3d, 4s, and 4p atomic orbitals of Ni from a Herman-Skillman atomic calculation of the neutral Ni $3d^{8.5} 4s^1 4p^{0.5}$ configuration. Results for the 5-layer slab show an excess 0.03 electrons per atom on the surface layer. This almost neutral result contrasts sharply with the non-self-consistent results of CARUTHERS et al. for Fe(001) [4.17], which they found to have a 1.5 electron per atom deficit on the surface layer. CARUTHERS et al. attributed their unphysical result to deficiencies of the LCAO basis set and expect it to be a feature of *all* LCAO style calculations. WANG and FREEMAN disagreed, ascribing the difficulty to the lack of self-consistency. They expect the result to change with changing parameters of the surface potential, and emphasize that a unique answer requires a self-consistent solution.

They found a work function of 5.52 eV for the 5-layer case in reasonable agreement with experiment (5.15-5.27 eV) [4.55]. The non-self-consistent value was 9.14 eV. The one- and three-layer model give almost the same value.

The self-consistent Ni(001) bands agree in general with the 35-layer non-self-consistent results of DEMPSEY and KLEINMAN [4.56]. The three- and five-layer densities of states are quite similar, as seen in Fig.4.14, providing some evidence that a 5-layer film adequately represents the surface. For both cases the surface layer DOS is markedly different from the central layer or layers; but for the 5-layer

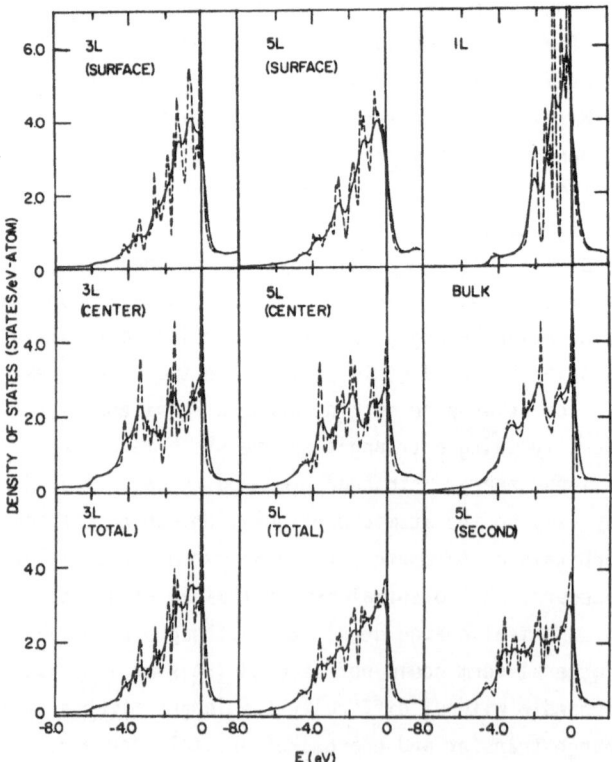

Fig.4.14. Density of states and projected local density of states for the one-, three-, and five-layer Ni(001) films in units of electron states/atom-eV. The dashed curves are the raw curves smoothed by a Gaussian broadening function of 0.1 eV FWHM; the solid curves are the result of a similar broadening of 0.5 eV. The bulk DOS for paramagnetic Ni of WANG and CALLAWAY is shown in the right-hand center panel

case, the second layer DOS is very similar to that of the central layer. Here again is evidence that the disturbance due to the presence of the surface seriously affects only the surface layer and that all other layers are very much bulk-like. This result seems to be common to all self-consistent calculations. It is evidence that one must be wary of drawing conclusions from any but self-consistent results.

For the Ni(001) monolayer, the computed d bands are very much narrower, presumably because of the drastic reduction in the number of nearest neighbors. This was also true of an earlier Cu(001) monolayer calculation of WANG and FREEMAN [4.57] and of the copper monolayer calculation of ARLINGHAUS et al. [4.58]. A similar but smaller narrowing effect is seen in the surface layer DOS of the 3- and 5-layer Ni(001) results.

WANG and FREEMAN have also calculated electronic structure and surface states for a nine-layer ferromagnetic Ni(100), using a spin-polarized version of their technique [4.59]. They found a d-like surface state in the majority-spin band which lies above the Fermi energy, thus reducing the surface layer magnetic moment to 0.44 μ_B, a reduction of about 20% relative to the bulk. They did not find the sharp structure near the Fermi energy that was computed by DEMPSEY and KLEINMAN [4.56].

WANG and FREEMAN have also used this discrete variational technique to calculate the self-consistent electronic structure of a c(2 × 2) overlayer of oxygen on Ni(001) [4.60]. They used a model consisting of a 3-layer Ni(001) slab upon which is placed a layer of oxygen atoms in the c(2 × 2) arrangement which is found experimentally [4.61]. The numerical basis set consists of the 3d, 4s, and 4p orbitals of Ni plus the 2s and 2p orbitals of oxygen, in each case orthogonalized to the frozen Ni or O core wave functions. Calculations are done at 10 values of $\underline{k}$ in the irreducible $1/8^{th}$ of the two-dimensional Brillouin zone. This calculation is paramagnetic.

The oxygen atoms are placed in alternate fourfold sites at 1.7 a.u. above the surface nickel plane. There are thus two different Ni sites, which leads to a secular equation twice the size that would arise for a (1 × 1) monolayer adsorbate. At self-consistency, the central Ni plane is almost neutral, while each of the two inequivalent surface Ni atoms have lost ~0.3 electron, and each oxygen has gained ~ 0.6 electron.

Figure 4.15 compares the self-consistent density of states for the 3-layer Ni(001) slab with and without the c(2 × 2) oxygen layer chemisorbed on it. The surface layer d-band peak is much reduced in intensity, while new peaks appear above and below the d band. The more striking is the double peak at 5.5 eV below E_F and ~ 2 eV wide. This is clearly seen in the top panel of Fig.4.15 to be an oxygen 2p state with considerable admixture of Ni surface-layer states. The Ni density has both sp and 3d character. There is considerable change to the central Ni plane density as well. There is a less striking peak induced just above E_F, which is apparently an oxygen antibonding state.

Since the calculation is carried to self-consistency, the method is quite able to deal with charge transfer and orbital occupation changes which comprise the

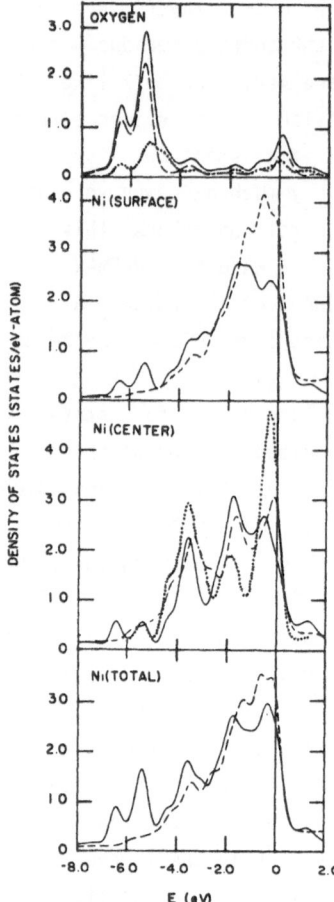

Fig.4.15. Total and projected DOS for a three-layer Ni(001) slab (dashed lines) and c(2 × 2)O on Ni(001) (solid lines), including the oxygen e (long dashes) and a_1 (dotted) orbital DOS

large scale effects of the presence of a surface or of an adsorbate. The method has actually been applied and carried through for these several interesting cases, and useful results have appeared.

Unfortunately, the various other restrictive approximations lessen the value of these results. The basis set is less general than that of APPELBAUM and HAMANN [4.43]. Charge polarization is not allowed. Most serious of all, the restriction of the form of the charge density to a sum of overlapping spherical charge densities, and the restriction of the self-consistency iteration to variation of the respective s, p, and d occupancies, places severe constraints on the possible self-consistent adjustments, thus negating a large part of the advantage. These calculations must therefore be judged only partly self-consistent.

4.6 Fully Self-Consistent Methods

The preceding two methods impose restrictions on the form of the potential which limit the possible variations to specified types. While self-consistent, these calculations are self-consistent only within these imposed constraints. In this final section, we go on to discuss two methods which reduce such restrictions to the level of the various numerical approximations and therefore impose no serious restrictions. These methods, which represent the best available at the time of this writing, we choose to style *fully* self-consistent.

4.6.1 Pseudopotential Method

The first point-by-point self-consistent calculation for a d-band metal was a pseudopotential calculation for niobium [4.62]. A nine-layer slab with (100) sur-faces was used. The potential consists of nonlocal pseudopotentials for the Nb^{+5} case which are self-consistently screened by the valence pseudo wave functions. The calculation employs in large part the methods successfully used by COHEN and coworkers in calculating the surface electronic structure of semiconductors and simple metals [4.63]. In addition to the periodicity in the plane of the slab, an artificial periodicity in the perpendicular direction is introduced by repeating the slab indefinitely with a separation equivalent to six layers of Nb atoms. This device permits the use of standard bulk pseudopotential methods albeit with a much increased density of reciprocal lattice vectors because of the large unit cell repeat distance in the perpendicular direction.

Figure 4.16 shows the local density of states for the central and surface layers of Nb(001), as well as a difference curve. The increased surface density between 0 and 2 eV reflects the presence of surface states. A large number of surface states were found in this energy region. There is unfortunately no experimental data on Nb(001) to compare. The calculated work function, 3.6 eV, does agree fairly well with the measured 4.0 eV [4.64].

More recently, KERKER et al. [4.65,66] have done similar calculations for Mo(001). However, in this case they abandoned the simple pseudopotential formalism used in the Nb(001) calculation in favor of a mixed basis set pseudopotential formalism, in which the usual plane-wave basis is augmented with localized Gaussian orbitals to represent the more localized d functions. This method, due to LOUIE et al. [4.68], will be discussed in detail further on. Only one Gaussian per site was used in the Mo(001) calculation.

A wide variety of surface states and surface resonances were found in this cal-culation. The results are essentially summarized by Fig.4.17, which shows the Mo(001) surface density of states. The surface excess density (i.e., the difference between surface layer and central layer densities of states) is also shown in this figure, as the shaded areas. There is excellent agreement between these results and angle-resolved photoemission experiments [4.67].

Self-consistency is important in these calculations, especially with regard to surface states, some of which are not found at all in non-self-consistent cal-culations. An interesting observation is that rigid-band ideas are of little use

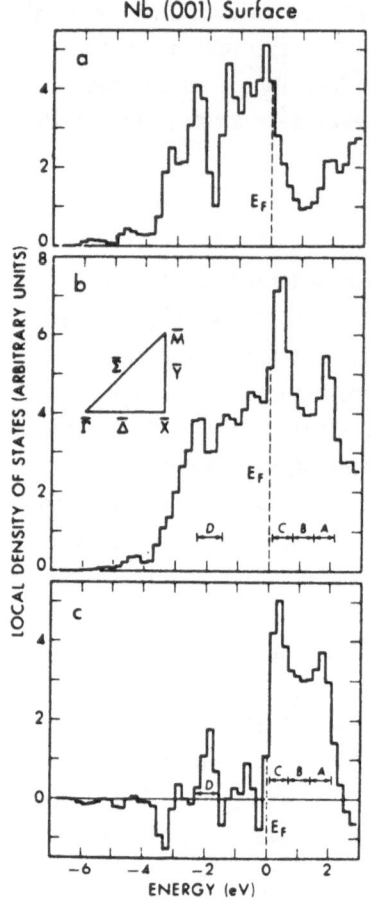

Nb (001) Surface

LOCAL DENSITY OF STATES (ARBITRARY UNITS)

ENERGY (eV)

◄ Fig.4.16a-c. Calculated local density of states for the Nb(001) ideal surface. (a) LDOS for the fifth layer from the surface. (b) LDOS for the surface layer (first layer). Inset, the irreducible part of the 2D surface Brillouin zone. (c) Difference curve (see text). The calculated work function is 3.6 eV; the measured value is 4.0 eV; (001) surface; [4.64]

Fig:4.17. The local density of states for the Mo(001) surface layer. The shaded areas represent the excess density of states (see text). Energies are measured with respect to the Fermi level. The positions of the principal bulk peaks are marked by open arrows

▼

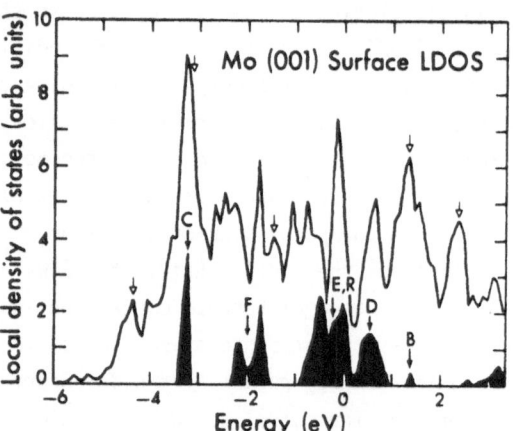

in predicting surface states — the Nb(001) and Mo(001) surface states are radically different.

The mixed-basis pseudopotential [4.68] theory of LOUIE has been applied to the clean palladium (111) surface [4.69] and to hydrogen chemisorbed on palladium [4.70].

LOUIE used a mixed-basis set to achieve the flexibility needed in dealing with transition metals which contain both highly extended, delocalized electrons and the more localized d electrons. No constraints were made on the form of the potential. The wave functions are expanded as

$$\Psi_{\underline{k}}(\underline{r}) = \sum_{\underline{G}} \alpha_{\underline{G}}(\underline{k}) \exp[i(\underline{k} + \underline{G}) \cdot \underline{r}] + \sum_{j} \beta_{j}(\underline{k}) \phi_{j}(\underline{k},\underline{r}) \qquad (4.4)$$

that is, in a mixed basis consisting of both plane waves, as in the usual pseudopotential methods, and of Bloch sums of localized Gaussian orbitals. The $\underline{G}$'s are reciprocal lattice vectors and the ϕ_{j}'s are the Bloch sums of Gaussians. The

Gaussians have the form $r^{\ell}\exp(-\alpha r^2)$ times an angular part. A fixed ionic pseudo-potential is used for the core. Matrix elements involving the Gaussians are evaluated numerically using fast-Fourier-transform techniques.

The surface is modeled in the slab geometry, using several atomic layer thick slabs. These slabs have complete lateral periodicity in two dimensions but are repeated artificially in the direction normal to the surface (at some distance apart) in order to maintain three-dimensional periodicity for computational purposes.

LOUIE has investigated the Pd(111) surface using a seven-layer slab [4.69]. The slab is repeated periodically in the normal direction, with slabs four layer spacings apart. The Pd^{+10} core is replaced by an ℓ-dependent pseudopotential fitted to computed neutral Pd atomic energies and wave functions. This core potential used with a basis set of 30 plane waves and 5 Gaussians per atom gives a self-consistent bulk Pd band structure that agrees well with experiment and with previous calculations [4.71]. The equivalent basis set was used for the surface calculations — 300 plane waves and 35 Gaussians (5 per atom times 7 layers).

The surface state bands which result are shown in Fig.4.18. The 5.8 eV computed work function agrees well with the measured value of 5.6 eV [4.72]. The slab is almost neutral layer-by-layer, but there is considerable charge rearrangement in going to self-consistency, so that the constraint of atom-by-atom neutrality used in many non-self-consistent calculations is likely too rigid.

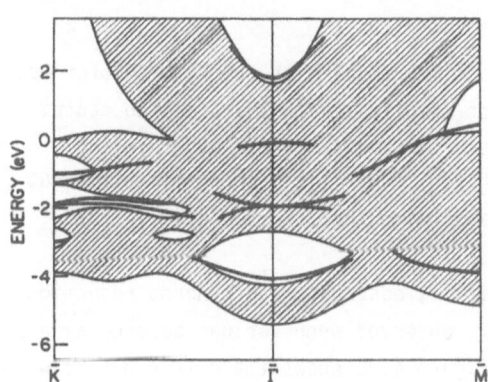

Fig.4.18. Surface state bands (solid curves) and the projected band structure (shaded areas) in the $\bar{K} = 2\pi/a$ (0,-2/3,2/3) and $\bar{M} = 2\pi/a(1/3,-1/3,2/3)$ directions for the Pd(111) surface. (E_F is 0 eV)

Figure 4.18 shows 11 surface state bands, most of them d-like. An interesting exception is the surface state ~2 eV above E_F, which is sp-like. This state penetrates some distance into the surface. A similar state has been seen experimentally on the (111) surfaces of Cu, Ag, Au [4.46,47], and Ni [4.73]. The other d-like states are quite sharply localized on the first or second layer.

Figure 4.19 shows the local densities of states for the central plane of the 7-plane slab, the second plane from the surface, and the surface plane. The central

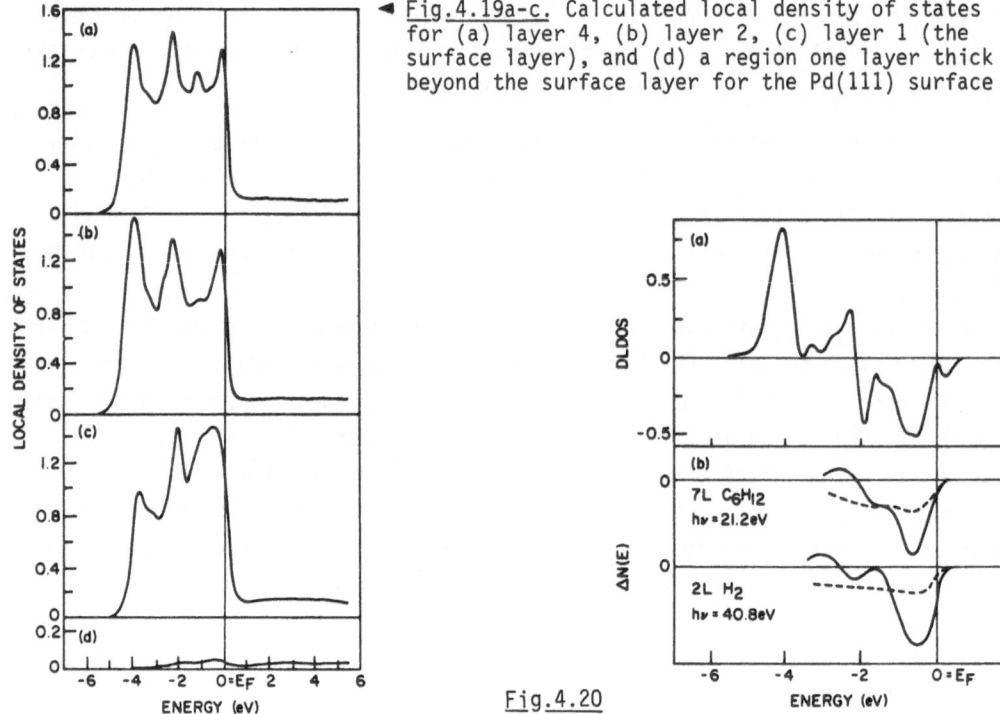

Fig.4.19a-c. Calculated local density of states for (a) layer 4, (b) layer 2, (c) layer 1 (the surface layer), and (d) a region one layer thick beyond the surface layer for the Pd(111) surface

Fig.4.20

Fig.4.20. (a) Calculated difference in LDOS between layer 4 and the surface layer. (b) Adsorbate-induced differences in photoemission intensities from [4.72]. The dashed curves indicate the estimated attenuation if uniform attenuation of the d-band were to occur

plane planar density of states (PDS) is almost identical with the bulk DOS for Pd, and the PDS for the next two planes is almost identical with that for the central plane. The surface layer PDS, however, is very different. It is larger in the region from 0 to 2 eV below E_F, and noticeably narrower. The second moment is 16% smaller than that of the other planes (compared with the 25% predicted by naive tight-binding theory).

Further, the results explain a characteristic, adsorbate-independent reduction of photoemission signal from 0-2 eV below E_F, observed when various adsorbates are put down on the Pd(111) surface [4.72,74]. Figure 4.20 shows the difference between the central and surface layer PDS. The negative region of this difference curve corresponds to excess electron states in the surface region. The lower curves of Fig.4.20 show the differential photoemission signal between clean and adsorbate covered Pd(111) [4.72]. The deficit in the 2 eV region below E_F can clearly be associated with surface states or rather with their removal to lower energies; they are the states sensitive to adsorbates, since they exist only near the surface. The bulk states are only mildly affected. The effect is adsorbate-independent because the states involved are characteristic of the *substrate*.

LOUIE also discussed the chemisorption of hydrogen on Pd(111) [4.70]. The model used is exactly that of the Pd(111) clean surface calculation, augmented to account for the addition of a (1 × 1) layer of atomic hydrogen. A regular monolayer of H atoms is added to each side of the 7-layer Pd(111) slab. An H^{+1} pseudopotential represents the hydrogen core (nucleus). The same 300 plane waves and 35 Gaussians used for clean Pd(111) calculations were used for the adsorbate calculations.

Three possible adsorption sites are considered — H directly above each surface atom, a threefold site directly above a second layer Pd atom, and a threefold site above a third layer Pd atom [an fcc(111) surface has hexagonal symmetry]. The H-Pd distance assumed is the sum of the 1.37 Å Pd metallic radius and the 0.32 Å H covalent radius. The results can be used to distinguish these three possibilities, all of which give a (1 × 1) surface geometry. The first possibility is clearly at variance with experiment, but the two possible threefold sites agree well with experiment and are essentially indistinguishable from one another.

The surface band structure for the third case is shown in Fig.4.21. The intrinsic Pd(111) surface states are strongly affected by H adsorption — some change character, some move to lower energies, and some disappear and presumably become part of adsorbate states. The two most striking new features are the narrow H-Pd bonding band which appears about 2 eV below the Pd d bands, and the 4 eV wide antibonding band above E_F near the $\bar{K}$ point in the surface Brillouin zone.

The H-Pd adsorbate states are almost completely confined to the H layer and the surface Pd layer, the Pd contribution being mostly d-like. Figure 4.22 shows the bulk Pd density of states, plus the surface-layer density of states for the Pd(111) surface before and after H adsorption (in the third case site). The bonding Pd-H band appears as a distinct peak at ~-6.5 eV, and there is an impressive reduction in density near E_F which is due to the removal of intrinsic surface states (some of which, of course, have reappeared in the -6.5 eV peak).

A very important result is the absence of any peak at all corresponding to antibonding Pd-H states, as would be predicted by naive chemisorption models [4.75, 76]. These states do exist; we see them in Fig.4.21. But they appear as a wide band above E_F, so spread out in energy that they are invisible in the DOS curves. This is a clear-cut rationale for calculating surface band structure. Cluster or other simple models which downplay the role of the periodic substrate just will not make the correct prediction.

Finally, Fig.4.23 compares the theoretical difference curves for density of states (clean vs dirty) for the three possible chemisorption sites with the photoemission difference spectra [4.72]. All three possibilities show the measured decrease near E_F, which comes from intrinsic surface state removal and is thus nonspecific, as already discussed. The peak at -6.5 eV, however, is moved all the way to -4.5 eV for case A (the onefold site), and clearly serves to rule out that possibility. Cases B and C cannot be distinguished on the basis of this peak or any

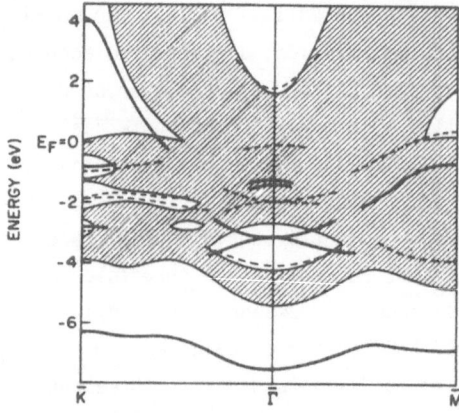

Fig.4.21. Localized states (solid curves) and the projected band structure (shaded areas) in the $\bar{K} = 2\pi/a$ $(0,-2/3,2/3)$ and $\bar{M} = 2\pi/a(-1/3,-1/3, 2/3)$ directions for the H-chemisorbed Pd(111) surface with H in the C-site. Also indicated are the surface states (dashed curves) for the clean surface

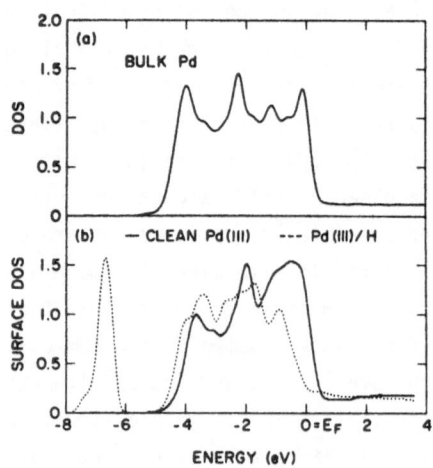

Fig.4.22a,b. Calculated LDOS for (a) Pd and (b) Pd(111) surface with and without a monolayer of H in the C-site

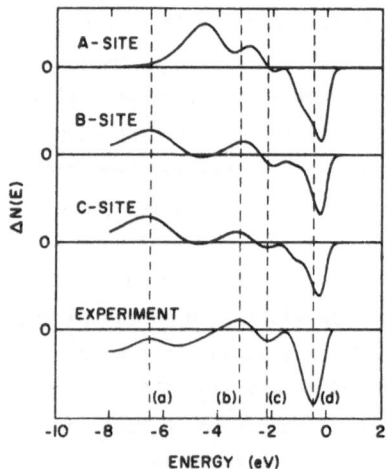

Fig.4.23. Comparison of the calculated H-induced photoemission difference spectra with the experimental difference curve. The experimental curve is from [Ref.4.72, Fig.2]

of the others. These two cases both have hydrogen in threefold sites, differing only in second layer geometry. Since surface effects do not appear to penetrate beyond the first layer, as we have seen, there is no reason to expect B and C to give rise to different results; they remain unresolvable here.

These calculations are fully self-consistent and use a very flexible basis set, including Gaussians as well as plane waves. Excellent results have been obtained. The palladium calculations are particularly complete and revealing. With the clean and adsorbate-covered surfaces done in exactly the same way, the changes upon chemisorption are easy to follow. The existence of surface states, their removal or modification upon chemisorption, and the appearance of adsorbate states both bonding and anti-bonding — these add up to a very complete description.

The only possible drawbacks are that, because of the use of pseudopotentials, core states cannot be calculated in this formalism, total energies are likely to be unreliable, and the 3d series of transition metals may not yield to the method (although an extensive enough set of Gaussian orbitals may well obviate this last difficulty).

4.6.2 Self-Consistent Local Orbital Method

We have developed an ab initio self-consistent procedure for calculating surface electronic structure which can be used for both clean d-band metal surfaces and for surfaces with chemisorbed atoms [4.77-79]. We give here a brief description of the method, followed by a discussion of our results for nitrogen chemisorbed on Cu(100). For a complete description of the computational procedure we refer to a recent publication [4.80].

Our self-consistent local orbital (SCLO) technique employs a local density approximation for exchange and correlation [4.81], linear combinations of Gaussian orbitals as basis functions, and a slab geometry with several layers of atoms arranged to represent the desired crystal face.

For maximum flexibility, the Gaussian basis should consist of many linearly independent orbitals of each angular momentum value. Such a basis set would be prohibitively large. Consequently we choose to "contract" the basis, by selecting as basis functions linear combinations of Gaussian orbitals chosen to approximate the atomic wave functions found from a Hartree-Fock-Slater calculation [4.34] for the atoms of the slab. These functions can be augmented as necessary, by the inclusion of "free" Gaussians for added variational freedom.

There are important reasons for using such a local orbital basis. A negative reason is that a plane-wave basis — such as used with pseudopotentials, or more sophisticated techniques — is not a practical approximation for the case of d orbitals. As already discussed, the d orbitals tend to preserve their identities and remain localized in the solid. The plane-wave basis is more appropriate to a very delocalized case. The atomic orbitals are correct in the limit of weak interactions, which for d orbitals seems much closer to reality. Furthermore, they form a very accurate representation of the core orbitals without further difficulty. Thus we are able to treat both core and valence orbitals within the confines of the same representation, which leads to welcome simplifications in the formulation and computational work.

More important still are the conceptual advantages of this atomic orbital basis, in that one obtains a natural or "chemical" picture of the interaction involved in chemisorption. A plane-wave representation is the handier one when interest centers on k space, on the energy bands, on Fermi surface properties. But our interest is on real-space properties, on the local properties of the interaction. Chemisorption occurs at a surface, a very localized concept; an adsorbate is a local thing—*where* it is, is very important. Everything, then, of interest to us in chemisorption is localized and must be handled in a formalism which deals properly with such local properties.

And we are interested in photoemission and in the initial density of states, which, we are going to demonstrate, is basically what photoemission measures, in the absence of strong matrix element effects. Now in studying photoemission and the density of states, it is of paramount importance to identify the causes, the sources of changes in the DOS curve which occur in the chemisorption process. In this context, the idea of a *local* density of states is essential to our understanding. This LDOS measures the contribution to the total density of various regions or orbitals. In an atomic orbital basis, this vital LDOS information is easily and naturally obtained.

Periodicity in the plane of the slab is used to reduce the Schrödinger equation to a number of finite matrix eigenvalue problems with dimension on the order of the number of electrons per atom, times the number of layers in the slab. We model the crystal in the slab geometry so as to retain this two-dimensional periodicity parallel to the surface. This allows us to retain most of the practical advantage of plane-wave methods and allows us to use a model containing a realistically large number of atoms at reasonable cost, where, e.g., a cluster model, which by nature is without such periodicity, would become prohibitively expensive for a relatively small number of atoms.

To further facilitate the calculation of the changes in matrix elements during the iteration to self-consistency, an artificial periodicity is introduced in the direction perpendicular to the slab by repeating the slab periodically. The gap between slabs is made large enough to insure negligible interaction between them. This artifice permits the use of reciprocal-space Fourier transform techniques developed for bulk matrix element calculations [4.82,83].

Finally, and probably most important of all, we are performing a fully self-consistent calculation. The changes involved in going from first guess to self-consistent answer are subtle and cannot be guessed a priori or sensibly approximated by extrapolation from available starting data. The changes in the electronic potential near the surface, just within and just outside, involve a complicated interaction back and forth. Small changes may have dramatic consequences, and it is hopeless to guess. Moreover, there is charge transfer between adsorbate and substrate, and the adsorbate is in a highly asymmetric environment. These changes are more profound, and even less easy to anticipate. Because we are interested in the chemical bonds formed and in the charge densities of this involved system, we must require self-consistency.

Our SCLO method thus proceeds as follows. Overlapping atomic potential matrix elements are computed by direct space integration. These are then converted into matrix elements of a potential between overlapping atomic charge densities by computing corrections in reciprocal space by Fourier transform techniques. In both calculations the expansion of the atomic orbitals as linear combinations of Gaussians allows all integrals to be done in closed form.

These potential matrix elements are used to initiate an iterative process in which the slab Hamiltonian is solved using a special points scheme [4.35] in the two-dimensional Brillouin zone of the slab, followed by the calculation of corrections to the potential matrix elements obtained from the difference between the charge density of the current solution and the charge density which gave the potential from which it was calculated.

A significant feature of the SCLO procedure is that the self-consistency corrections are computed entirely in reciprocal space, using Fourier transforms. As a consequence there are no constraints on the shape of the final self-consistent potential, other than the inherent symmetry of the slab. This contrasts with other methods where the potential is constrained to be of a specified form (muffin-tin, sum of spherical potentials, sum of s-type Gaussians, etc.).

These correction matrix elements, which are calculated in reciprocal space, are initially quite large and require careful handling if the iteration process is to converge. This difficulty is particularly acute in chemisorption calculations where atoms of different electron affinities are allowed to interact. To illustrate the nature of the difficulty consider the case of the Cu:N calculation [4.78]. In the starting potential (derived from overlapping atomic charge densities), the more electro-negative nitrogen attracts electrons. The effect is so strong that there are actually holes in the Cu d bands of the solution to the slab Hamiltonian [4.78]. The charge density associated with this solution thus assigns an excess of electrons to the nitrogen and a deficiency to the copper. Consequently, in the potential due to this charge density, the nitrogen now repels electrons and the copper attracts them. In the next Hamiltonian solution the nitrogen is stripped of electrons with the copper acquiring an excess. This "sloshing" of electrons is generally unstable and will grow rather than diminish. The standard counter measure is to add only a small fraction of the difference charge density to the old charge density at each iteration, but often more subtle tricks are necessary for ultimate convergence [4.80].

We now discuss the results for nitrogen chemisorbed on Cu(100). These involve a monolayer of nitrogen on either side of a 3-layer Cu(100) slab. In presenting these results, we begin by describing the 3-layer clean Cu(100) calculation which is used as the basis for comparison in discussion of before-and-after features of the nitrogen chemisorption.

The clean calculation was done for a 3-layer slab with (100) surfaces [4.77]. The slab is shown in Fig.4.24. The Herman-Skillman program [4.34] was used to obtain 1s through 4s atomic orbitals which were then approximated by even-tempered Gaussian fits obtained using the prescription of BARDO and RUEDENBERG [4.84]. These were used to generate a starting Hamiltonian which was iterated to self-consistency as described above. Kohn-Sham exchange, i.e., $X\alpha$ with $\alpha = 2/3$, was used throughout for the exchange potential.

In Fig.4.25a is shown the total density of states (TDS) for bulk copper as calculated by MUELLER [4.85] (who fit to BURDICK's [4.86] APW bands). In Fig.4.25b is our TDS for the thin film computed using 45 $k_{\shortparallel}$ points in the $1/8^{th}$ irreducible wedge of the surface Brillouin zone. Monte Carlo averaging and smoothing techniques were used. Note that there is *detailed* agreement between our TDS and the bulk TDS. The d bands have the same width, the same principal peaks, and are located at the same place relative to the Fermi level. This is a remarkable result for it suggests, barring monumental coincidence, that a three-layer Cu film looks very much like bulk Cu. Second, our atomic orbital calculation and an APW calculation—using entirely different methods—lead to similar results.

The curve in Fig.4.15c is the experimental photoemission spectrum for 21.2 eV photons incident on Cu(100) as taken by BURKSTRAND et al. [4.87].

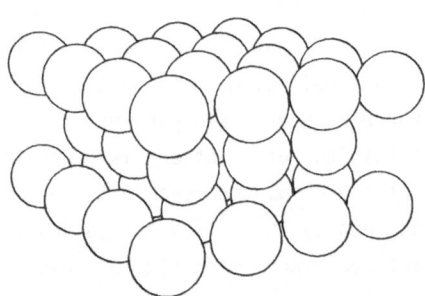

Fig.4.24. Computer drawn plot of Cu slab

Fig.4.25. (a) Bulk copper density of states as calculated by MUELLER [4.85] who fit to BURDICK's [4.86] bands. (b) TDS for thin film. (c) 21.2 eV photocurrent from [4.87]. (d) Central plane LDS. (e) Surface plane LDS

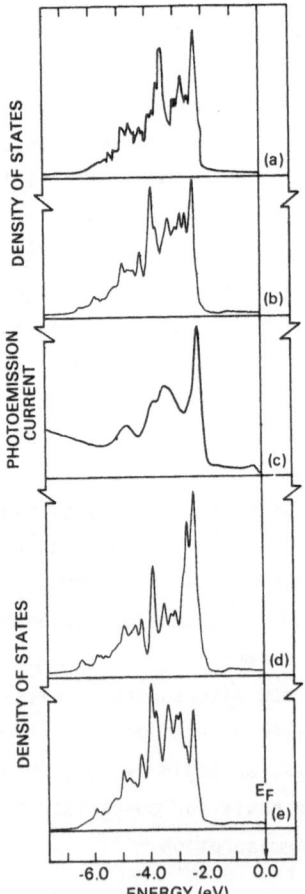

Note that it has essentially the same d-band width and energy interval between the Fermi level and the d-band edge as in Fig.4.25b. Now the peaks in the photoemission spectrum come primarily from electrons which have not suffered inelastic collisions. For copper in the 20-40 eV kinetic energy range, the electron mean free path is approximately [4.88] 6 Å. The distance between planes is 1.81 Å.

Thus it is of interest to compute the local densities of states (LDS). These can be computed easily for each plane in the localized basis that we use. The result for the central plane, the second plane from the surface, is shown in Fig.4.25d. That for the surface plane is in Fig.4.25e. Note that the second moment of the central plane PDS is larger than that for the surface. This is to be expected from the larger number of near neighbors for the second plane in. The most important conclusion one can draw from the comparison of Figs.4.25d and e is that the PDS varies quite markedly between the first and second plane from the surface. Such a marked variation has been seen [4.89] for alkali and alkaline earth metals, at rather short mean free paths. It also is suggested in the angular dependent photo-emission data from Cu(100) of FADLEY et al. [4.90].

In Figure 4.26 we show the self-consistent electronic charge density for the conduction band. Note that by the central plane, the charge contours are nearly spherical. This shape is very much like what one finds in bulk copper (see, e.g., [Ref.4.91, Fig.10]). In the surface plane, the charge spreads into the vacuum and smooths parallel to the surface, as originally postulated by SMOLUCHOWSKI [4.92]. It is this spreading and smoothing which leads to the surface dipole barrier.

We now turn to a discussion of the calculation for a monolayer of N atoms chemisorbed on Cu(100). The calculation is for a five-layer slab consisting of a three-layer Cu slab with a monolayer of N added to either side. The slab is shown in Fig.4.27. This Cu:N system was chosen because of the extensive experimental data which has been accumulated [4.87,93,94]. We had available as input to the calculation, the N layer registry [the N atoms occupy fourfold sites directly above the centers of the squares formed by Cu atoms in the Cu(100) surface] and the Cu-N interlayer spacing of 1.45 Å, which were determined by analysis of low energy diffraction data [4.93]. Ultraviolet photoemission spectroscopy (UPS) data on Cu(100):N shows striking changes relative to clean Cu(100) [4.87,94]. This pro-vides an excellent test for the theoretical calculation since a successful cal-culation must reproduce these changes to the extent that they are caused by changes in the electronic density of states.

The calculation discussed here was performed exactly as described for the clean Cu(100) calculation except that the Cu atomic orbital basis was augmented by N 1s, 2s, and 2p atomic orbitals located on the N atoms. This was the first self-con-sistent chemisorption calculation performed on a d-band metal. The experimental UPS spectrum of Cu(100):N is shown in Fig.4.28a in comparison with that for clean

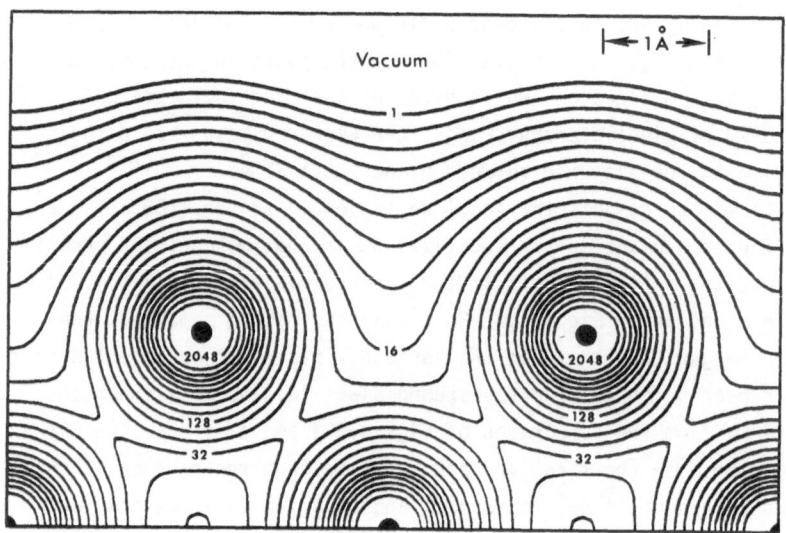

Fig.4.26. Electronic charge density at a copper (100) surface plotted on a plane perpendicular to the surface and passing through a line connecting a surface atom with one of its nearest neighbors in the second plane of atoms. The units of charge density are 1.15×10^{-3} a.u. Charge densities on successive contours are in the ratio $\sqrt{2}$. The range of contours shown is 1 to 2048

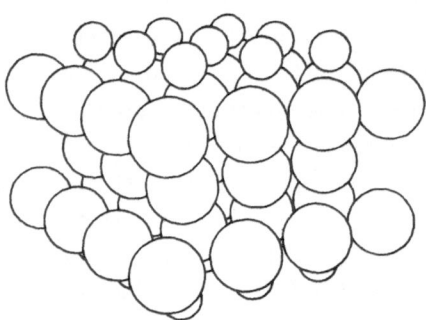

Fig.4.27. Computer drawn plot of Cu:N slab

Fig.4.28. (a) UPS data of N covered and ▶ clean Cu(100). (b) Theoretical TDS for the CuN slab and a clean Cu(100) slab

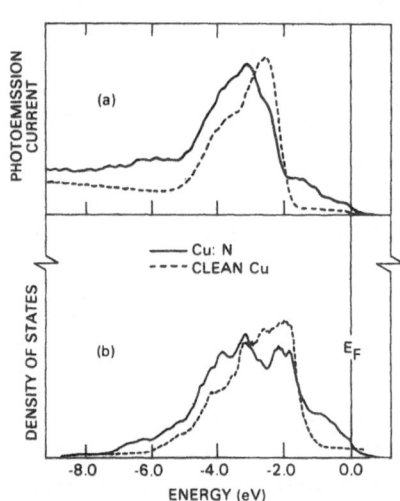

Cu(100)[1]. The adsorption of N is accompanied by substantial changes in the UPS spectrum which may be described as follows: 1) the intensity is increased in a region beginning somewhat below the bottom of the Cu d band and extending well into the d band, 2) the intensity is diminished in a region within and near the top of

[1]The UPS data was obtained as part of the research reported in [4.94]. It was obtained with 40.8 eV photons incident on a Cu(100) surface with 70% of a monolayer of N adsorbed. We thank Drs. Burkstrand and Tibbetts for supplying this data.

the d band, and 3) the intensity is increased in a region running from the top of
the d band to the Fermi level. In Fig.4.28b we show the calculated total density
of states (TDS) of the Cu(100):N slab in comparison with that for the clean
Cu(100) slab. The TDS were folded with a Gaussian of 0.4 eV width at half maximum.
This is the halfwidth of the electron spectrometer of [4.94] for electrons photoemitted by 40.8 eV photons. It is clear that each of the changes of the experimental spectrum described above is reproduced with sensibly the correct magnitude.
This is notable in view of the fact that the UPS data depend not only the initial
density of states (Fig.4.28b), but also on final state and matrix element effects[2].
However, we believe the ability of the TDS to reproduce the qualitative features
of the UPS spectra is strong evidence for the correctness and accuracy of our calculation.

It is necessary to look more closely at these results in order to determine the
detailed nature of the electronic interaction which causes these changes to take
place.

The PDS for each plane of the Cu:N slab are shown in Fig.4.29 along with the
TDS. The PDS of the N plane is normalized to 3/11 of the PDS of the Cu planes to
give the same weighting per valence electron. The PDS on the planes of the clean
Cu slab are shown for comparison.

The clean Cu densities of states were located relative to the N covered densities of states (in Fig.4.28 as well as Fig.4.29) by bringing the central plane
PDS into coincidence.

We may use these PDS to determine the origin of the features in the theoretical
TDS which lead to the good agreement with the UPS data discussed above. Figure
4.29 breaks down the copper-nitrogen results plane by plane. At the top is the
total density of states for the nitrogen-covered copper (100) slab. The second
panel shows the PDS for the central Cu plane (the solid curve) compared with that
same central plane PDS for clean copper. These curves are quite similar, in striking contrast with the case for the two Cu surface plane PDS shown next. Here the
copper surface 3d band has shifted sharply to lower energy, and broadened out as
well, upon the adsorption of nitrogen. Finally, the PDS for the nitrogen layer
is at the bottom. Nitrogen levels contribute throughout the energy range of the
d band, of course, but their maximum contribution clearly occurs above the d band.

---

[2]The initial TDS is only one of a number of sources of structure in UPS spectra
and it is only in special circumstances that the initial TDS is rigorously mir-
rored in the UPS spectrum [4.95]. Further the contribution of a plane of atoms to
the UPS signal should decrease exponentially with depth because of the short photoelectron escape depth. The slab TDS accounts for this only crudely by having two
N planes and two outer Cu planes

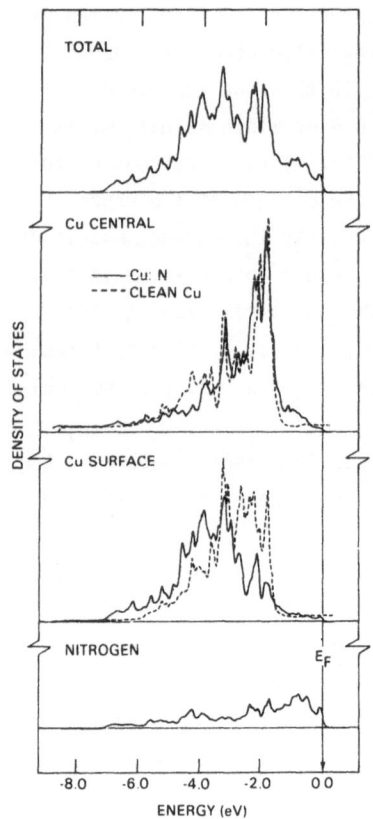

Fig.4.29. (a) Theoretical TDS of the Cu-N slab. (b) Central plane Cu PDS for the Cu-N slab (solid) and for the clean Cu slab (dashed). (c) Surface plane Cu PDS for the Cu-N slab (solid) and for the clean Cu slab (dashed). (d) N plane PDS

We are thus able to identify the features of the photoemission spectrum for nitrogen on copper with specific aspects of the electronic structure. As might be expected the principal contribution to the enhanced structure above the Cu d band is seen from Fig.4.29 to come from the 2p electrons on the N plane (see footnote 2). On the other hand, by comparison of the clean and N covered PDS of Fig.4.28, we see that the changes in and below the d band occur primarily because the Cu d-electron energies in the surface Cu plane are shifted downward and spread by interaction with the N electrons. Thus the increased structure near the bottom of the d band is not directly due to N electrons. It is emphatically not due to nitrogen 2s levels, which lie at much lower energy, off the scale of the figure.

We have previously remarked on the bulk-like quality of the Cu potential and charge density in the central plane. Additional evidence for this appears in Fig.4.29c where we see that the central plane PDS is largely unaffected by the chemisorption of N, indicating that the central plane atoms are not much perturbed by the presence of an adsorbate layer two planes away.

In Fig.4.30 we show the self-consistent charge density for the conduction band. The bulk-like nature of the central plane is again evident and, when this plot is

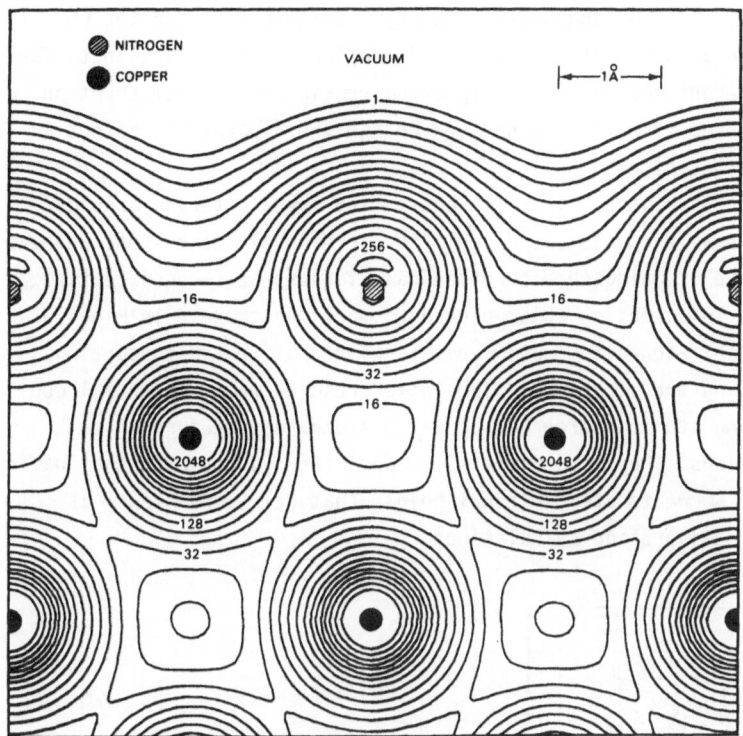

Fig.4.30. Self-consistent electronic charge density of the Cu:N conduction band plotted in a plane perpendicular to the surface and oriented to pass through the centers of neighboring atoms in each plane. The units are 1.5×10^{-3} a.u. Successive contours differ by a factor of $\sqrt{2}$ and range from 1 to 2048. Note that the maximum contour near the Cu nuclei is almost a factor of 10 larger than the maximum contour near the N nuclei

compared with the corresponding figure for clean Cu(100), it is clear that the central plane densities are virtually identical.

These facts show dramatically how rapidly metals heal from the effects of a perturbation. This rapid healing provides further confirmation that our results for the five-layer Cu:N slab can be meaningfully related to N chemisorbed on Cu(100).

We can summarize these results in a few words. First, the local density of states varies rapidly in the region near the surface; the charge density becomes remarkably like the bulk even at the second layer in from the surface. Chemisorption significantly alters the *surface* d-band density of states, with much less effect on planes further into the interior of the solid. The charge density is only locally changed upon chemisorption; the metal heals rapidly from this perturbation.

The photoemission spectra agree well with the initial densities of states, at least in this case. More calculations are needed before we can judge the degree to which this is true in general. There is an interesting contrast between these Cu:N results and the Ni:CO results discussed previously [4.32,39]. The latter displayed two well-separated features: a d band unperturbed by chemisorption, and the CO molecular levels, broadened and shifted upon chemisorption but still distinct. The Cu:N case is quite different. No distinct N spectrum remains after chemisorption. The N levels broaden, merge with the d band, and change the d band. The effect is a blend instead of distinct pieces.

Our Cu:N calculations have been fully self-consistent ones, offering considerable advantages over non-self-consistent calculations previously described. We have emphasized the importance of doing chemisorption calculations self-consistently throughout this review. To dramatize this point once more, we compare in Fig.4.31 the TDS from our final self-consistent Cu:N results with that from our starting potential, before starting the iterations. The latter result from a potential constructed from overlapping atomic charge densities, a standard procedure in bulk band structure calculations. They are completely wrong, even to the extent of having holes in the Cu d band. We reiterate our final point, that self-consistency is essential to reliable chemisorption calculations.

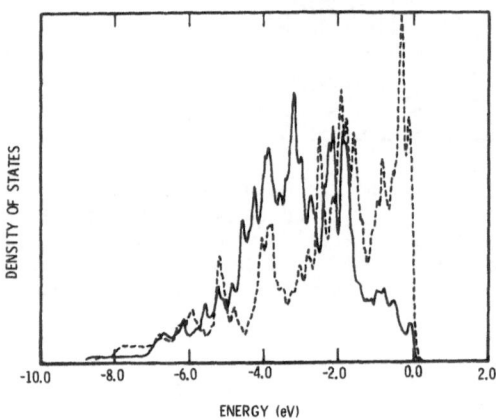

Fig.4.31. Comparison of the TDS of our self-consistent Cu:N calculation (solid) with that obtained (dashed) from our starting potential constructed by overlapping atomic charge densities. The Fermi level is at 0.0

Recently the authors have used a nine-layer slab model to investigate the self-consistent electronic structure of Cu(100) [4.96] and Ni(100) [4.97] surfaces. Our earlier Cu(100) calculation [4.77] was self-consistent and agreed well with experiment, using 3-layer-thick slabs; but a 3-layer slab is not thick enough to make apparent the localization of a state in the surface region, and thus does not suffice for the search for surface states. These nine-layer calculations were specifically motivated by the desire to determine the surface states/surface resonances of Cu(100) and Ni(100) surfaces.

We use as basis functions all the occupied core and valence atomic orbitals of Cu or Ni, augmented by 4p, 4d, and 5s orbitals to provide additional variational freedom. This provides a basis set of equivalent quality to a "double zeta plus polarization" basis set used in molecular calculations [4.98]. While this basis

set is not complete, of course, it does overlap both the bound and continuum eigenfunction subspaces [4.99] of the solid.

Since self-consistency corrections are determined entirely by Fourier transform
techniques, there are no constraints on the shape of the final self-consistent
potential. Once the iterative calculation achieves self-consistency, detailed
charge densities and electronic densities of states are computed, and a thorough
search is made for those states which can be called surface states or resonances.

Within the frame of reference of a slab model, surface states may be defined for
practical purposes as those states which exist primarily on the outer planes of the
slab and avoid the central planes. Accordingly, we accept an electronic state of
the slab as a surface state or resonance if more than 80% of the probability density
of its wave function is located on the outer two planes of the slab.

For the Cu(100) case, a number of such states are found, including a previously
unreported set of bands located at the upper edge of the copper d band. The two-
dimensional bands for Cu(100) are shown in Fig.4.32, computed on a mesh of 45 uniformly spaced points in the irreducible one-eighth of the two-dimensional Brillouin
zone. The occupied states on this mesh which satisfy the 80% criterion and which
lie along the high symmetry directions are shown in Fig.4.32 by the open circles.
The selected states lying about -6.5 eV correlate reasonably well with surface
state bands found in [4.22,100]. But the extensive group of states centered at
about -6 eV and running along the top edge of the d bands have not been observed
previously. We conclude that these edge bands, which are predominantly d-like, are
a distinctive feature of the *self-consistent* Cu(100) potential. Previous calculations
show no evidence of such a large surface state density near the top of the d band.

The planar densities of states (PDS) arising from the surface plane, second plane,
and central plane of the nine-plane slab are shown in Fig.4.33, as well as the total
density of states (DOS) for the slab. From the close resemblance of the central plane
PDS and the total DOS, it is apparent that the central region of the film is bulk-
like. But while the second plane in from the surface is very much like the central
plane, there is a striking difference in the surface plane. There is a definite
narrowing of the d band at the surface, with the second moment or mean square width
being reduced to only two-thirds of its value for the other planes.

The most interesting feature of Fig.4.33 is the large enhancement of the surface
PDS relative to the bulk (as exemplified by the central plane PDS) in the approximately 1 eV range at and above the d-band edge. Clearly this is principally due
to the surface states of Fig.4.32, which are heavily concentrated in the region at
-6 eV or just at the top edge of the d band.

We now discuss the experimental evidence for the existence of these surface edge
bands in Cu(100). Several studies have been performed which examined the changes in
photoelectron spectra which occur when various atomic species are chemisorbed on
Cu(100). The experiments have involved chemisorption of nitrogen [4.87], nitrogen,

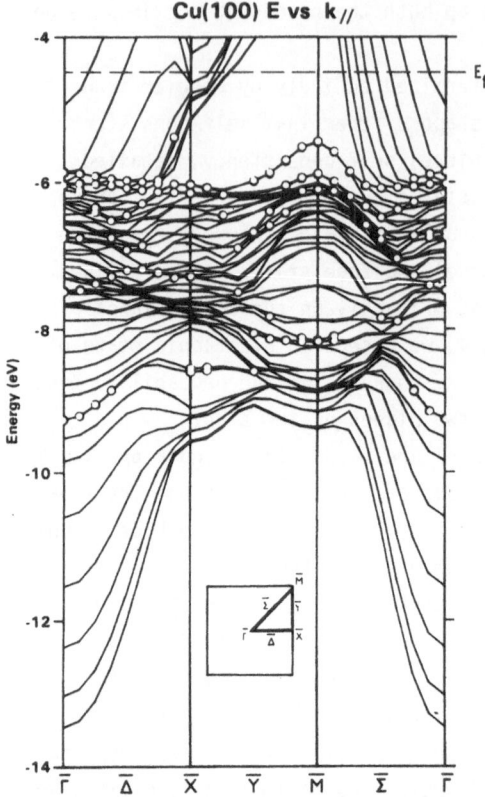

Cu(100) E vs k∥

Energy (eV)

Fig.4.32. Two-dimensional energy bands along high symmetry directions for the 9-plane Cu(100) slab. The open circles represent states on a 45-k point mesh (see text) which are highly localized in the surface of the slab

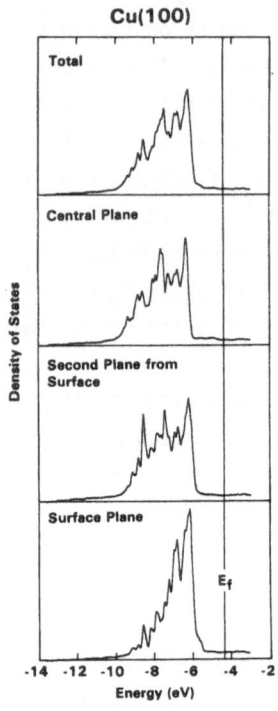

Cu(100)

Total

Central Plane

Second Plane from Surface

Surface Plane

Density of States

Energy (eV)

Fig.4.33. Calculated density of states curves for the 9-plane Cu(100) slab

oxygen and sulfur [4.94], and oxygen [4.101]. A universal feature of the data from these experiments is that a sharp peak in the Cu(100) spectrum at the top edge of the band is substantially attenuated when the atomic species are chemisorbed. This is illustrated in Fig.4.34, taken from [4.94], where the differences between the clean Cu(100) spectrum of 40.8 eV photon energy and the spectrum with fractional monolayers of nitrogen, oxygen, and sulfur are shown. The dominant feature in each difference spectrum is the sharp dip at the d-band edge which arises from attenuation of a large peak in the clean Cu(100) spectrum. In addition to appearing for all three adsorbates, the attenuation is also present in nitrogen chemisorption experiments at 16.7 and 21.2 eV photon energy [4.87], and in angular resolved oxygen chemisorption experiments at 13 and 19 eV [4.101].

Figure 4.34 also shows the theoretical difference curve, the difference between the planar densities of states of the central and surface planes of our film. The

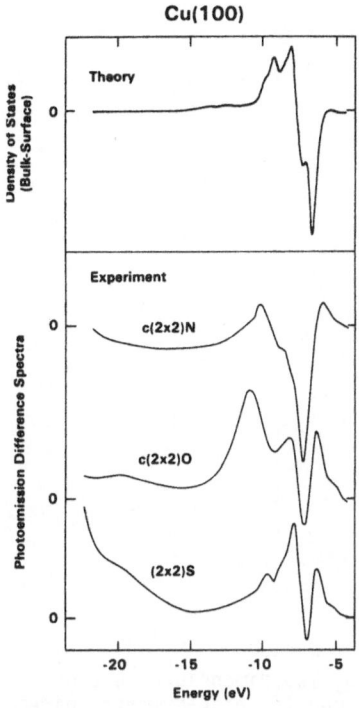

Fig.4.34. (a) Theoretical difference density of states between central and surface planes; (b) experimental difference spectra for Cu(100) covered with partial monolayers of various adsorbates

correlation with the experimental curves is striking. The fact that the large attenuation of the photoemission signal at the top of the d band is quite insensitive to photon energy and adsorbate shows that it can be traced to a property of the Cu(100) substrate. This large attenuation appears to be due to a high density of surface states, which are removed by chemisorption through bonding to the chemisorbed species.

The surface state band which peaks at $\bar{M}$ at about -5.5 eV (see Fig.4.32) has actually been observed in angular photoemission [4.102]. This surface state band is easiest to observe because it has the largest separation from the d band. Both the peak location and dispersion observed experimentally agree well with that predicted in Fig.4.32. To our knowledge this is the first time a surface state has been predicted theoretically prior to being detected experimentally.

The bands of Ni(100) are shown [4.97] in Fig.4.35. Again surface states are indicated by open circles. The Fermi energy for nickel lies within the d bands, near the top edge, unlike copper where E_F was well above the d bands. The striking contrast with the Cu(100) results is the relative paucity of surface states for Ni(100); almost none, as compared with copper.

Experimentally, the situation for Ni(100) is not so clear. However, the recent work of PLUMMER and EBERHARDT [4.103] does give evidence of a surface state at the Fermi energy. We see a few states in that vicinity, in the appropriate part of the surface Brillouin zone. PLUMMER and EBERHARDT proposed, however, that theirs is a

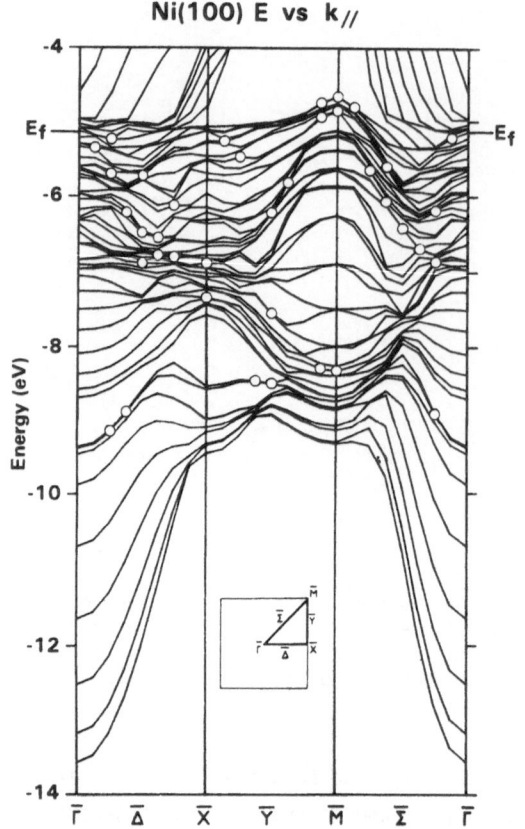

Ni(100) E vs k$_{//}$

Energy (eV)

E_f

$\bar{\Gamma}$ $\bar{\Delta}$ $\bar{X}$ $\bar{Y}$ $\bar{M}$ $\bar{\Sigma}$ $\bar{\Gamma}$

Fig.4.35. Two-dimensional energy bands along the high-symmetry directions for the 9-plane Ni(100) slab. The open circles represent states on a 45 k-point mesh which are highly localized in the surface of the slab

"magnetic" surface state, allowed because of a spin gap in the nickel bands. Our calculation is strictly paramagnetic, so that if the nickel surface state is truly a magnetic effect, it would not appear in our results.

The striking difference between the Cu(100) and Ni(100) predictions as to surface states remains perhaps the most thought-provoking feature of these results. One can only conjecture as to the source of this difference. On the one hand, it may be that a spin-polarized Ni calculation would give quite another result, as just mentioned. On the other hand, it may be that the position of the Fermi energy is the relevant variable. The Cu(100) d band is completely filled, with $E_F \sim 1.3$ eV above the d-band edge while of course the Ni(100) d band is not quite filled, with E_F within the d band, very near the top edge. This may be the relevant distinction. As results become available for more metals, this very intriguing point should become clearer; for now, the issue remains clouded.

4.7 Summary

This chapter has set forth a variety of techniques for calculating the effect of chemisorption on the electronic structure of d-band metals. A wide variety of models and non-self-consistent methods have demonstrated the cleverness of the modelers and the tremendous variety of conceptually interesting approaches to the problem. Many of these methods are quite useful in explaining experimental results.

As these methods developed, however, it began to appear that something better was needed. It is just too difficult to predict a priori the charge rearrangements and concomitant changes to the electronic potential which take place at either clean or adsorbate-covered surfaces. It is necessary to treat the problem self-consistently in order to understand the details. We outlined two methods, due to APPELBAUM and HAMANN [4.43], and to WANG and FREEMAN [4.54], which go some distance toward solving this problem.

Finally, we outlined the mixed-basis pseudopotential treatment of LOUIE [4.68] as well as the authors' self-consistent local orbital method [4.77-80], which are fully self-consistent computational methods proceeding from two different points of view. Both methods give results in excellent agreement with experiment and appear capable of application to a general class of chemisorption problems.

The methods work; it remains to apply them widely. We confidently expect the next few years to see the proliferation of self-consistent calculations on chemisorption systems, using these methods and perhaps others.

References

4.1   F. Ducastelle, F. Cyrot-Lackmann: J. Phys. Chem. Sol. *31*, 1295 (1970)
4.2   J.P. Gaspard, F. Cyrot-Lackmann: J. Phys. C *6*, 3077 (1973)
4.3   R. Haydock, M.J. Kelly: Surface Sci. *38*, 139 (1973)
4.4   M.C. Desjonquères, F. Cyrot-Lackmann: J. Phys. F *6*, 567 (1976)
4.5   K. Sturm, R. Feder: Solid State Commun. *14*, 1317 (1974)
      R. Feder, K. Sturm: Phys. Rev. B *12*, 537 (1975)
4.6   M.C. Desjonquères, F. Cyrot-Lackmann: J. Phys. F *5*, 1368 (1975)
4.7   M.C. Desjonquères, F. Cyrot-Lackmann: Solid State Commun. *20*, 855 (1976)
4.8   F. Cyrot-Lackmann: J. Vac. Sci. Technol. *9*, 1045 (1972)
4.9   E.W. Plummer, T.N. Rhodin: J. Chem. Phys. *49*, 3479 (1968)
4.10 W. Kohn: Phys. Rev. B *10*, 382 (1974)
4.11 J.C. Slater: *Symmetry and Energy Bands in Solids* (Dover Publications, New York 1972)
4.12 B.R. Cooper: Phys. Rev. Lett. *30*, 1316 (1973)
4.13 W. Kohn: Phys. Rev. B *11*, 3756 (1974)
4.14 N. Kar, P. Soven: Phys. Rev. B *11*, 3761 (1974)
4.15 S.J. Gurman: Surf. Sci. *55*, 93 (1976)
4.16 S.J. Gurman, J.B. Pendry: Phys. Rev. Lett. *31*, 637 (1973)
4.17 E. Caruthers, D.G. Dempsey, L. Kleinman: Phys. Rev. B *14*, 288 (1976)
4.18 J.C. Slater, G.F. Koster: Phys. Rev. *94*, 1498 (1954)

4.19 D.G. Dempsey, L. Kleinman, E. Caruthers: Phys. Rev. B *12*, 2932 (1975)
4.20 D.G. Dempsey, L. Kleinman, E. Caruthers: Phys. Rev. B *13*, 1489 (1976)
4.21 D.G. Dempsey, L. Kleinman, E. Caruthers: Phys. Rev. B *14*, 279 (1976)
4.22 K.S. Sohn, D.G. Dempsey, L. Kleinman, E. Caruthers: Phys. Rev. B *13*, 1489 (1976)
4.23 K.S. Sohn, D.G. Dempsey, L. Kleinman, E. Caruthers: Phys. Rev. B *14*, 3185 (1976)
4.24 K.S. Sohn, D.G. Dempsey, L. Kleinman, E. Caruthers: Phys. Rev. B *14*, 3193 (1976)
4.25 L. Kleinman: *Transition Metals – 1977*, The Institute of Physics (London) Conference Series No. 39, ed. by M.J.G. Lee, J.M. Perz, E. Fawcett (The Institute of Physics, Bristol and London 1978) Chap.5, p.314
4.26 D.J.M. Fassaert, A. van der Avoird: Surf. Sci. *55*, 291 (1976)
4.27 H. Basch, A. Viste, H.B. Gray: Theoret. Chim. Acta *3*, 458 (1965)
4.28 N.V. Smith, L.F. Mattheiss: Phys. Rev. Lett. *37*, 1494 (1976)
4.29 P.J. Estrup, J. Anderson: J. Chem. Phys. *45*, 2254 (1966)
4.30 B. Chakraborty, W.E. Pickett, P.B. Allen: Phys. Rev. B *14*, 3227 (1976); Phys. Lett. *48*A, 91 (1974); private communication to N.V. Smith and L.F. Mattheiss
4.31 B. Feuerbacher, R. Wills: Phys. Rev. Lett. *36*, 1339 (1976); private communication to N.V. Smith and L.F. Mattheiss
4.32 D.W. Bullett, M.L. Cohen: Solid State Commun. *21*, 157 (1977)
4.33 P.W. Anderson: Phys. Rev. *181*, 25 (1969)
4.34 F. Herman, S. Skillman: *Atomic Structure Calculations* (Prentice Hall, New York 1963)
4.35 D.J. Chadi, M.L. Cohen: Phys. Rev. B *8*, 5747 (1973)
      S.L. Cunningham: Phys. Rev. B *10*, 4988 (1974)
4.36 O.K. Andersen, R.V. Kasowski: Phys. Rev. B *4*, 1064 (1971)
4.37 R.V. Kasowski: Solid State Commun. *17*, 179 (1975)
4.38 R.V. Kasowski: Phys. Rev. Lett. *33*, 1147 (1974)
4.39 R.V. Kasowski: Phys. Rev. Lett. *37*, 219 (1976)
4.40 D.E. Eastman: Phys. Rev. B *3*, 1769 (1971)
4.41 H.D. Hagstrum, G.E. Becker: Phys. Rev. Lett. *22*, 1054 (1969)
4.42 T. Gustafsson, E.W. Plummer, D.E. Eastman, J.L. Freeouf: Solid State Commun. *17*, 391 (1975)
4.43 J.A. Appelbaum, D.R. Hamann: Solid State Commun. *27*, 881 (1978)
4.44 H. Sambe, R.H. Felton: J. Chem. Phys. *62*, 1122 (1975)
4.45 B.C. Allen: Trans. Met. Soc. AIME *227*, 1175 (1963)
4.46 P.O. Gartland, B.J. Slagsvold: Phys. Rev. B *12*, 4047 (1975)
4.47 P. Hermann, H. Neddermeyer, H.F. Roloff: J. Phys. C *10*, L17 (1977)
4.48 D.E. Ellis, G.S. Painter: Phys. Rev. B *2*, 7887 (1971)
4.49 A. Zunger, A.J. Freeman: Phys. Rev. B *15*, 4716 (1977)
4.50 W. Kohn, L.J. Sham: Phys. Rev. A *140*, 1133 (1965)
4.51 W.Y. Ching, C.C. Lin: Phys. Rev. B *12*, 5536 (1975)
4.52 C.B. Haselgrove: Math. Compt. *15*, 373 (1951)
4.53 O. Jepsen, O.K. Anderson: Solid State Commun. *9*, 1763 (1971)
      G. Lehman, M. Tant: Phys. Status Solidi *54*, 469 (1972)
      C.S. Wang, J. Callaway: Comp. Phys. Commun. *14*, 327 (1978)
      J. Rath, A.J. Freeman: Phys. Rev. B *11*, 2019 (1975)
4.54 C.S. Wang, A.J. Freeman: Phys. Rev. B *19*, 793 (1979)
4.55 W. Eib, S.F. Alvarado: Phys. Rev. Lett. *37*, 444 (1976)
      D.E. Eastman: Phys. Rev. B *2*, 1 (1970)
      Y. Fukuda, W.T. Elam, R.L. Park: Phys. Rev. B *16*, 3322 (1977)
4.56 D.G. Dempsey, L. Kleinman: Phys. Rev. Lett. *39*, 1297 (1977)
4.57 C.S. Wang, A.J. Freeman: Phys. Rev. B *18*, 1714 (1978)
4.58 F.J. Arlinghaus, J.G. Gay, J.R. Smith: Phys. Rev. B *20*, 1332 (1979)
4.59 C.S. Wang, A.J. Freeman: J. Appl. Phys. *50*, 1940 (1979)
4.60 C.S. Wang, A.J. Freeman: Phys. Rev. B *19*, 4930 (1979)
4.61 J.E. Demuth, D.W. Jepsen, P.M. Marcus: Phys. Rev. Lett. *31*, 540 (1973)
4.62 S.G. Louie, K.M. Ho, J.R. Chelikowsky, M.L. Cohen: Phys. Rev. Lett. *37*, 1289 (1976); Phys. Rev. B *15*, 5627 (1977)

4.63 See for example M. Schlüter, J.R. Chelikowsky, S.G. Louie, M.L. Cohen: Phys. Rev. B 12, 4200 (1975)

4.64 R.P. Leblanc, B.C. Vanbrugghe, F.E. Girouard: Can. J. Phys. 52, 1589 (1974)

4.65 G.P. Kerker, K.M. Ho, M.L. Cohen: Phys. Rev. Lett. 40, 1593 (1978)

4.66 G.P. Kerker, K.M. Ho, M.L. Cohen: Phys. Rev. B 18, 5473 (1978)

4.67 Shang-Lin Weng, T. Gustafsson, E.W. Plummer: Phys. Rev. Lett. 39, 822 (1977)

4.68 S.G. Louie, K.M. Ho, M.L. Cohen: Phys. Rev. B 19, 1774 (1979)

4.69 S.G. Louie: Phys. Rev. Lett. 40, 1525 (1978)

4.70 S.G. Louie: Phys. Rev. Lett. 42, 476 (1979)

4.71 J.F. Janak, D.E. Eastman, A.R. Williams: Solid State Commun. 8, 271 (1970)

4.72 J.E. Demuth: Surf. Sci. 65, 369 (1977)

4.73 F.J. Himpsel, D.E. Eastman: Phys. Rev. Lett. 41, 507 (1978)

4.74 H. Conrad, G. Ertl, J. Kuppers, E.E. Latta: Surf. Sci. 58, 578 (1976)

4.75 D.M. Newns: Phys. Rev. 178, 1123 (1959)

4.76 K. Schönhammer: Solid State Commun. 22, 51 (1977)

4.77 J.G. Gay, J.R. Smith, F.J. Arlinghaus: Phys. Rev. Lett. 38, 561 (1977)

4.78 J.R. Smith, F.J. Arlinghaus, J.G. Gay: Solid State Commun. 24, 279 (1977)

4.79 J.R. Smith, F.J. Arlinghaus, J.G. Gay: *Transition Metals – 1977*, The Institute of Physics (London) Conference Series No. 39, ed. by M.J.G. Lee, J.M. Perz, E. Fawcett (The Institute of Physics, Bristol and London 1978) Chap.5, p.303

4.80 J.R. Smith, J.G. Gay, F.J. Arlinghaus: Phys. Rev. B (to be published)

4.81 J.C. Slater: *The Self-Consistent Field for Molecules and Solids* (McGraw-Hill, New York 1974)

4.82 E. Lafon, C. Lin: Phys. Rev. 152, 579 (1966)

4.83 J. Rath, J. Callaway: Phys. Rev. B 8, 5398 (1973)

4.84 R.D. Bardo, K. Ruedenberg: J. Chem. Phys. 59, 5956 (1973)

4.85 F.M. Mueller: Phys. Rev. 153, 659 (1967)

4.86 G.A. Burdick: Phys. Rev. 129, 138 (1963)

4.87 J.M. Burkstrand, G.G. Kleiman, G.G. Tibbetts, J.C. Tracy: J. Vac. Sci. Technol. 13, 291 (1976)

4.88 C.J. Powell: Surf. Sci. 44, 29 (1974)

4.89 C.R. Helms, W.E. Spicer: Phys. Rev. Lett. 57A, 369 (1976)

4.90 L.F. Wagner, Z. Hussain, C.S. Fadley: Solid State Commun. 21, 257 (1976)

4.91 C.Y. Fong, J.P. Walter, M.L. Cohen: Phys. Rev. B 11, 2759 (1975)

4.92 R. Smoluchowski: Phys. Rev. 60, 661 (1941)

4.93 G.G. Kleinman, J.M. Burkstrand: Solid State Commun. 21, 5 (1977)

4.94 G.G. Tibbetts, J.M. Burkstrand, J.C. Tracy: Phys. Rev. B 15, 3652 (1977)

4.95 B. Feuerbacher, R.F. Willis: J. Phys. C 9, 169 (1976)

4.96 J.G. Gay, J.R. Smith, F.J. Arlinghaus: Phys. Rev. Lett. 42, 332 (1979)

4.97 F.J. Arlinghaus, J.R. Smith, J.G. Gay: Phys. Rev. B (to be published)

4.98 Henry F. Schaefer III: *The Electronic Structure of Atoms and Molecules* (Addison-Wesley Publishing Co., Reading, Mass. 1972) p.77

4.99 H. Shull, P. Lowdin: J. Chem. Phys. 30, 617 (1959)

4.100 C.M. Bertoni, O. Bisi, C. Calandra, F. Nizzoli, G. Santors: J. Phys. F 6, L41 (1976)

4.101 D.T. Ling, J.N. Miller, P.A. Pianetta, D.L. Weissman, I. Lindau, W.E. Spicer: J. Vac. Sci. 15, 495 (1978)

4.102 P. Heimann, J. Hermanson, H. Miosga, H. Neddermeyer: Phys. Rev. Lett. 42, 1782 (1979); Phys. Rev. B 20, 3059 (1979)

4.103 E.W. Plummer, W. Eberhardt: Phys. Rev. B 30, 1444 (1979)

5. Cluster Chemisorption

A. B. Kunz

With 9 Figures

In the past several years, considerable emphasis has been placed on the study of solid surfaces and on the interaction of adsorbates with such surfaces. This interest has arisen for a number of reasons. The chief reasons are: surface problems are of considerable scientific and technical interest of themselves; and that the sensitivity of experimental procedures, and the power of theoretical techniques combined with the increase in availability of high-speed digital computers have finally made possible surface science studies to a wide variety of investigators. An example of a problem of considerable scientific as well as technological interest, which is amenable to cluster study, would be the relative change in position of energy levels in a semiconductor due to an impurity as a function of whether that impurity is substitutional, interstitial, in the bulk or at a surface. Techniques are evolving to where this question may be answered in the near future. A second example which is also of scientific and technological interest concerns the influence of surface detects on chemisorption. This question is currently under intensive study by several groups at the University of Illinois and elsewhere. The other paramount reason for the interest in surface science has been brought about by the "energy crisis". Considerable interest has been focused on the use of more efficient catalysts for coal gasification, or fuel cells for example, or producing better converters for solar generation of electricity. The net result of all this has been an enormous effort in understanding the various properties of surfaces. In this chapter we will concentrate upon a subset of the theoretical techniques used for such a study. This subset is the use of finite clusters of atoms/ions to simulate a solid surface or even the bulk solid.

In the context of this chapter a cluster will be defined as a collection of *one* or more atoms, ions, or molecules with *appropriate boundary conditions* which are studied in some detailed sense for one or more physical properties and is considered either by virtue of size of system studied or its boundary conditions to represent this behavior for a larger system such as that of an infinite solid or a semi-infinite surface problem. This definition is problematic in that the nature of the imposed boundary conditions and the size of the cluster is left to the discretion of the person studying the problem. In this chapter we shall briefly consider some of the formal aspects of these questions. A second problem to be faced in the application of cluster techniques is that certain systems and certain problems are far more amenable to such an approach than are others. Furthermore, certain useful methods are more directly applicable to studies of certain specific phenomena than are other methods and vice versa. This complicates further the choice of technique since one may wish to match techniques to phenomena. It is this question which we shall address first. In this chapter we shall discuss the question of specific methods of solving cluster problems, the question of boundary conditions, and finally consider in some detail recent specific examples in which cluster techniques are used to solve problems of scientific interest.

5.1 Ideal Choice of Problem for Cluster Simulation

The cluster by virtue of its finite physical size is from the outset clearly un-suited to describe any physical phenomenon for which the spatial extent of that phenomenon is greater than the physical extent of the cluster in any significant way. On this basis, a study of phenomena involving, say, electrical conductivity or the energy bands of a metal would require a cluster so large as to be impracti-cal from a computational standpoint. In any case such long-range phenomena are better studied using the techniques of energy band theory and are excluded from further consideration in this chapter. On the other hand, phenomena in which the significant physical extent of such phenomena are smaller than the spatial extent of the cluster may be ideal subjects for a cluster simulation. Examples of such phenomena include: localized excitations in solids (a tightly bound exciton is a prime example); the formation of self-trapped quasiparticles in a solid (the self-trapped hole or V_k center in an alkali halide is one such system); various properties of point defects or impurities in solids (the F center or electron trapped at a halogen vacancy in an alkali halide is the usual text book example for such a system) [5.1]; studies related to the formation and breaking of the chemical bond (chemisorption forming our most important example of this) [5.2]. In this chapter we concentrate our examples wholly upon this last example. It is a matter of some importance, however, to acknowledge that the formulae presented and the techniques described are simply and directly applicable to the entire range of scientific problems for which cluster techniques are applicable.

The question of chemisorption and the formation of chemical bonds is amenable to study by cluster techniques for several reasons. These include the fact that the chemical bond and in particular covalent types of bonds are normally regarded as being of very short range in character (typically of order one internuclear separa-tion) and therefore, a cluster large than the bond length is easily studied. In many real world cases the adsorbate during chemisorption is not deposited in a uniform, ordered layer on a surface and hence long-range order in the plane of the surface is absent and traditional solid state techniques using translational periodicity are of little use, and here a cluster model is useful and efficient. It is acknowledged that traditional techniques of solid state physics may be patched up in order to extend them to the case of an aperiodic surface problem. In doing this one constructs an artificially large unit cell including the solid and some adsorbate and then replicates this system in some periodic way. If this is to be a successful technique, it is necessary to make the unit cell so large so that the effects of the adsorbate superlattice formed in this way are negligible [5.3]. Unfortunately such a system may be even less economical to solve than is a comparable cluster and the degree of difficulty is greater. Therefore, the author is of the opinion that for these problems in which long-range order doesn't occur naturally, cluster models are definitely preferable.

Finally it is worth stating: that due to the inherent size limitations in a cluster simulation, and always assuming, of course, that an accurate model is used to solve for the physical properties, questions of a chemical nature such as bond strength, bond length and geometrical data in general should be most accurately given by a cluster model. Physical properties such as absorption spectra, photoemission data may be very poorly represented by a finite cluster unless the optical absorption is a local excitation or in the case of photoemission the final-state hole remains localized spatially. In this respect, the proper test of whether or not a cluster model is valid for describing nonlocalized electronic states is *not* whether the cluster has a set of eigenvalues which bear some resemblence to some measured spectrum, but whether or not the cluster electronic structure accurately reproduces the same electronic states calculated using the techniques of energy band theory assuming, as is necessary, that the same model is used to describe the cluster and the energy band problem.

It is fortunate that there are at least two available models today for which one has obtained solutions in the band theory limit and in the cluster limit. These are the Hartree-Fock model (HF) [5.4] and the scattered wave X-α model (MSX-α) [5.5]. The basis of these models is discussed in the next section. We may nonetheless see how these models make a transition from the atomic systems to solid state properties for a real material, namely solid metallic lithium [5.6] (a nearly free electron case) and also for the case of NiO a tightly bound insulator [5.7,8]. Some relevant physical quantities are shown in Tables 5.1 and 2.

There is one immediate and obvious lesson to be learned from these tables. This is, for small clusters the convergence of cluster properties to the infinite solid limit seems to be far more rapid for the rather tightly bound system such as NiO than for simple metallic systems such as Li metal clusters. There is one additional lesson to be gleaned from these examples. In cases such as NiO (here there is only 1 Ni atom in the cluster) crystal field effects and spin polarization

Table 5.1. The Fermi energy, ε_F, the work function, ε_ω, the nearest neighbor distance, nn, are shown for clusters of from 1 to 13 Li atoms and for bulk lithium. Results in the Unrestricted Hartree-Fock (UHF) approximation and in the X-α limit are given as are pertinent experimental results when available. The X-α value for ε_ω is given using both the X-α eigenvalue and also the theoretically more accurate transition state method. Energies are in eV and lengths are in atomic units (1 a.u. ≈ 0.53 Å). Data is from [5.6]

Property / No. of atoms	ε_F *1			ε_ω				nn		
	Exp.	UHF	X-α	Exp.	UHF	Eigenvalue X-α	Transition State X-α	Exp.	UHF	X-α
1	0	0	0	5.39	5.34	2.6	5.4	-	-	-
2	0	0	0	n.a.	4.71	3.2	5.2	5.04	5.60	5.08
8	n.a.	2.67	1.4	n.a.	4.36	3.8	4.8	n.a.	6.08	5.59
9	n.a.	4.31	2.8	n.a.	4.31	2.7	3.9	n.a.	5.66	5.25
13 *2	n.a.	n.a.	2.8	n.a.	n.a.	3.7	4.3	n.a.	n.a.	6.10
∞	4.	9.2	3.5	2.28	(3)	(3)	n.a.	5.65	n.a.	n.a.

1. The Fermi energy is the energy difference of the least bound valence orbital and the most bound valence orbital.
2. The 13 atom cluster is in an icosahedral geometry.
3. These values are not given because the solid work function is a composite of the ∞ solid band energies and of a surface dipole moment. Due to inadequate detail in the calculated band structure potential information, it is not possible to correct these results for the surface dipole.

have a very large role in determining the energy level structure whereas overlap and the number of neighbors seem to dominate electronic structure determination of Li. We would infer from these studies that properly handled small clusters of only a few atoms may yield good qualitative and fair quantitative understanding of spatially local properties such as the formation of chemical bonds.

5.2 Methods of Obtaining a Solution

Methods used for cluster studies fall into two basic types, empirical methods and nonempirical methods. Within each type there is a near infinity of subdivision of methods available. In this section we briefly discuss only a few of the more popular types of models. We acknowledge at the outset that most methods have some good points and some bad points; when used by careful workers results of significant utility are produced. Inside the category of nonempirical methods (also often termed ab initio or first-principle methods) one can further subdivide into two general categories; variationally determined models (such as Hartree-Fock) which this author admittedly favors, and local density models which have a conceptual basis in the Hohenberg-Kohn Theorem [5.11]. We shall briefly consider both methods

Table 5.2. Cluster calculations for NiO_6^{10-} exist in both the restricted Hartree-Fock (RHF) limit, the unrestricted Hartree-Fock (UHF) limit and the X-α limit. Band calculations for this system exist in the restricted Hartree-Fock limit and in a non-self-consistent Slater exchange limit, and both self-consistent Slater ($\alpha = 1$) and Kohn-Sham ($\alpha = 2/3$) exchanges. The essential properties are similar in the Slater and Kohn-Sham limits and since the X-α limit lies between these extremes we give the Slater results by means of comparison here. Energies are in eV here and in the case of the X-α cluster calculation transition state energies are used. A negative gap is equivalent here to a band overlap of that amount. The X-α results are from [5.7], the UHF results from [5.8], the X-α band results from [5.9] and the HF band results from [5.10].

Property System	Width of Ni 3d energy band [eV]	Width of O 2p energy band [eV]	Gap between Ni 3d and O 2p band[2] [eV]
NiO_6^{10-} UHF [1]	2.2	4.4	- 0.5
NiO_6^{10-} RHF [1]	1.5	4.0	- 1.1
NiO_6^{10-} X-α [1]	2.5	6.3	0.5
correlated HF band	2.6	6.5	- 1.7
Slater band nonself-cons.	2.0	2.4	4.22
Slater band self-cons.	2.3	3.1	0.7
exp. on o solid	~ 3.5	~ 5	~ - 2

1. The notation 10- implies the 6-O plus 1-Ni atoms in the cluster have a net charge of -10 electron charges. The surrounding point ion field has a net charge of +10 electron charges to force charge neutrality.
2. A negative band gap implies an overlap of the bands.

below. The empirical models are of seemingly infinite variety and of varying sophis-tication. Of these models we shall consider briefly the philosophy behind only the xtended Hückel (EHT) model [5.12]. The EHT model is that empirical method which s most widely used for cluster chemisorption studies. In all these studies we shall gnore the kinetic energy of the nuclei and therefore employ the conventional Born-ppenheimer approximation [5.13].

The starting place for such studies is the n-electron, N-nuclei Hamiltonian, H. n this case we use lower case letters for electron properties and upper case etters for nuclear properties. H is given as

$$H = -\sum_{i=1}^{n} \frac{\hbar^2}{2m} \nabla_i^2 - \sum_{i=1}^{n} \sum_{I=1}^{N} \frac{e^2 Z_I}{|r_i - R_I|}$$

= h/2π (normalized Planck's constant)

$$+ \frac{1}{2} \sum_{\substack{i=1 \\ i \neq j}}^{n} \sum_{j=1}^{n} \frac{e^2}{|r_i - r_j|} + \frac{1}{2} \sum_{\substack{I=1 \\ I \neq J}}^{N} \sum_{J=1}^{N} \frac{e^2 Z_I Z_J}{|R_I - R_J|} \tag{5.1}$$

where R_I is the position of the Ith nucleus, Z_I is its charge, e is the electron charge, m is its mass and r_i is the position of the ith electron. Ideally one wishes to solve an n-electron problem in which the n-electron wave function $\bar{\psi}$ is dependent upon the x_i, the space and spin coordinator of each electron, and parameterically depends upon the coordinates of each nucleus. Practically, this is impossible and we shall at once specialize to the one-electron type of approximation. (This is often called the independent particle model).

We may require that $\bar{\psi}$ be approximated as an antisymmetrized product of one-electron orbitals $\psi_i(x_j)$ [5.14]. This is conveniently represented as a Slater determinant,

$$\bar{\psi}_{HF}(x_1 \cdots x_n) \simeq \frac{1}{\sqrt{N!}} \det || \psi_i(x_j) || \ . \tag{5.2}$$

In what follows we will require that the one-electron orbitals be orthonormal and that these be determined by use of the variational method. Therefore, if we use

$$\delta_{\psi_k}^{\dagger} \left[\langle \psi | H | \psi \rangle + \sum_{i,j} \pi_{ij} \left(\delta_{ij} - \int \psi_i^* \psi_j d\tau \right) \right] = 0 \quad , \tag{5.3}$$

one obtains an equation defining $\psi_i(x)$. This is the Hartree-Fock equation, and is (after diagonalizing the Lagrange multiplier matrix),

$$\left[-\frac{\hbar^2}{2m} \nabla^2 - \sum_I \frac{e^2 Z_I}{|r-R_I|} + e^2 \sum_j \int \frac{\psi_j(x')^2}{|r-r'|} dx' \right] \psi_i(x)$$

$$e^2 \sum_j \psi_j(x) \int \frac{\psi_j^*(x') \psi_i(x')}{|r-r'|} dx' = \varepsilon_i \psi_i(x) \quad . \tag{5.4}$$

In obtaining (5.4) the only constraint which has been placed on the wave function is that the orbitals, ψ, be functions of the space-spin coordinates of a single particle and that these orbitals form an orthonormal set. Other than the requirement that energy minimization occur, no additional constraint has been placed upon the orbitals. In this unconstrained limit, we term this the generalized Hartree-Fock (GHF) approximation. This level of approximation is not used in practical calculations since by not having orbitals which are eigenfunctions of z component of spin, s_z, computer time and storage will be increased. The author is of the opinion that this method should be further considered to see if this objection is outweighed by any advantage gained from the flexibility of the one-electron orbital.

The simplest constraint which one can usefully employ is to require that $\psi_i(\underline{x})$ be an eigenfunction of s_z. That is, we require,

$$\psi_i(\underline{x}) = \psi_i(\underline{r}) \begin{matrix} \alpha \\ \beta \end{matrix} \quad . \tag{5.5}$$

Here α is the Pauli eigenfunction of $s_z = 1/2$ and β for $s_z = -1/2$. In doing this the functional form of (5.4) is not modified and the form of $\psi_i(\underline{r})$ is unconstrained. This is the unrestricted Hartree-Fock approximation (UHF). This method is used for practical calculations [5.4,6].

For the purposes of solution and simplicity, (5.4) is often written in terms of the Fock-Dirac first-order density matrix, $\rho(\underline{x}\,\underline{x}')$,

$$\rho(\underline{x}\,\underline{x}') = \sum_{i=1}^{n} \psi_i(\underline{x})\psi_i^{\dagger}(\underline{x}') = \sum_{i=1}^{n} |i(\underline{x})|><|i(\underline{x}')| \quad . \tag{5.6}$$

Note here this form permits $\rho(\underline{x}\,\underline{x}')$ to act as an operator. The Fock equation is simply

$$F\,\psi_i(\underline{x}) = \varepsilon_i\psi_i(\underline{x}) \quad ,$$

$$F = -\frac{\hbar^2}{2m} \nabla_i^2 - \sum_I \frac{e^2 Z_I}{|\underline{r}-\underline{R}_I|} + e^2 \int \frac{\rho(\underline{x}'-\underline{x}')}{|\underline{r}-\underline{r}'|}\,d\underline{x}$$

$$- e^2\rho(\underline{x}\,\underline{x}')/|\underline{r}-\underline{r}'| \cdot \hat{P}(\underline{x}'\underline{x}) \quad . \tag{5.7}$$

$\hat{P}$ is the operator which interchanges coordinate $\underline{x}$ with $\underline{x}'$. In addition to computational utility, this change of notation is useful in that it makes clear the Hermiticity of the GHF or UHF equation and hence shows how the orthonomality constraint is achieved.

There are many other levels of approximation possible in the Hartree-Fock case and the interested reader is referred to SCHAEFER's book to see their discussion [5.4]. There is one other level which must concern us at this time, that is the restricted Hartree-Fock method (RHF). This method is normally called simply the Hartree-Fock method (HF) in the literature and should not be confused with the GHF or UHF model previously discussed. It is with respect to an RHF solution, historically, that WIGNER defined the concept of correlation energy [5.15]. The constraints are as follows: electrons are assumed to occur in spin up, spin down pairs as far as possible and that the spatial part of the orbitals for such a spin up, spin down pair be identical. Secondly the spatial part of the orbitals are constrained to transform according to one of the irreducible representations of the space group for the nuclear geometry. In all except filled shell cases the RHF solutions are not obtained from (5.7) but from (5.7) plus a set of off-diagonal Lagrange multipliers chosen to impose the constraints. In addition, one often actually constructs a multideterminant solution from the RHF solution such that the multideterminant solution will be an eigenfunction of operators such as spin and spatial symmetry. In this latter case it is important to decide if one projects out the desired eigenfunctions before or after performing the variation. This point is considered fully by SCHAEFER.

There is one very important point to be considered in the use of the RHF model, which has been the subject of a detailed review [5.16], and which is that the imposed symmetry constraint is sufficiently strong so that the process of energy minimization is greatly hindered, in some cases. Therefore estimates of total energy may be found better in some cases by the UHF method. This point is quite well illustrated for the case of H_2.

Physically one believes that if two hydrogen atoms are far apart from each other, they exist as two atoms not as an H_2 molecule. The symmetry constraint of the RHF model contradicts this fact and the Σ_g^1 ground state of the H_2 molecule doesn't dissociate properly into hydrogen atoms. By contrast this is not the UHF case. In the UHF solution, the orbitals are the same as the RHF case for close separation of nuclei (ca. 2 a.u. or so) but dissociates into two H atoms. In the case of the lowest lying Σ_u^1 excited state both the RHF and the UHF solutions dissociate properly. The potential energy curve in which the zero of energy is the energy of two isolated H atoms is seen in Fig.5.1 for the Σ_g^1 state of H_2 in both the RHF and UHF case. The up sin one-electron orbital, $\psi_{1\sigma_g}\alpha$, is shown at large separation and at small separation for both the RHF and UHF cases. Because of this difficulty with the RHF solution, it is often asserted that the Hartree-Fock method is invalid for describing the formation and dissociation of chemical bonds. Here we have seen that this problem is in fact due to constraints imposed upon the Hartree-Fock method and not essential to the method itself. It is also seen that the UHF orbital solution may not transform according to an irreducible representation of the group of the nucleus, while of course the exact many-body wave-function solution would. This is also fully discussed by LÖWDIN [5.16]. The author does not consider the absence of the symmetry of the exact solution in the UHF approximation to be a severe disadvantage in that one doesn't find an exact eigenfunction of energy in any of the methods and it seems a bit inconsistent to insist on exact eigenfunctions of other operators when approximations to energy are accepted. The author also notes that in most cases the UHF solution is nearly a correct eigenfunction of the symmetry operators, so that symmetry information is often retained. One may always project out correct symmetry states from the UHF solution if necessary.

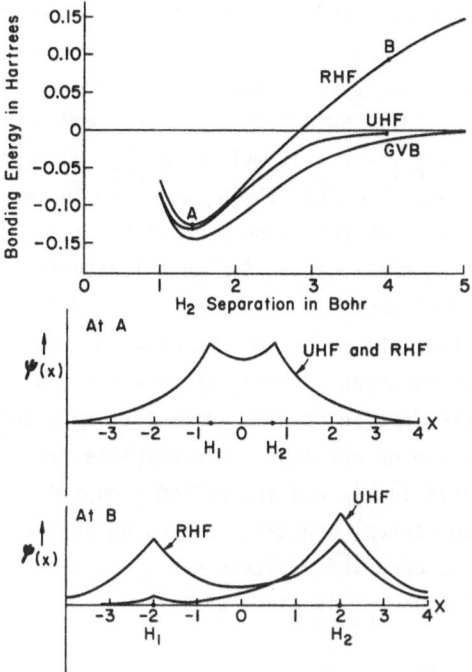

Fig.5.1. The binding energy for the H_2 molecule is shown for the RHF solution, the UHF solution and the GVB solution as a function of internuclear separation. In all cases the lowest energy state (the Σ_g^1) is shown. Energies are in atomic units (1 a.u. = 1 Hy = 2 Ry = 27.2 eV). Lengths are in atomic units (1 a.u. = 0.53 Å). The spin up orbital is shown for both UHF and RHF as a function of position along the inter-nuclear axis for positions A and B on the binding curve

A second criticism leveled at the Hartree-Fock method is that Fermi statistics are not obeyed in that one may have empty energy levels whose orbital eigenvalue is less than that of some filled levels. Largely this problem illustrates the inadequacy of Koopman's theorem in identifying eigenvalues with energies. More importantly it illustrates the troubles of the RHF model. In RHF one constraints a given subshell to have the same radial dependence and also eigenvalue even though the subshell is partly empty. Thus the use of constraints causes a problem here too. The UHF model, and presumably the GHF model, avoids this trouble by allowing each electron to have its own orbital and its own energy. In such an event Fermi statistics seem to be obeyed even using Koopman's theorem.

There is one method in current use which retains the symmetry information of an RHF solution and still permits the proper dissociation and formation of chemical bonds. In its simplest form, this is the perfect pairing generalized valence bond (GVB) model [5.17]. This model is simple in principle although rather complicated in use. The idea is to form a particular two determinant wave function from the RHF solution. In this limit one has

$$\Psi_{GVB}(\underline{x}_1 \cdots \underline{x}_n) = \frac{1}{\sqrt{N!}}[\tilde{A} \ \psi_1(\underline{r}_1)\alpha_1\psi_1(\underline{r}_2)\beta_2 \cdots \psi_k(\underline{r}_{2k-1})\alpha_{2k-1}\psi_k(\underline{r}_{2k})\beta_{2k} \cdots \psi_{n/2}(\underline{r}_n)\beta_n]$$

$$+ \ B \ \frac{1}{\sqrt{N!}} \ \tilde{A}[\psi_1(\underline{r}_1)\alpha_1\psi_1(\underline{r}_2)\beta_2 \cdots \hat{\psi}_k(\underline{r}_{2k-1})\alpha_{2k-1}\hat{\psi}_k(\underline{r}_{2k})\beta_{2k}$$

$$\cdots \ \psi_n/2(\underline{r}_n)\beta_n] \quad . \tag{5.8}$$

In (5.8) A is the antisymmetriser and both the ψ's, $\hat{\psi}$ and B are varied independently subject to constraints of symmetry and orthonormality of orbitals. This method may be used for any number of orbital pairs simultaneously even though (5.8) shows only a single pair being so treated. This wave function is an explicitly correlated wave function, in that it is clearly a function of $\underline{x}_{2k-1} - \underline{x}_{2k}$ as well as $\underline{x}_{2k-1}$ and $\underline{x}_{2k}$. Therefore, it goes well beyond a simple Hartree-Fock model. The H_2 solution in the GVB limit is seen in Fig.5.1. It agrees here with UHF for large separation of nuclei and is superior for small separation. The small separation superiority is often but not always found for other molecular systems.

Additional correlations may be added to either the HF or the GVB model by use of configuration interaction (CI). The CI wave function is most easily expressed in terms of the HF wave function. Let α^+ be the fermion creation operator and α the fermion annihilation operator. Let us recall, the HF operator has a complete set of one-particle orbitals, some of which are used in Ψ_{HF} and are called occupied orbitals and the remainder of which are called virtual orbitals. Let us adopt the convention that subscripts i, j, etc., refer to occupied orbitals and a, b, etc., refer to virtual orbitals. Then the exact solution Ψ_{CI} is given as

$$|\Psi_{CI}\rangle = |\Psi_{HF}\rangle + \sum_i \sum_a c_i^a \alpha_a^+ \alpha_i |\Psi_{HF}\rangle + \sum_{i,j} \sum_{a,b} c_{ij}^{ab} \alpha_a^+ \alpha_b^+ \alpha_i \alpha_j |\Psi_{HF}\rangle + \cdots$$

$$+ c_{12 \cdots n}^{ab \cdots d} \alpha_a^+ \alpha_b^+ \cdots \alpha_d^+ \alpha_1 \alpha_2 \cdots \alpha_n |\Psi_{HF}\rangle \quad . \tag{5.9}$$

In this case the c's are determined variationally. In practice only a subset of (5.9) is ever used for real studies, and by careful choice of subset, results of great numerical precision may be obtained. There are many other ways of going beyond Hartree-Fock and a variety of these are discussed by SCHAEFER [5.4]. In actual application to cluster calculations, such methods are not currently used to any great extent. In part, the use of such techniques could well be inconsistent in that correlation potentials are often long-range and the finite termination could easily introduce significant boundary effects.

The methods are based upon explicit application of the variational method. There is one other fundamental starting procedure which is used to develop models for cluster calculations. The X-α method results from such a procedure. This starting point is the theorem of HOHENBERG and KOHN [5.11]. This theorem states that the ground state energy of a many-fermion system, if the ground state is nondegenerate, is a unique functional of the fermion density, $n(\underline{r})$. This implies the functional exists. Unfortunately, the exact functional relationship has not been determined except for a high density electron gas. Theories based conceptually upon the Hohenberg-Kohn theorem are termed local density models. There are at least two well-defined local density limits which are in use. These are due to SLATER [5.18] and to KOHN and SHAM 5.19 . In both cases one determined $n(\underline{r})$ from one-particle orbital $\psi_i(\underline{r})$ by $n(\underline{r}) = \sum_{i=1}^{n} \psi_i^2(\underline{r})$.

In the Slater limit one starts to develop the functional using the HF expression as a starting point and recognizes that all terms except the exchange terms are already a function of simple one-body operators or of the particle density. One then looks at the exchange term which is just

$$\sum_j \psi_j(\underline{x}) \int \frac{\psi_j^*(\underline{x}')\psi_i(\underline{x}')}{|\underline{r}-\underline{r}'|} d\underline{x}' = \sum_j \psi_j(\underline{x})/\psi_i(\underline{x}) \int \frac{\psi_j^*(\underline{x}')\psi_i(\underline{x}')}{|\underline{r}-\underline{r}'|} d\underline{x}' \times \psi_i(\underline{x})$$

$$\equiv V_{ex}^i \psi_i(\underline{x}_i) \quad . \tag{5.10}$$

The exchange operator V_{ex}^i is simply then,

$$\sum_j \frac{\psi_j(\underline{x})}{\psi_i(\underline{x})} \int \frac{\psi_j^*(\underline{x}')\psi_i(\underline{x}')}{|\underline{r}-\underline{r}'|} d\underline{x}' \quad .$$

In proceeding one makes a weighted average of this operator and one uses the partial density ρ as a weighting factor and plane waves for the ψ's. This results in the Slater exchange approximation

$$V_{ex}^S(\underline{r}) = -3[3/8\pi n(\underline{r})]^{1/3} \quad . \tag{5.11}$$

If one substitutes this into the HF equation one may determine the orbitals needed for $n(\underline{r})$. The approach of Kohn-Sham is to start with the exchange contribution to the total energy in the HF case and to make the weighted replacement prior to performing the energy variation. The exchange part of the total energy is simply

$$E_{ex} = -\frac{3}{2} (3/8\pi)^{1/3} \int n^{4/3}(\underline{r})d\underline{r} \quad . \tag{5.12}$$

This then yields an exchange operator for determining a set of one-electron orbitals, of the form

$$V_{ex}^{KS}(\underline{r}) = -2[3/8\pi n(\underline{r})]^{1/3} \quad , \tag{5.13}$$

which is exactly 2/3 the Slater value.

The important point is both limits yield the same functional form for V_{ex} and only the constant changes. This has led SLATER to postulate a constant α which varies between 2/3 and 1 for different atoms and yields an exchange potential,

$$V_{ex}^{X\alpha} = \alpha V_{ex}^S \quad . \tag{5.14}$$

The constant α is determined empirically for each atom in the periodic table [5.20]. This then is the basis for the much used X-α model.

One may ask if the one-electron eigenvalues, ε_i from either an X-α model or an HF model have any meaning. In the HF case the meaning is provided by Koopman's theorem. Here, approximately, ε_i is the negative of the energy needed to remove an electron from state i in the system if in so doing the other electron orbital remains undisturbed while ε_a is the energy needed to put an electron into state a of the system [5.21,22]. In the X-α case the energy eigenvalue is simply the derivative of the total energy with respect to particle number [5.5].

The X-α model seems to permit bonds to dissociate properly unlike the RHF model. The X-α method may be criticized because the method only produces a particle density $n(\underline{r})$, not a true wave function Ψ, and hence much symmetry data is unavailable unless additional approximations are made. This is the N-representability problem. In short the X-α model is not really able to answer questions uniquely for which detailed wave-function knowledge is needed such as the Compton profiles. In addition strictly speaking only nondegenerate ground states are given but in actual practice the method is used for finding approximate solutions for degenerate states and excited states [5.23].

Ideally, the methods based upon either variational techniques or upon the Hohenberg-Kohn theorem are able to provide full information about: chemical bond strength; geometric data including bond length or bond angle; electronic charge distribution; optical/X-ray absorption and/or emission spectra; etc. This ideal is not always achieved in practice due to the methods employed to solve the equations. In the case of the X-α method SLATER and JOHNSON have developed effective and efficient techniques of solving the eigenvalue equations using a multiple scattering formalism (MSX-α) similar in concept to the KKR formalism used in solid state physics [5.5]. In order to obtain full efficiency in this model approximations to the local potential are made.

The most widely used approximation is probably the muffin-tin approximation. As this is applied to molecules, one surrounds each nucleus with a sphere of a radius chosen such that spheres about external nuclei may be tangent but not overlapping. Inside this sphere the potential about the nucleus is assumed to be spherically symmetric. The entire molecule is then surrounded by a large sphere which encloses all the spheres about the nuclei and may be tangent to them. This sphere is called the Watson sphere and the potential inside it is set equal to a constant. Outside the Watson sphere the potential is set to zero. This potential is illustrated graphically in Fig.5.2 for a simple diatomic molecule. In more recent works the muffin-tin potential has been modified to permit regions in which the spheres surrounding the nuclei are allowed to overlap. In this model the charge density in the overlapping sphere region is treated approximately to avoid multicenter integral calculation. Some of the difficulties with this are obvious. One need question

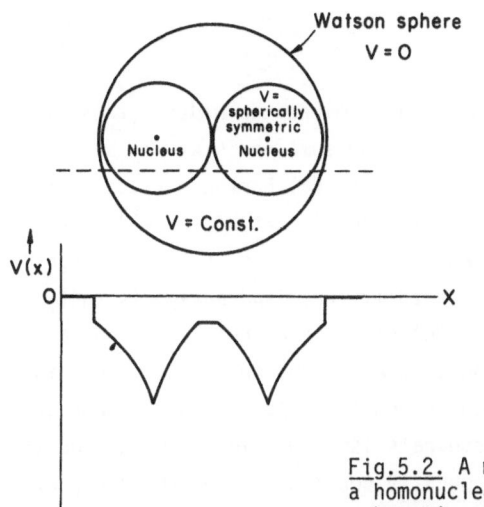

Fig.5.2. A muffin-tin potential is illustrated here for a homonuclear diatomic molecule. In the top panel a schematic of the whole potential is illustrated and in the bottom panel the potential on the dashed line in the top panel is shown

for each problem the suitability of a spherical potential approximation or the adequacy of a constant potential in a Watson sphere. There are more subtle prob- lems as well. These are, if one wishes to perform studies relating to bond length and angle, 1) how does one adjust the radius of the muffin tins or of the Watson sphere and 2) how is the constant potential chosen so that the resulting answers reflect the physical situation rather than the parameters chosen? No unambiguous answers to these questions are available and hence most users of this approximation restrict their studies to a single nuclear geometry for a given system. Thus one may obtain useful spectroscopic information but much chemical data is lost. We note at once the X-α method is not limited to this approximation and furthermore it is quite possible to do HF studies efficiently using a muffin-tin approximation as DAGENS and PERROT have shown in several solid state calculations [5.24].

Recently ELLIS and PAINTER have developed a self-consistent technique, the dis- crete variational method (DVM), for solving without any use of muffin tins the X-α equations for large or small molecules. This method employs sampling techniques to numerically integrate the equations and should ultimately be able to provide chemical as well as spectroscopic data. This technique seems to be more costly of computer time than is in MSX-α method, but the removal of the ambiguities inherent to the muffin-tin methods makes the additional cost worthwhile in this author's opinion [5.25].

Current HF or GVB or CI methods used for cluster chemisorption studies revolve about expanding the one-electron orbital, ψ_i in a linear combination of Gaussian type orbitals. This is, one assumes

$$\psi_i(\underline{r}) = \sum_A C_A^i P_A(\underline{r}) \, \exp -\alpha_A(\underline{r} - \underline{R}_A)^2 \quad . \tag{5.15}$$

The $\underline{R}_A$ are the origins of the Gaussian and the α_A determines the width, (these may be obtained variationally but are most often taken from someones's atomic basis set), $P_A(\underline{r})$ is a polynomial which acts much like a spherical harmonic (i.e., x/r, xy/r^2, etc.) and C_A^i is the variationally determined coefficient of the Ath Gaussian in the ith orbital. The advantage of a Gaussian expansion set is that all integrals may be performed easily in an essentially closed form. The disadvantage is that a Gaussian basis function does not behave asymptotically at either large separation from the nuclei or at short distance from a nucleus as a true one-electron orbital would and thus a large number of Gaussians need be employed to accurately represent a wave function. The time to do all needed integrals is proportional to the fourth power of the number and hence one is usually greatly constrained in the number of Gaussians employed and great care is needed in their choice if useful answers are to be obtained. Even so it is unlikey that one can obtain very accurate information on quantities such as electron charge or spin density at the nucleus by this technique, thus as an example, hyperfine interaction studies would be suspect.

Finally, there is a class of technique used for some cluster studies which are empirical in nature. This set of techniques is of course limited in utility since differing choices of empiricism can clearly yield answers which make radically different predictions for the same system. Nonetheless, in the hands of a careful worker, given a good set of physical insights, useful studies with such models can be performed. The basic models in use for cluster studies are the extended Hückel method (EHT) and various levels of the complete neglect of differential overlap method (CNDO) or the bond energy, bond order method (BEBO). As an example of how such methods develp we will discuss only EHT [5.12,26]. The other methods are essentially variations or extensions of this theme.

The EHT can be visualized by constructing a solution to the HF equation using a basis set of atomic orbitals $X_i(\underline{r} - \underline{R}_A)$. Thus,

$$\psi_j(\underline{r}) = \sum_i \sum_A C_{iA}^j X_i(\underline{r} - \underline{R}_A) \quad . \tag{5.16}$$

The coefficients C_{iA}^j and the orbital eigenvalues ε_j may be determined by solving the matrix equation,

$$[F - \varepsilon S][\psi] = 0 \quad . \tag{5.17}$$

The matrix elements of F and S are of course,

$$F_{iAjB} = [X_i(\underline{r} - \underline{R}_A), F \, X_j(\underline{r} - \underline{R}_B)] \quad , \tag{5.18}$$

and

$$S_{iAjB} = [X_i(\underline{r} - \underline{R}_A), X_j(\underline{r} - \underline{R}_B)] \quad .$$

Clearly if these elements were evaluated exactly and if a sufficient set of X's were used one would have a solution to the HF equation. The empiricism comes from attempts to guess the matrix elements F_{iAjB} rather than evaluating them. In the EHT, one uses normally the simple rules

$$F_{iAjA} = \delta_{ij}\varepsilon_{iA} \quad , \tag{5.19}$$

and

$$F_{iAjB} = (K/2)S_{iAjB}(\varepsilon_{iA} + \varepsilon_{jB}) \quad , \quad A \neq B \quad .$$

Here the ε_{iA}'s are the negative of the energy needed to remove an electron from state i on atom A and may be determined by either experiment or calculation. The overlap element S_{iAjB} is normally calculated and the constant K is an empirical constant called the Wolfberg-Helmholtz parameter and is often set to 1.75. Many refinements of this model are possible and we will not consider if further. There have been several attempts to justify the EHT on the basis of some derivation from first-principle models but these derivatives are at best applicable to certain special cases. Since the EHT seems to be useful for many cases beyond those which are derived it remains best to treat this model as an empirical one. The reader is referred to [5.26,27] for various first-principle discussions of the EHT.

5.3 Interaction of Cluster and Environment

Recently, MARSHALL et al. have proposed the use of the unrestricted-Hatree-Fock model (UHF) for solid state cluster simulations. We shall in this section develop the UHF technique to include the question of interaction of the cluster with its environment [5.6]. It is understood from the outset that the development applies equally well to any other theory which uses an energy functional based on the first-order density matrix or its trace formed from one-particle orbitals. Thus, this derivation is valid for the X-α model or the restricted Hartree-Fock model (RHF) which most authors term the Hartree-Fock model. In this model the many-electron Hamiltonian, H, is given by (5.1). In the UHF model the n-electron wave function is approximated as (5.2) where the one-electron orbital is subject the constraints of (5.3) and (5.5). The form of $\psi_i(r_i)$ is otherwise unconstrained. If ψ_i is varied to minimize the expectation value of H, the ϕ_i are determined by solving

$$F(\rho)\phi_i = \varepsilon_i\psi_i \quad ,$$

$$\rho = \sum_{i=1}^{n} \psi_i(\underline{x})\psi_i^+(\underline{x}') \quad ,$$

$$F(\rho) = -\frac{\hbar^2}{2m}\,\nabla^2 - \sum_I \frac{e^2 Z_I}{|\underline{r}-\underline{R}_I|} - e^2 \int \frac{\rho(\underline{x}',\underline{x}')}{|\underline{r}-\underline{r}'|'}\,d\underline{x}'$$

$$-e^2\rho(\underline{x},\underline{x})\,|\underline{r}-\underline{r}'|^{-1}\hat{P}(\underline{x}',\underline{x}) \quad .$$

In this formula, $\hat{P}(\underline{x}',\underline{x})$ is the operator which exchanges coordinate $\underline{x}$ with $\underline{x}'$. It is this equation which we would like to solve for the solid as a whole.

Solution of this system of equations for an entire solid is often impossible, therefore, we simulate the entire system by a finite cluster. This region of the cluster is termed A here and the remainder of the system is termed the environment of A. The issue at hand is how to partition the system rigorously and what potential or boundary condition to place on A in order to do this rigorously. To effectively carry this out we have recourse to the method of local orbitals of ADAMS-GILBERT-KUNZ [5.28,29].

The question of boundary conditions on clusters in general is quite difficult and is not yet completey solved, in fact, we find no general condition at all for metallic systems. Since clusters are smaller than the system represented one need determine, somehow, which conditions to place on the cluster. As a first try, one might try free space conditions; however, for reasons which are quite obvious, this is likely to require a very large cluster to achieve accuracy for either a metal or an ionic solid (the first is due to the diffuse valence orbitals and the second is due to long-range ionic potentials). In the case of a covalent solid, artifacts such as dangling surface bonds may distort results unless a large cluster is used. Therefore, one may expect that a use of free space boundary conditions is best left to cases where very large clusters may be studied or for studies of molecular crystals, where clusters may be chosen such that neither appreciable covalency, nor long-range potentials occur across the cluster-environment boundary. The appropriate boundary condition to use on a metal is not well defined and one need often use large clusters or be satisfied with qualitative rather than quantitative answers. The cases of molecular bonded systems, ionic systems or a covalent system are subject to mathematical derivation.

Let us consider the local orbital formalism of ADAMS and GILBERT. Let us use F to represent either the Fock operator for the entire system or a local-density operator. Let A be the region which our cluster occupies, and let E be the remainder (environment of A). Then assign M electrons to A (this is chosen to be physically reasonable for the appropriate case, such as the number of electrons on the ions inside A if one has an ionic system). Let

$$F = F_A + U_A \quad , \tag{5.20}$$

F_A being that part of F which includes kinetic energy, nuclear attraction of the electrons and nuclei inside A and the electron-electron potential including Coulomb and exchange parts for electrons assigned to A. We would like to solve the UHF equation, but since we wish to study only part of the system, let us solve instead

$$[F_A + U_A - \rho W \rho] \phi_i = \pi_i \phi_i \quad , \tag{5.21}$$

where W is an arbitrary hermitian operator. This is the Adams-Gilbert equation. As they have shown, provided a common W is used for all electrons,

$$\rho = \sum_{i=occ} \psi_i \psi_i^+ = \sum_{i=occ} \phi_i \phi_j^+$$

and the ϕ's are orthonormal.

First consider an ionic or molecular crystal. We divide U_A into two parts, V_A^M is an ionic (Madelung) contribution and is long range, and V_A^S is the remainder and is short range. Of course for the molecular system $V^M = 0$. Let $W = V_A^S$. Therefore one solves (using completeness) for the orbitals ($\rho \phi_i = \phi_i$).

$$[F_A + V_A^M] \phi_i = \pi_i \phi_i - V_A^S \phi_i + \rho V_A^S \phi_i \quad . \tag{5.22}$$

Now we are only concerned with the m orbitals of (5.22) which lie in A. Provided the appropriate number of electrons is assigned to A, the solutions found for these electrons should only weakly penetrate E. If, for these orbitals, one finds as is reasonable for ionic cases that they don't appreciably penetrate E and since V_A^S doesn't penetrate A appreciably and in the limit of self-consistency $V_A^S \phi_i$ is cancelled by $\rho V_A^S \phi_i$ on the average (this is true since the eigenvalue of ρ for an occupied orbital is 1), one finds the appropriate approximate equation including interaction with the remainder of the system to be

$$[F_A + V_A^M] \psi_i = \pi_i \psi_i \quad . \tag{5.23}$$

From these solutions one may calculate the approximate potential energy surface for an entity wholly in A by evaluating the energy of system A as if it were an isolated system A in a potential field V_A^M. One sees that for a molecular bonded system (e.g., Ar) the approximate equation is just the free space cluster, and for an ionic system the equation includes a Madelung field. In most muffin-tin calculations no such field is used and since at best the Watson sphere potential can simulate the correct Madelung potential at either a cation or an anion, but not both, the information about relative separation of one-electron energy levels on differing sites may be poorly given.

This treatment extends directly to a covalent system. Just as before, we envision the system as being composed of ions or molecules; here we envision the system as being composed of electrons in bond pairs and ions. Thus for diamond we have a system of C^{4+} ion and groups of two-electrons bonds, and on the average, two two-electron bonds per C^{4+} ion. We proceed as before except we assume the boundary between A and E passes through n-two-electron bonds. To achieve localization in the Adams-Gilbert sense, we cannot break these bonds, therefore, we must include a more (fewer) electrons in A than a simple atomic view of the system would require and thus U_A has a long-range part due to n more (fewer) protons than the atomic view would require and these long-range parts are concentrated near the A-E interface. This long-range potential is entirely due to surface charge at the interface and not to any long-range Madelung field. From here one solves (5.22) for the n more (fewer) electrons, or approximately (5.23) for this system using a V_A^M due to the n more (fewer) protons at the boundary. One may recognize here the common trick of using H atoms to tie off the dangling bonds in a covalent cluster simulation.

This transformation cannot be done in general for metals since the bond behavior of them is such that one cannot simply rotate into a localized representation in a one-particle theory. Of course, if the phenomena studied are local in nature despite the metallic host, then one is free to proceed. The forming of local bonds to a metal or local magnetism are examples of such phenomena [5.30].

The eigenvalues π_i of (5.22) or the approximate (5.23) represent the Koopmans' theorem eigenvalues of the infinite solid, ε_i, only in the limit that the orbital ψ_i in question is localized in A so that in fact ψ_i and ϕ_i are identical. Alternately, here the eigenvalue π_i may also correspond to an ε_i in the case where the energy band of which ε_i is a part has infinitesimal width as is the case for a core level. For levels ε_i which have finite width, at best the eigenvalues π_i represent some state in that bond most often near the center of gravity of the band. This property can be easily deduced by using the ϕ_i corresponding to the ε_i to generate a Bloch basis set for use in a LCAO band model. Clearly here the one site matrix element of F is π_i and using the invariance of the trace of the F matrix under diagonalization, if the overlap of ϕ_i into E is small as we assumed, π_i must lie near the center of gravity of the band containing ε_i.

Similar considerations may be performed to see that chemical data can be realized for atoms included in A. Consider the total energy expression. We will partition ρ into two parts, those ψ_i in A and those in E, or:

$$\rho(\underline{x},\underline{x}') = \sum_i^A \psi_i(\underline{x})\psi_i^+(\underline{x}^1) + \sum_j^E \psi_j(\underline{x})\psi_j^+(\underline{x}^1) \quad .$$

The Hamiltonian can be broken apart as:

$$H = \sum_i^A \frac{h^2}{2m}\nabla_i^2 + \sum_i^E \frac{h^2}{2m}\nabla_i^2 - e^2\left(\sum_I^A\sum_i^A \frac{Z_I}{|\underline{r}_i-\underline{R}_I|} + \sum_I^E\sum_i^E \frac{Z_I}{|\underline{r}_i-\underline{R}_I|}\right.$$

$$\left.+ \sum_I^A\sum_i^E \frac{Z_I}{|\underline{r}_i-\underline{R}_I|} + \sum_I^E\sum_i^A \frac{Z_I}{|\underline{r}_i-\underline{R}_I|}\right) + e/2^2\left(\sum_i^A\sum_j^A \frac{1}{|\underline{r}_i-\underline{r}_j|}\right.$$

$$\left.+ \sum_i^E\sum_j^E \frac{1}{|\underline{r}_i-\underline{r}_j|} + 2\sum_i^E\sum_j^A \frac{1}{|\underline{r}_i-\underline{r}_j|}\right) \quad .$$

From here one may proceed in the manner originally derived by LÖWDIN for ionic solids cohesive energy calculations, to obtain an expression for E, the total energy, for our wave function. Here, however, all our orbitals are orthogonal because of our solving a common Hermitian equation and therefore Löwdin's S energy will be absent. The total energy, E, schematically is a sum of E_A, E_E, E_{AE}, where E_A is the energy of the system of orbitals in A with respect to that part of the Hamiltonian which refers only to A, E_E is the energy of the system of orbitals in E with respect to that part of the Hamiltonian which refers only to E, and I_{AE} is the interaction energy of orbitals in A with those in E. If only nuclei in A move and A is large so that orbitals in E are unchanged by this, the term E_E is constant throughout and since only energy changes are necessary, we neglect E_E hereafter. E_A is the energy of the pseudomolecules in A in the usual Hartree-Fock approximation. Thus only I_{EA} need be evaluated. We have

$$I_{EA} = \sum_I^A\sum_J^E \frac{e^2 Z_I Z_J}{|R_I-R_J|} + \sum_i^A\left\langle i\left|-\sum_I^E \frac{Z_I e^2}{|r-R_I|}\right|i\right\rangle + \sum_i^E\left\langle i\left|-\sum_I^A \frac{Z_I e^2}{|r-R_I|}\right|i\right\rangle$$

$$+ \sum_i^A\sum_j^E\left(\left\langle ij\left|\frac{e^2}{|r_1-r_2|}\right|ij\right\rangle - \left\langle ij\left|\frac{e^2}{|r_1-r_2|}\right|ji\right\rangle\right) \quad . \tag{5.24a}$$

One may greatly simplify this if one assumes orbitals in E weakly penetrate into A and vice versa. In the extreme limit of no overlap this reduces to a very simple result. Let there be G_I electrons associated with the Ith nucleus. Then the ionicity, I_I is $G_I - Z_I$. Then approximately, (5.24a) becomes

$$I_{EA} \cong \sum_{I}^{A} \sum_{J}^{E} \frac{Z_I Z_J e^2}{|R_I - R_J|} - \sum_{I}^{A} \sum_{J}^{E} \frac{G_I Z_J e^2}{|R_I - R_J|}$$

$$- \sum_{I}^{A} \sum_{J}^{E} \frac{G_J Z_I e^2}{|R_I - R_J|} + \sum_{I}^{A} \sum_{J}^{E} \frac{G_I G_J e^2}{|R_I - R_J|} + 0 \quad .$$

This immediately reduces to

$$I_{EA} \cong \sum_{I}^{A} \sum_{J}^{E} \frac{I_I I_J e^2}{|R_I - R_J|} \quad . \tag{5.24b}$$

Thus in the lowest approximation the total energy charge for moving a nucleus about in A is given by the energy of a pseudomolecule in A plus a "Madelung" contribution of the atoms in A with those in E.

The importance of such an external potential to simulate the solid can be demonstrated for LiF. LiF is an ionic solid in a fcc structure. The system is essentially a collection of F^- and Li^+ ions. Here we chose a cluster of 6 Li ions and 1 F ion in the center. The Li are octahedrally coordinated with the F ion and thus the cluster is $Li_6 F^{5+}$. We solve (5.23) here. As a comparison, two other studies are performed. The first is to surround the cluster with a Watson sphere of charge -5e to effect charge neutrality as is often done in MSX-α calculations, and the third is to employ simple free space conditions for the cluster. In forming the Watson sphere, we use a radius of 5.5095 a.u. The results are relatively insensitive with small changes in variation of change or radius of the Watson sphere.

The results of this calculation for the one-electron eigenvalues and their differences are seen in Table 5.3. There are three salient points to be made from this table. First, the eigenvalues differ greatly from calculation to calculation as one expects due to differing imposed potential. Second, the difference of eigenvalues on a given ion, (e.g., F^- ϵ 1s - ϵ 2s) is very insensitive to boundary condition, and finally the difference in eigenvalues on two different ions (e.g., Li^+ ϵ 1s - F^- ϵ 2s) is greatly effected by external potential. In the case of two eigenvalues on the same ion the greatest difference between two calculations is about 0.005 eV whereas for two eigenvalues on different ions the deviation is always greater than 1 eV between the current theory and the other two methods.

One may conclude that if one wishes accurate values for energy differences for orbitals on different ions a simple Watson sphere is inadequate for ionic system and the error is proportional to the degree of ionicity as seen. Hence for a doubly ionized solid such as TiO or NiO say, the relative error of levels on the anion compared to the cation may be of order 3 eV if a Watson sphere or free space boundary conditions are used.

<u>Table 5.3.</u> The one-electron eigenvalues computed for the $(Li_6F)^{5+*}$ cluster are given. The calculations are performed using the ab initio external potential discussed in the text and shown in column I; the Watson sphere potential which yields charge neutrality and shown in column II; and the free space boundary which is shown in column III. Also shown are the differences in energy between each pair of eigenvalue. Results are in eV

	I	II	III
F ϵ 1s	-712.6993	-718.3818	-743.0681
F ϵ 2s	- 39.9698	- 45.6561	- 70.3407
F ϵ 2p	- 15.7505	- 21.4378	- 46.1225
Li ϵ 1s	- 63.6852	- 68.0178	- 92.8108
|Li ϵ 1s - F ϵ 1s|	649.0141	650.3640	650.2573
|Li ϵ 1s - F ϵ 2p|	23.7154	22.3617	22.4701
|Li ϵ 1s - F ϵ 2p|	47.9347	46.5800	46.6883
|F ϵ 2s - F ϵ 2s|	672.7295	672.7257	672.7274
|F ϵ 1s - F ϵ 2p|	696.9488	696.9440	696.9456
|F ϵ 2s - F ϵ 2p|	24.2193	24.2183	24.2182

*The superscript implies that the net charge on the ions considered explicitly in the cluster is +5 electron charges.

5.4 Examples of Cluster Calculations

In recent years a great deal of use has been made of cluster technqiues for studying chemisorption. All the techniques discussed in this chapter and some not discussed have been used for these studies. Studies have been performed on all classes of solid such as the solid rare gases, [5.31], ionic insulators [5.32], semiconductors [5.33] and simple and transition metals [5.34,35].Of all available systems, the one most widely studied is probably Ni. Ni is an important example for study in that it is a useful catalyst. Therefore, one of the principal focuses of Ni studies has been to determine the nature and strength of the bond joining an adsorbate to Ni.

The thrust of many early studies has been colored by a vast lore which has grown up about the transition metals in general. Ni is a ferromagnetic solid and in this sense the unfilled 3d shell of the Ni atom plays a dominant role. In fact, the magnetic properties of the entire transition metal series are dominated by the presence of an unfilled d shell. Therefore, due to the fact that the d shell is unfilled for any of the transition metals, and since our knowledge of chemistry inclines one to believe that bonds form most readily to an unfilled shell, it was a natural assumption that the catalytic activity and the nature of the chemisorbed bond would also be due to the unfilled d shell. Attempts to correlate catalytic properties of transition metals with known solid state properties or atomic properties have been made [5.36]. This is both a simple and natural assumption, but recent studies both theoretical and experimental show that this view is incorrect for Ni for at least some adsorbates and that the principal ingredient of the chemisorbed bond may be the occupied Ni 4s orbital, hybridized mostly with the Ni virtual 4p orbital.

Several empirical studies on Ni exist. In these studies a cluster of Ni atoms with free space boundary conditions are used as is the EHT method. The basis used for example by FASSAERT et al. [5.37] was an atomic basis set for Ni and did not include any Ni 4p polarization function. As a test particle for chemisorption a predissociated H atom was chosen. The essential results of this calculation may be summarized briefly. The H atom was found to bond to the Ni cluster and the bonding was to the partly filled 3d levels of the Ni cluster with some 4s participation. At the time it was performed this empirical calculation was in a good agreement with the commonly held beliefs regarding transition metal catalysis. Later calculations [5.38] including the 4p functions also found some 4p participation. Furthermore, the H-Ni cluster bond looked quite similar to a diatomic NiH calculation [5.39,40].

There was a more sophisticated attempt to study chemisorption of H by Ni and this was due to BLYHOLDER [5.41,42]; he also used an empirical model. In this case the CNDO method was used. A cluster of 8 Ni atoms was employed and Blyholder's basis set included virtual p polarization functions. Free space boundary conditions were employed. This calculation also found that H was bound to the Ni cluster. However, this calculation produced an unexpected result. The dominant bond was found to be into the Ni 4s levels and these were highly hybridized with the virtual 4p levels. This unexpected result was not tested experimentally and was politely ignored for a short while.

The essential lack of impact of these calculations may be largely due to the fact that differing empirical methods were employed and each yielded somewhat different results. Even though the CNDO model should be more accurate than the EHT method, it was the EHT calculation which agreed better with the common prejudices.

The next tpye of model employed to the Ni system was the MSX-α method. The initial calculations were for Ni 8 and 13 atoms clusters [5.43]. In part these studies were hindered by the difficulties of obtaining accurate potential energy surfaces when using a muffin-tin potential. The MSX-α equation was solved only for a single choice of nuclear separation for a given geometry. In doing this one loses all detail about actual bonding curves and geometries. The data most easily obtained from such a calculation relates to the one-electron eigenvalues and the one-electron orbitals themselves. The MSX-α cluster was found to have one-electron eigenvalues in reasonable agreement with the existing photoemission data for Ni. The next step was to use the available orbital data to discuss the possible mechanisms of chemisorption. Charge density maps of several of the valence orbitals were presented and these showed a marked presence of Ni 3d character in the valence orbitals. An example of an orbital density map (A_{2g}) is shown in Fig.5.3 for a 13 atom Ni cluster, and the one-electron energy scheme for an 8 atom Ni cluster is seen in Fig.5.4. These results were in accord with the preconceived ideas about catalysis. There are two criticisms to be made about this interpretation. The first is that no actual calculation of bonding curves and geometries were made and hence no demonstration of bonding to the d orbitals was shown. The second is more subtle. This is, the Ni 4s orbital is much more diffuse than is the 3d orbital (the radial maximum for the 4s is about 2.5 times greater than that for the 3d orbital) and since the wave function is a volume type quantity, the sensitivity to the 4s orbital in a charge density map study is about an order of magnitude less; therefore in this simple case the 3d behavior can obscure the 4s behavior. More recent X-α studies have found important sp hybridization contributions to the chemical properties of Ni [5.44].

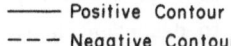
—— Positive Contour
— — Negative Contour

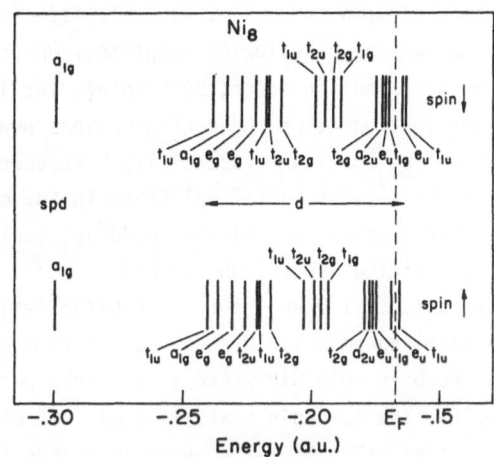

Energy (a.u.)

Fig.5.3. Equimagnitude contour lines are shown for the A_{2g} surface orbital on a square face of a 13 atom cubo-octahedral cluster of Ni. [5.44]

Fig.5.4. The spin dependent one-electron energy levels and approximate orbital percentage are seen for an 8 atom cluster of Ni. Energies in a.u. (1 a.u. = 2 Ry = 1 Hy = 27.2 eV). These are the results of [5.44]

The next study complete is by GUSE et al. [5.45] and uses both the UHF and GVB method. In addition this calculation makes a further digression from the previous work in the nature of the cluster used. The reason here is that in actual catalysts one rarely uses a bulk sample of the metal but rather a fine particle (5-100 or even more atoms) or the metal bonded to a substrate such as alumina or silica. In this case the clusters included a single Ni and a single H atom, on up to a cluster of 4 H atoms, 2 Ni atoms and 2 O atoms. This latter cluster simulated a pair of Ni atoms adsorbed to the O on the surface of silica. The bonds formed by the O to the rest of the silica are tied off by a pair of H atoms as we discussed previously and an H_2 molecule is chemisorbed. One important and useful result was that the quantitative behavior of the NiH bond was unaffected by the nature of the substrate (whether it was an H atom, part of silica or free space). The remainder of the results are also worth summarizing in some detail here. In doing so only the NiH and the H Ni H cases are discussed because of the high degree of agreement with the more complicated clusters.

The ground state of a nickel atom is ^3F with the configuration $1s^2 2s^2 2p^2 3s^2 3p^6 3d^8 4s^2$ and is nearly degenerate with its first excited state ^3D with the configuration $1s^2 2s^2 2p^6 3s^2 2p^6 3d^9 4s$ which is experimentally only about 0.07 eV higher in energy. For large Ni-H separations where the two states are nearly degenerate the configuration interaction matrix elements between these configurations vanish due to symmetry and the neglect of this interaction is justified. In the HF calculations, one finds the $3d^9 4s$ state to be about 1.7 eV higher than the $3d^8 4s^2$ state. One must consider NiH and NiH_2 states resulting from both of these configurations. The nickel $^3F(3d^8 4s^2)$ state, which requires several configurations to describe an eigenfunction of spin and space, was calculated using the UHF method which has the distinct advantage of allowing completely unrestricted adjustment of the two d holes to the polarization effect of bonding. The interaction between the principal bonding configuration and configurations needed to produce a spin-space eigenfunction will be small and essentially R independent. Also it is worth noting that this method allows radial splitting in the core orbitals which is not permitted in the other methods. The nickel $^3D(3d^9 4s)$ excited state was treated using both the perfect pairing GVB and UHF methods.

The potential energy curves of $Ni(3d^8 4s^2)H$ are shown in Fig.5.5. From the spin 1 Ni atom and spin 1/2 H atom one can form NiH with spin 3/2 or 1/2. One finds that the high spin NiH state is a nearly pure quartet with an $\hat{S}^2$ expectation value of 3.765. The low spin state is a mixture of a quartet and a doublet with an $\hat{S}^2$ expectation value of 1.766 which indicates that the state is $0.81|1/2> + 0.58|3/2>$. One deduces that the energy of the pure doublet state at the minimum is raised above that of our low spin state by 0.0025 hartrees (0.07 eV) with the UHF orbitals indicating approximately the energetic importance of the space-spin restrictions. Both high and low spin states are bound with an energy of about 1 eV with, at most, a very small energy barrier. The equilibrium distance for each state is 3.25 bohr.

In order to determine the nature of the binding in NiH the author examined the valence orbitals of the molecule for large and small internuclear separations. The orbitals for the high spin state are almost identical to those of the low spin state, which are shown in Fig.5.6. It is immediately clear that the nickel 3d orbital does

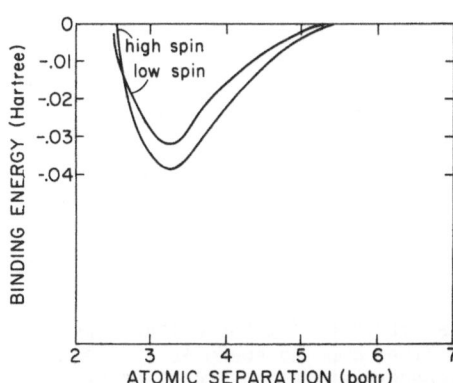

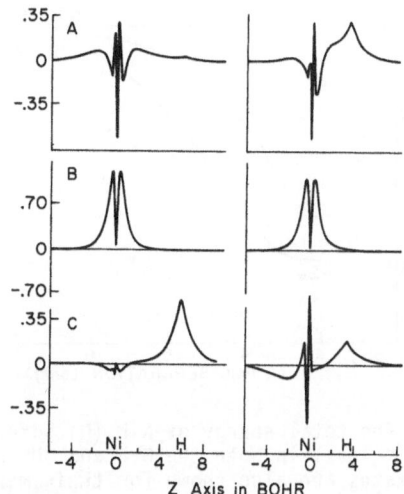

Fig.5.5. The potential energy curves for the low and high spin states of NiH $(3d^8 4s^2)$. The results are in a.u. (1 a.u. = 1 Hy = 2 Ry = 27.2 eV). Lengths are in a.u. (1 a.u. = 0.53 Å). [5.45]

Fig.5.6. Amplitude of the three valence orbitals as a function of internuclear separation for NiH $(3d^8 4s^2)$. A shows the doubly occupied Ni 4s orbital, B the Ni $3d_z^2$ orbital and C the hydrogen 1s orbital. The origin is at the Ni nucleus and the plane includes the H nucleus. [5.45]

not contribute significantly to the bonding. This is consistent with the findings of BLYHOLDER for the adsorption of hydrogen on nickel. As the internuclear separation decreases the doubly occupied nickel $4s^2$ orbital splits into two essentially sp hybrid orbitals. One of these forms a covalent bond with the hydrogen atom while the other becomes a singly occupied nonbonding orbital directed away from the hydrogen.

Both the UHF and GVB methods are compared for the Ni$(3d^9 4s)$ H spin doublet in Fig.5.7 where the UHF state has an $\hat{S}^2$ expectation value of 0.761. We calculated the potential energy curve of the $^2\Sigma^+$ state using GVB with the split pair being the Ni 4s orbital and the H 1s orbital. Also shown are the energies of the $^2\Pi$ and $^2\Delta$ states at 3.0 bohr using the GVB method. We conclude that the GVB method is more suitable for describing the $3d^9 4s$ state of Ni since it produces a lower total energy for the system than does UHF. Taking into account the experimental fact that the asymptotic nickel $3d^9 4s$ and $3d^8 4s^4$ state are nearly degenerate, GUSE et al. estimated the ground state of NiH to be the $^2\Delta$ state arising from Ni $3d^9 4s$ configuration. In each of these states the bond involves the Ni 4s and H 1s orbitals, leaving a single hole in the nickel d shell.

Initially the hydrogen atom is far from the nickel atom which is in the ^3F state with the $3d^8 4s^2$ configuration. As the hydrogen approaches the nickel atom the $3d^9 4s$ configuration begins to mix in until a point is reached where the $3d^8 4s^2$

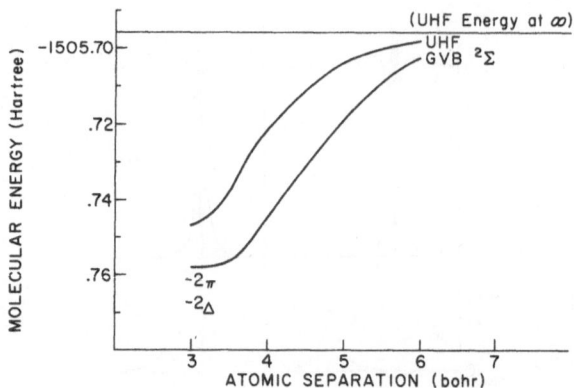

Fig.5.7. The total energy of NiH (Ni $3d^9 4s$) as a function of internuclear separation for the $^2\Sigma$ level in both the GVB and UHF scheme. The computed energies of the 2π and 2Δ states are also shown for their minimum energy separation. Energies are in a.u. (1 a.u. = 1 Hy = 2 Ry = 27.2 eV). Lengths are in a.u. (1 a.u. = 0.53 Å). [5.45]

and $3d^9 4s$ configuration mix strongly. As the hydrogen approaches the potential minimum the $3d^9 4s$ configuration becomes dominant.

A hydrogen atom adsorbing on the surface of a supported catalyst does not interact with a free nickel atom but rather with one already bound to a surface such as SiO_2 or $Al_2 O_3$ or Ni. In order to investigate this process we discuss where the nickel-surface bond by NiH is approximated by an H at the potential minimum for NiH and calculate the potential energy curve of a second hydrogen atom binding to this system to form a linear H (NiH) molecule.

One begins with the UHF Ni($3d^8 4s^2$) H system and brings in a second hydrogen atom which forms triplet H (NiH) with an $\hat{S}^2$ expectation value of 2.007 (see Fig.5.8). Examination of the wave function indicates that the triplet coupling comes from the d orbital combinations. The hydrogen 1s electron combines with the electron in the nonbonding Ni 4sp hybrid orbital to form a deep (2.2 eV), long-range covalent bond. The energy required to bend this second bond while keeping the nickel-hydrogen distance fixed at its equilibrium value is shown in Fig.5.8.

The H Ni($3d^9 4s$)H potential energy curve was calculated in two ways by GUSE et al for the NiH $^2\Sigma^+$ state with the bonding orbitals being doubly occupied, restricted Hartree-Fock orbitals rather than a GVB pair. The second hydrogen atom is brought in with the GVB pair being the second H1s and the unpaired nickel $3d_{z^2}$ electron forming $^1\Sigma^+$HNiH. The second calculation is identical to the first except that the first NiH bonding orbitals are left a GVB pair, consequently lowering the total energy of the system. It may be seen that hydrogen interacts very weakly with the 3d electron, showing that the ground state of HNiH is derived from the $3d^8 4s^2$ configuration of nickel. GUSE et al. are unable to find solutions to H-NiH arising from the $3d^9 4s$ state of Ni for H-NiH separations less than 4 a.u. because the solution completely utilizes the $3d^8 4s^2$ Ni configuration which has lower energy.

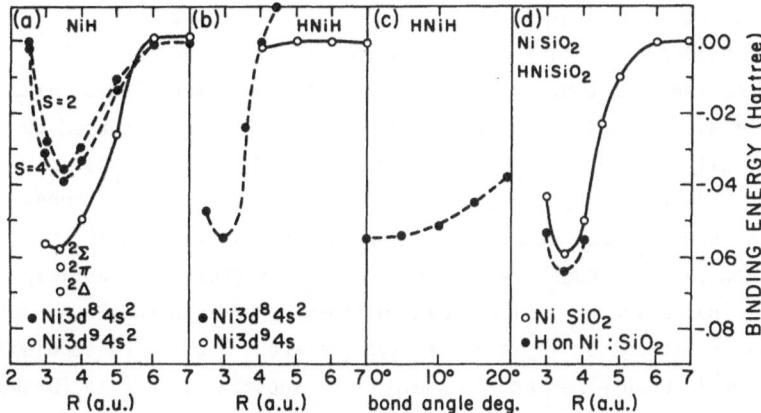

<u>Fig.5.8.</u> The composite potential energy curve for NiH (Ni $3d^8 4s^2$ and $3d^9 4s$) is shown in (a). In (b), the potential energy curve for linear HNiH molecule is shown. In (c), the potential energy curve for bending the linear HNiH molecule is shown. Finally, in (d), the potential energy curve for bonding a Ni atom to a silica surface—directly above an O ion—(the Ni-O distance is varied here) is shown along with the potential energy curve to bond a H atom to the Ni atom bonded to silica (the H-Ni distance is varied here) [5.45]. Energies are in a.u. (1 a.u. = 1 Hy = 2 Ry = 27.2 eV). Lengths are in a.u. (1 a.u. = 0.53 Å)

The studies of H adsorption by Ni bonded to a surface of SiO_2 show the chemisorption behavior in these cases to be similar to that for the H-NiH system.

In this calculation it seems clear that the 3d electrons are not significantly involved in the chemisorption bond other than as donors of electrons into bonding orbitals. Rather, the hydrogen atom forms a covalent bond with an electron which is ultimately in a nickel 4s or 4 sp hybrid orbital. An important point of the NiH calculations is the existence of a highly reactive low-lying excited state which has implications for both chemisorption and catalysis. A nickel atom on a surface will probably behave as in NiH. That is, it will have the $3d^9 4s$ configuration with the 4s orbital participating in the atom surface bond. As a hydrogen atom approaches it will begin to form a covalent bond with the singly occupied 3d orbital and be weakly attracted to the surface. As it nears, the $3d^8 4s^2$ configuration will begin to mix in until it becomes dominant at the potential minimum. Thus, although the hydrogen bonds with a 4s or 4sp orbital, the electron in this orbital may originate in the 3d shell.

The qualitative features of the calculations of GUSE et al. were confirmed by nearly simultaneous studies of chemisorption of H by a few Ni atoms by MELIUS [5.46] using CI techniques on top of a GVB calculation and also by studies of bondings CO to Ni by MELIUS et al. [5.47]. Subsequent to the studies of chemisorption by Ni cited above MESSMER et al. [5.48] using the X-α method have considered the interaction of atomic hydrogen with small clusters of Ni, Pb, and Pt. In these studies, which finally include an adatom or interstitial atom as well as the host metal, one finds strikingly different behavior for the case of Ni-H bonding as com-

pared to Pd-H bonding or Pt-H bonding. In these studies the Ni-H bond is essentially 4s derived, the Ni3d contribution being only 35% of the bonding orbital,
whereas the Pd-H bond or the Pt-H bond is mostly metal d in character. The studies
performed by MESSMER et al. are not specifically for the case of H chemisorbed to
the metal, but are actually intended to simulate the case of H interstitials in the
metal lattice. Clearly this X-α study found greater 3d participation in the bonding of H to Ni than is found by GUSE et al. who found 3d participation in the bonding orbital of less than 20%. One question arises here. Is the greater d participation in the X-α model indicative of a significant difference between the X-α model
and the Hartree-Fock-like methods or is it due to the different geometries studied?
At present this question is unanswered and it would be of substantial use to further
study this question.

More recently still, large cluster calculations for the adsorption of atomic
hydrogen on Ni have been performed by UPTON and GODDARD [5.49]. In these studies
twenty Ni atoms are used and a single H atom. The atoms are arranged to simulate
various surface planes such as (001), (110), (112), (111). Furthermore, the large
number of Ni atoms permit UPTON and GODDARD to study sites in which the H atom is
atop of Ni atom, bridged between two Ni atoms or in an open site among three or
four Ni atoms. These calculations are performed using the Hartree-Fock technique
and by replacing the electrons in the Ni core (1s, 2s, 3s, 2p, 3p) with an effective potential. Additionally, they also do studies in which the Ni3d shell is also
replaced with an effective potential. In this latter case, of course, any contribution to the chemical bond by the Ni3d is absolutely prevented. To find the
effect of the Ni3d orbital on bonding a bridged site for H was chosen. When the
Ni 3d orbitals were permitted to participate a bond strength of 2.86 eV is obtained
and a NiH distance of 1.61 Å is found. When the Ni 3d orbitals were excluded from
participation a bond energy of 2.73 eV is found and a NiH distance of 1.59 Å is
obtained. Thus even in the large cluster, the participation of the Ni3d orbital in
bonding to H is found small in the Hartree-Fock method.

UPTON and GODDARD have many other results of interest. They found the open sites
to bind H more strongly than the bridged sites, which in turn bind H more strongly
than the atop sites. The strongest atop site bond is for the (100) surface and is
1.56 eV with a bond length of 1.50 Å. It is interesting to contrast this with the
value obtained by GUSE et al. for bonding of H to a single Ni atom interbonded to
SiO_2. GUSE et al. found a bond strength of 1.65 eV and a bond length of 1.7 Å. Thus
the large bulk cluster of Ni differs little in bond strength for H atop adsorption
than does a single Ni atom supported on SiO_2. In the case of a bridged site the
strongest bond was found by UPTON and GODDARD to be 2.73 eV and is to a (001) surface. Finally in the case of an open site the strongest bond is 3.21 eV and is to
a (111) surface. In the event of the open site on the (001) surface the bond
strength is 3.04 eV. The results of UPTON and GODDARD suggest the degree of co-

ordination of H to Ni has a much greater effect on bond energy than does the face adsorbed to.

It is gratifying to note that despite some degree of quantitative doubt as to the participation of the Ni 3d in the chemisorptive bond among X-α, Hartree-Fock and semi-empirical studies, there is an emerging qualitative view that the 4s participation is far greater than the 3d participation. This is an exciting result in that this contradicts the simplistic view of transition metal chemistry which had prevailed for a long period. Fortunately, recent experimental results support the theoretical studies of H on Ni. DE MUTH [5.50] at IBM, and SPICER [5.51] at Stanford are able to support the view of H chemisorbed to Ni primarily by the Ni 4s orbital. These conclusions are obtained by performing photoemission experiments on clean Ni surfaces and Ni surfaces after H is chemisorbed. By comparison of the two spectra one infers that the Ni-H bond is largely due to the Ni 4s shell.

Another question which one may wish to answer is: Is there some solid state case in which the unfilled Ni 3d shell forms a strong covalent bond to an adsorbate? This cannot be answered generally at this time. However, systems such as NiO may be informative. In NiO the Ni 4s level is completely empty and therefore, the rather diffuse Ni 4s electron no longer would act as a potential barrier to an adsorbate forming a bond into the Ni 3d shell. Very recent UHF and GVB studies have been performed on a perfect [100] surface of NiO in order to answer this question. If one ignores the size of the O^{--} orbitals, one might expect this to be a favorable case for forming bonds to the 3d shell.

NiO is one of a class of oxides (both transition metal or other oxides) which in some cases are able to act as a catalyst. In particular NiO is useful as an oxidation catalyst. Recent experimental data of IWAMOTO et al. [5.52] for chemisorption of CO can only be understood if only about 5% of all surface sites are chemically active.

SURRATT and KUNZ have performed a series of studies applying both the ab initio unrestricted Hartree-Fock and perfect pairing generalized valence bond methods to a cluster approximation of the interaction of predissociated hydrogen with a NiO surface [5.53]. In this study, 100 or more atoms are treated explicitly in some degree of accuracy. These atoms located close to the adsorbate and the adsorbate are treated exactly in the context of the model, and the remainder by their appropriate point ion potential (pseudopotential) as described in Sect.5.4.3.

Due to the unfilled nature of the Ni 3d band, one might anticipate covalent bonding of H with a Ni atom at the surface. This is *not found*, possible due to electron-electron repulsion from the diffuse electron cloud about the O^{2-} ions, thus the Ni 3d orbitals remain essentially chemically inert in these studies. It is found that the H adsorbs at best weakly (bond energy less than 0.5 eV) to a perfect NiO [100] surface. If, however, a defect on the [100] surface is created by creating a Ni vacancy, a bond of considerable strength is found between the de-

144

fect and H. The bond is largely ionic and is about 7 eV in strength. The H atoms sits below the surface and directly above a O-ion in the first atom plane beneath the surface. In creating the defect, the electronic structure of the O^{--} ions about the defect changes so that this defect is electron deficient and this is what permits the formation of the strong ionic bond.

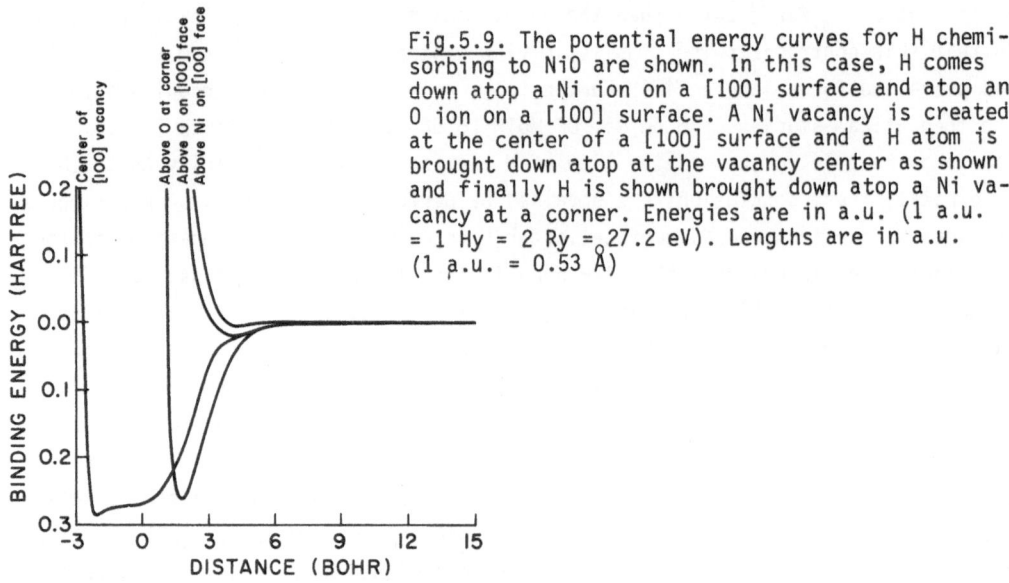

Fig.5.9. The potential energy curves for H chemisorbing to NiO are shown. In this case, H comes down atop a Ni ion on a [100] surface and atop an O ion on a [100] surface. A Ni vacancy is created at the center of a [100] surface and a H atom is brought down atop at the vacancy center as shown and finally H is shown brought down atop a Ni vacancy at a corner. Energies are in a.u. (1 a.u. = 1 Hy = 2 Ry = 27.2 eV). Lengths are in a.u. (1 a.u. = 0.53 Å)

This set of results is summarized in Fig.5.9, where we see the potential energy curves for bonding a predissociated H atom to a Ni ion on a [100] face of NiO, to an O ion on a [100] face, to a Ni vacancy on a [100] face and to a Ni vacancy at a cube corner. This calculation is in strong qualitative agreement with those for chemisorption by Ni in predicting the low activity for the unfilled Ni 3d orbitals.

Finally, it should be mentioned that the conclusion from the bonding study of H on Ni or on NiO need not be expected to apply to all other systems. It is true that other first transition elements (Sc, Mn, and Cu) were found to have an sp bonding character for H [5.45]. It is also true that the same experiment which demonstrated the essential s-p hybrid nature of the bond between H and a Ni surface has also shown that there is substantial or even dominant d bonding in the cases of H on Pd or Pt (these are second and third transition period elements) [5.50]. It is quite nice to see a substantial qualitative change in bonding character between the first transition period elements and the other transition periods in that the quantitative behavior of these periods for such catalytic processes as hydrogenolysis of ethane is totally different [5.36].

Additionally there is one case in which some substantial d bonding involving first transition elements may occur. This is for the low Z side such as Sc, Ti, and V. These metals form oxide systems and also form into perovskites of the form A B O_3. In these cases the T.M. element is octahedrally coordinated with the oxygens as is true for NiO. There is one important difference between these systems and NiO. The ions of Sc, Ti or V are much larger than Ni but in the monoxide form NiO or CoO or FeO have similar or larger lattice constants than does TiO or VO. This implies that substantial covalent involvement of the 3d orbitals on the Sc, Ti or V occurs. This may be understood further since the radial extent and polar-izability of the Sc, Ti or V 3d orbitals is much greater than is the case for Ni, Fe or Co. This has been investigated carefully for the band structure properties of TiO [5.54], and the covalent involvement of the Ti d's established. Alternate aspects of the d involvement in bonding are given for the perovskite structures in a series of studies by T. Wolfram using BEBO techniques and also perturbed band theoretic methods. Further discussion of their results is found in Chap.6.

Acknowledgments. The author is deeply indebted to his associates and students (both past and present) for assistance in preparing this manuscript. He is most particularly indebted to Dr. M.P. Guse and Professor G.T. Surratt for their help in performing many of the calculations upon which this chapter is based.

Research is supported in part by National Science Foundation under Grant DMR-76-01058 and by the U.S. Air Force Office of Scientific Research under Grant AFOSR-76-2989.

References

5.1 W.B. Fowler: *The Physics of Color Centers* (Academic Press, New York 1968)

5.2 J.C. Slater, K.H. Johnson: Phys. Today *27*, 34 (1974)

5.3 R.V. Kasawski: Phys. Rev. Lett. *32*, 83 (1974)

5.4 H.F. Schaefer III: *The Electronic Structure of Atoms and Molecules* (Addison-Wesley, Reading, Mass. 1972)

5.5 J.C. Slater: *Quantum Theory of Molecules and Solids,* Vol.4 (McGraw-Hill, New York 1974)

5.6 R.F. Marshall, R.J. Blint, A.B. Kunz: Solid State Commun. *8*, 731 (1976); Phys. Rev. B *13*, 3333 (1976)

5.7 K.H. Johnson, R.P. Messmer, J.W.D. Connolly: Solid State Commun. *12*, 313 (1973)
R.P. Messmer, C.W. Tucker, Jr., K.H. Johnson: Surf. Sci. *42*, 341 (1974)

5.8 G.T. Surratt, A.B. Kunz: Solid State Commun. *23*, 555 (1977)

5.9 L.F. Mattheiss: Phys. Rev. B *5*, 290 (1972)

5.10 A.B. Kunz, G.T. Surratt: Solid State Commun. *25*, 9 (1978)

5.11 P. Hohenberg, W. Kohn: Phys. Rev. *136*, B 864 (1964)

5.12 R. McWeeny: In *Molecular Orbitals in Chemistry, Physics and Biology,* ed. by P.O. Löwdin, B. Pullman (Academic Press, New York 1964)

5.13 J.C. Slater: *Quantum Theory of Molecules and Solids,* Vol.1 (McGraw-Hill, New York 1963)

5.14   D.R. Hartree: *The Calculation of Atomic Structure* (John Wiley and Sons, London 1957)
5.15   E. Wigner: Phys. Rev. *46*, 1002 (1934); Trans. Faraday Soc. *34*, 678 (1938)
       E. Wigner, F. Seitz: Phys. Rev. *46*, 509 (1934)
5.16   P.O. Löwdin: In *Quantum Theory of Atoms, Molecules and the Solid State*, ed. by P.O. Löwdin (Academic Press, New York 1966)
5.17   W.J. Hunt, P.J. Hay, W.A. Goddard III: J. Chem. Phys. *57*, 738 (1972)
5.18   J.C. Slater: Phys. Rev. *81*, 385 (1951)
5.19   R. Gaspar: Acta Phys. Sci. Hung. *3*, 263 (1954)
       W. Kohn, L.J. Sham: Phys. Rev. A *140*, 1133 (1965)
5.20   K. Schwarz: Theoret. Chim. Acta *34*, 225 (1974)
5.21   T. Koopmans: Physica *1*, 104 (1933)
5.22   F. Seitz: *Modern Theory of Solids* (McGraw-Hill, New York 1940)
5.23   T.L. Gilbert: Phys. Rev. B *12*, 2111 (1975)
5.24   L. Dagens, F. Perrot: Phys. Rev. B *5*, 641 (1972)
5.25   D.E. Ellis, G.S. Painter: Phys. Rev. B *2*, 2887 (1970)
       D.E. Ellis, H. Adachi, F.W. Averill: Surf. Sci. *58*, 497 (1976)
5.26   T.L. Gilbert: In *Sigma Molecular Orbital Theory*, ed. by O. Sinanogler, K. Weiberg (Yale University Press, New Haven, Conn. 1969)
5.27   P.W. Anderson: Phys. Rev. Lett. *21*, 13 (1968)
5.28   A.B. Kunz, D.L. Klein: Phys. Rev. B *17*, 4614 (1978)
5.29   W. Adams: J. Chem. Phys. *34*, 89 (1961)
       T.L. Gilbert: In *Molecular Orbitals in Chemistry, Physics and Biology*, ed. by P.O. Löwdin, B. Pullman (Academic Press, New York 1964)
5.30   T.L. Gilbert: Phys. Rev. B *6*, 580 (1972)
5.31   M.P. Guse, A.B. Kunz: Phys. Status Solidi *71*, 631 (1975)
5.32   E.G. Derouane, J.G. Fripiat, J.M. André: Chem. Phys. Lett. *28*, 445 (1974)
       A.B. Kunz, M.P. Guse: Chem. Phys. Lett. *45*, 18 (1977)
5.33   A.J. Bennett, B. McCarroll, R.P. Messmer: Surf. Sci. *24*, 191 (1971)
       B. McCarroll, R.P. Messmer: Surf. Sci. *27*, 451 (1971)
5.34   J.G. Fripiat, K.T. Chous, M. Boudart, J.B. Diamond, K.H. Johnson: J. Mol. Catalysis (to be published)
5.35   J.P. Batra, P.S. Bogus: Solid State Commun. *16*, 1097 (1975)
5.36   J. Sinfeld: Adv. Catalysis *23*, 91 (1973)
5.37   D.J.M. Fassaert, H. Verbeck, A. Vander Avoird: Surf. Sci. *29*, 50 (1972)
5.38   D.J.M. Fassaert, A. Vander Avoird: Surf. Sci. *55*, 291 (1976)
5.39   R.C. Baetzold, R.E. Mack: J. Chem. Phys. *62*, 1513 (1975)
5.40   A.B. Anderson, R. Hoffman: J. Chem. Phys. *61*, 4595 (1974)
5.41   G. Blyholder: Surf. Sci. *42*, 249 (1974)
5.42   G. Blyholder: J. Chem. Phys. *62*, 3193 (1975)
5.43   R.P. Messmer, S.K. Knudson, J.B. Diamond, K.H. Johnson: Phys. Rev. B *13*, 1396 (1976)
       R.P. Messmer, C.W. Tucker, Jr., K.H. Johnson: Chem. Phys. Lett. *36*, 423 (1975)
5.44   R.P. Messmer, D.R. Salahub: In *Ludena*, ed. by L.V.N. Sabelli, A.C. Wahl (Plenum Press, New York) in press
       R.P. Messmer: Phys. Rev. B *15*, 1811 (1977)
       R.P. Messmer, S.K. Knudson, K.H. Johnson, J.B. Diamond, C.Y. Yang: Phys. Rev. B *13*, 1396 (1976)
5.45   A.B. Kunz, M.P. Guse, R.J. Blint: J. Phys. B *8*, L358 (1975)
       R.J. Blint, A.B. Kunz, M.P. Guse: Chem. Phys. Lett. *36*, 191 (1975)
       A.B. Kunz, M.P. Guse, R.J. Blint: Chem. Phys. Lett. *37*, 512 (1976)
       A.B. Kunz, M.P. Guse, R.J. Blint: Int. J. Quart. Chen. Synp. 10, *10*, 283 (1976)
       A.B. Kunz, M.P. Guse, R.J. Blint: In "Electrocatalysis on Non-Metallic Surfaces" (NBS Special Publication 455, 1976
       M.P. Guse, R.J. Blint, A.B. Kunz: Int. J. Quart. Chen. *11*, 725 (1977)
       M.P. Guse: Thesis, University of Illinois (1976) unpublished
5.46   C. Melius: Private communication (1975); Chem. Phys. Lett. *39*, 287 (1976)
5.47   C.F. Melius, J.W. Moskowitz, A.P. Mortola, M.B. Baille, M.A. Ratner: Surf. Sci. *59*, 279 (1976)

5.48 R.P. Messmer, D.R. Salahub, K.H. Johnson, C.Y. Yang: Chem. Phys. Lett. *51*, 84 (1977)

5.49 T.H. Upton, W.A. Goddard III.: Phys. Rev. Lett. *42*, 472 (1979)

5.50 J.E. Demuth: Surf. Sci. *65*, 369 (1977)

5.51 W. Spicer: Private communication (1977)

5.52 M. Iwamoto, Y. Yoda, M. Egashira, T. Selyama: J. Phys. Chem. *80*, 1989)1976)

5.53 G.T. Surratt, A.B. Kunz: Phys. Rev. Lett. *40*, 347 (1978)

5.44 D.R. Jennison, A.B. Kunz: Phys. Rev. Lett. *39*, 418 (1977)

6. Concepts of Surface States and Chemisorption on d-Band* Perovskites

T. Wolfram and Ş. Ellialtıoğlu

With 17 Figures

This chapter is principally concerned with the surface properties of the transition metal oxides with cubic perovskite structure. The topics discussed include: five- and sixfold coordinated transition metal-ion clusters and the bonding of molecules to such clusters, the bulk and surface electronic structure of the d-band perovskites and the nature of molecular bonding to the solid surface including band effects, the character of oxygen adsorption and the role of oxygen vacancies. Throughout the chapter we have attempted to show relationships between the electronic structure and simple, but general, concepts of importance in chemisorption and catalysis.

A review of recent photoemission experiments on $SrTiO_3$ surface states and oxygen adsorption is presented. Detailed comparisons of experimental results are made with theoretical calculations.

6.1 Introductary Remarks

6.1.1 General Comments

With the advent of high vacuum surface science involving a variety of photoelectron spectroscopies there is reason to believe that the nature of surface states and chemisorption on clean, single-crystal surfaces can be understood in a fundamental way.

Silicon and monatomic transition metal surfaces have been the subject of intense study using low energy electron diffraction, Auger spectroscopy, electron energy loss and ultraviolet photoemission measurements. It is now possible to compare the results of these experiments with theoretical calculations and a reasonably de-tailed description of these materials is emerging.

*Acknowledgement is made to the National Science Foundation for partial support of this work. Acknowledgement is made to the Donors of the Petroleum Research Fund, administered by the American Chemical Society, for the partial support of this research.

By contrast, relatively little experimental or theoretical information concerning the clean surface electronic structure of the transition metal oxides is available. This situation is beginning to change as more surface scientists are being attracted to the study of these interesting materials.

One of the objectives of this chapter is to call attention to some fascinating surface physical problems in the area of the transition metal oxides. From the point of view of "clean" surface physics, very little is known with certainty about surface states or chemisorption on these oxides. We shall present discussions of a number of concepts about the surface electronic properties of the class of transition metal oxides with the perovskite structure. Because of the early stage of development of this topic, some of the results must still be regarded as tentative.

The transition metal oxides include a wide variety of materials, such as insulators, semiconductors and metals. Many oxides possess magnetic properties and some undergo semiconductor-to-metal transitions. The magnetic metal oxides are particularly complex because electron correlation is sufficiently large to prevent delocalization of the d electrons. For such materials conventional energy band theory is not valid. As a result the bulk electronic structures are not well understood. Unfortunately, many of the catalytically active oxides such as NiO or LaCoO$_3$ belong to this class of highly correlated d-electron materials. Cluster models provide a qualitative description of the localized states of such materials but presently no satisfactory theoretical framework exists for investigating their surface electronic structure.

The transition metal oxides have long been of great technological importance because of their catalytic, electrocatalytic and photocatalytic behavior. Chemisorption and catalysis on the oxides has often been attributed to the presence of "active" surface sites. These sites are envisioned to interact with reactant molecules to form a "surface complex" either as the final state in chemisorption or as an intermediate state in catalysis. Active sites have been tentatively identified as "coordinatively unsaturated" transition metal surface ions or as surface defects such as oxygen vacancies. The precise nature of these active sites is of great importance but is not known in most cases. The microscopic details of the electronic mechanisms by which molecules adsorb on, and react with, an oxide substrate are often very obscure.

In the case of practical catalysis this cloudy situation is not likely to change in the short term because of the enourmously complex nature of actual catalyst systems. But, "hope springs eternal" that the study of well-characterized, clean, single-crystal surfaces may lead to some insight into the behavior of the catalysts. This proposition provided a great deal of motivation for the work described in this chapter.

6.1.2 The Perovskites

The perovskites with the ABO_3 (simple cubic) structure are a subgroup of the transition metal oxides that have attracted much interest recently. An excellent review of the qualitative properties of the perovskites has been given by GOODENOUGH [6.1]. Some typical perovskites are listed in Table 6.1 and the structure is shown in Fig.6.1. In the case of WO_3 and ReO_3 the A ions are absent from the structure.

Table 6.1. Some typical examples of ABO_3 perovskites

Insulators	Magnetic semiconductors	Metals
$SrTiO_3$	$LaCoO_3$	ReO_3
$BaTiO_3$	$LaMnO_3$	$LaTiO_3$
$KTaO_3$		$NaWO_3$
WO_3		$KMoO_3$

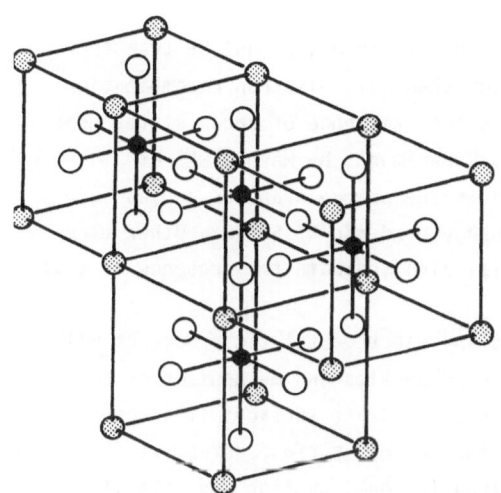

Fig.6.1. The ABO_3 perovskite structure. The black circles represent the transition metal ions (B ions), the A ions are represented by the shaded circles and the open circles are oxygen ions. The B-O distance is usually about 2 Å. Each B ion is surrounded by an octahedron of oxygen ions

PEDERSON and LIBBY [6.2], and MEADOWCRAFT [6.3] suggested the rare earth cobalt oxides, $RCoO_3$ (R = rare earth) as possible gas phase catalysts and more recently extensive studies have been carries out by VOORHOEVE et al. [6.4] using a variety of substituted perovskites such as $(La_{1-x}Pb_x)MnO_3$ and $(Pr_{1-x}Pb_x)MnO_3$. LIBBY and AEGERTER [6.5] found reduced $SrTiO_3$ and $BaTiO_3$ to be active in catalyzing isomerization, hydrogenation, synthesis and cracking reactions.

The perovskites have also been studied as electrocatalytic and photocatalytic electrodes in electrochemical applications. TSEUNG and BEVEN [6.6] have studied the behavior of $LaCoO_3$ and $(La_{0.5}Sr_{0.5})CoO_3$ electrodes in anodic and cathodic processes.

An interesting application of the perovskites is in the photoelectrolysis of water to produce hydrogen and oxygen. $SrTiO_3$, $BaTiO_3$, WO_3 and a large variety of other oxides have been employed as anodes in photoelectrolysis cells [6.7-10]. The highest cell efficiency, about 10%, was obtained using an n-type $SrTiO_3$ anode [6.7] subjected to band gap light (3.2 eV photons). These studies have demonstrated the feasibility of photoelectrolysis and have raised the exciting possibility of solar driven photolysis for the production of hydrogen fuel. The bulk band gap of the materials currently being studied is too large for efficient solar applications but excitations from band gap surface states to the conduction band are near in energy to the peak in the solar spectrum.

In the process of photoelectrolysis the photons adsorbed by the oxide anode create hole-electron pairs. These pairs must be separated so that rapid electron-hole recombination is avoided. In n-type $SrTiO_3$ the band bending at the electrolyte/solid interface provides an internal field for separating the hole-electron pairs.

To catalyze the anode reaction

$$2p^+ + 2OH^- \rightarrow \frac{1}{2} O_2 + H_2O \quad (p^+ = hole)$$

in electrolysis, photo holes in the valence band of the oxide must be transferred to the OH^- ions adsorbed on the surface faster than hole-electron recombination can occur. Efficient charge transfer requires the existence of anode states near in energy to the redox potential, $\varepsilon(OH^-/O_2)$. Experiments by MAVDROIDES et al. [6.7] suggest that the anode states involved in the charge transfer process are surface states which lie in the forbidden energy band gap of $SrTiO_3$. Other electrochemical experiments by FRANK et al. [6.11] also indicate the presence of such surface states.

Thorough understanding of these catalytic and electrocatalytic processes will require a knowledge of the surface electronic properties and mechanisms of adsorption and charge transfer which occur on the substrate surface. It is generally believed that the d-electron surface states and various surface defect states (such as a surface oxygen vacancy state) are involved in chemisorption and catalytic mechanisms. The surface of a perovskite provides an abundance of coordinatively unsaturated d orbitals. These "dangling" bonds provide geometrically and electronically favorable sites for chemisorption of molecules and for charge transfer between the solid and interacting molecules [6.12]. Surface oxygen vacancy states can also provide localized d-electron surface orbitals. Such vacancy states have wave functions that are combinations of d orbitals located on the cation sites adjacent to the oxygen vacancy. These states can act as a source or sink of electrons in surface chemical reactions and therefore can be catalytically active.

This chapter describes some of our recent work on the surface electronic structure of d-band perovskites including cluster model and solid state model calculations of surface states and surface energy bands. In Sect.6.2 we present a brief discussion of the electronic states of a simple transition metal ion cluster and how these states are altered when a ligand is missing. This allows a qualitative discussion of some possible chemisorption mechanisms. In Sect.6.3 we review the bulk electronic structure in terms of solid state energy bands and show how the bands are related to cluster states. We then describe the surface energy bands of d-band perovskites and consider the relation of the surface states to the cluster energy levels for a coordinatively unsaturated transition metal ion. As a final topic in Sect.6.3 we consider the effect of changes in the charges of surface ions on the position and occupation of surface bands.

In Sect.6.4 a discussion of the character of surface oxygen vacancy states is presented and the conditions for the occurrence of such states in the band gap are explored.

Section 6.5 is a review of recent photoemission studies of the (100) surface of SrTiO$_3$. Theoretical calculations of the electron energy distribution curves are described and compared with the experimental data. A brief discussion of oxygen adsorption on SrTiO$_3$ surfaces is presented and some tentative conclusions are described.

Finally, Sect.6.6 is a brief summary of the chapter.

6.2 Cluster Models of Transition Metal Oxides

The oxides are ionic materials with large Madelung potentials. In TiO$_2$ or SrTiO$_3$ the formal charges are +4 and -2 for the Ti and oxygen ions, respectively. Of course, the actual charges on the ions are considerably less because covalent bonding is significant. The dominant electronic structure is derived from the atomic orbitals of the transition metal ion and the oxygen ions.

A simple approach to understanding the electronic properties of the oxides is to study a representative group of atoms. Nearly all of the transition metal oxides possess a local structure consisting of the transition metal ion surrounded by six oxygen ligands. For the cubic perovskites the oxygen ions form a perfect octahedron as may be seen from Fig.6.1, which shows the ABO$_3$ structure. For oxides with other structures, TiO$_2$ (rutile) for example, the octahedron is distorted. Nevertheless, the central features of most of the oxides can be understood by considering a BO$_6$ cluster with cubic symmetry.

According to the LCAO (linear combinations of atomic orbitals) method the wave functions for the cluster states may be written as

$$\psi^j(\underline{r}) = \sum_\lambda a^j_\lambda \phi_\lambda(\underline{r}) + \sum_{i\lambda} b^j_{i\lambda} \phi(\underline{r} - \underline{r}_i) \quad , \tag{6.1}$$

where ψ^j, the jth cluster state wave function, is a sum of cation atomic orbitals, $\phi_\lambda(\underline{r})$, of the λ type (λ = 1s, 2s, 2p ...) and oxygen ion orbitals, $\phi_\lambda(\underline{r} - \underline{r}_i)$. The origin of coordinates is chosen at the cation site and the positions of the anions are specified by the set of vectors, $\underline{r}_i$. The coefficients, a_λ^j and $b_{i\lambda}^j$ are determined from the solutions of Schrödinger's equation for the cluster and specify the amplitudes of the cation and anion orbitals, respectively. A detailed discussion of the calculation of cluster states is given in Chap.5. Here we shall give only a brief qualitative discussion of the nature of the cluster states for the oxides.

The chemically active states of oxide clusters are derived principally from the oxygen 2p and cation nd orbitals (n = 3, 4, or 5). There are three distinct types of cluster states associated with the 2p and nd orbitals. Two of the types of cluster states involve significant cation and anion orbital amplitudes, while the third involves essentially only anion orbitals.

The first type, the *bonding* states, involve "in-phase" combinations of 2p and nd orbitals which concentrate bonding charge in the regions between the B and O ions. The second type, the *antibonding* states, involve "out-of-phase" combinations of orbitals which interfere destructively in the region between the B and O ions. These antibonding states are also referred to as crystal field states. The third type of cluster states are known as *nonbonding* states. They have wave functions that involve only the anion orbitals and do not contribute to the metal-ion-ligand bonding.

A schematic of the cluster levels of the BO_6 cluster is shown in Fig.6.2a. For insulating materials such as TiO_2, $SrTiO_3$, $BaTiO_3$ or $KTaO_3$ the highest occupied state is the $1t_{1g}$ and the lowest unoccupied state is the $2t_{2g}$. The energy separation, E_g, corresponds to the forbidden energy gap in the solid. For the oxides which are metals the $2t_{2g}$ levels are partially occupied. The crystal field (antibonding) states, $2t_{2g}$ and $3e_g$, correspond to the d bands (conduction bands of the solid). In the cluster model the separation of the $3e_g$ and $2t_{2g}$ levels is the crystal field splitting denoted by 10Dq. The crystal field (antibonding) states are admixtures of nd and 2p orbitals as illustrated in Fig.6.2b. The $3e_g$ consists of d_{z^2} (or $d_{x^2y^2}$) in combination with 2p orbitals directed toward the cation. The $2t_{2g}$ consists of the d_{xz} (or d_{yz},d_{xy}) in combination with 2p orbitals directed perpendicular to the BO axis. The oxygen admixture into these states is 20 to 30%. The corresponding bonding states are the $1t_{2g}$ and the $2e_g$ which have wave functions consisting roughly of 70 to 80% 2p and 20 to 30% nd orbitals.

The remaining levels shown in Fig.6.2a are the nonbonding states. They are composed almost entirely of combinations of oxygen 2p orbitals. The actual ordering of these levels depends upon the particular cluster in question.

In the calculation of the energy states of the BO_6 cluster the oxygen 1s and 2s orbitals as well as the cation (n + 1)s and (n + 1)p orbitals and core states must be included in the set of basis functions. Additional cluster levels are produced for this larger basis set including lower lying core-like states and un-

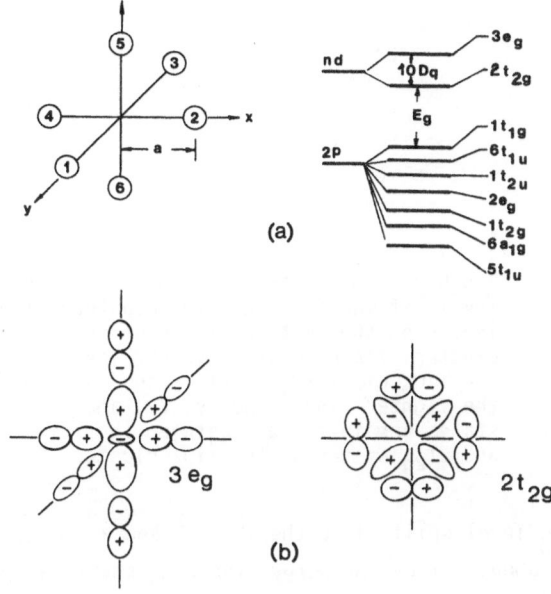

(a)

(b)

<u>Fig.6.2.</u> (a) A schematic of the BO_6 cluster and the cluster energy levels. The states are designated according to the appropriate irreducible representation of the cubic O_h group. The numerical prefix distinguishes different states of the same symmetry. (b) A schematic of the orbital combinations involved in the cluster wave functions for the $3e_g$ and $2t_{2g}$ states of the BO_6 cluster

occupied states above the $3e_g$. The qualitative features of the states in Fig.6.2, however, are not significantly altered.

Of principal interst in our discussion here are the crystal field states, $3e_g$ and $2t_{2g}$, which can play a central role in the surface physics and chemistry of the oxides.

At a surface of a solid the B ion may be missing one or more oxygen ligands. Such an ion is said to be "coordinatively unsaturated" [6.13], meaning that it has less than its normal complement of bonded ligands. This ion is expected to be chemically reactive since it may bond to a foreign molecule and charge transfer can occur.

The simplest model of an oxide surface is the TiO_5 cluster obtained by removing an oxygen atom from the TiO_6 cluster. The TiO_5 cluster approximates the environment of a fivefold coordinated Ti on a (100) surface of $SrTiO_3$ or on a (110) surface of TiO_2. In approximating the surface environment of a solid the effect of the Madelung potentials must be included.

Figure 6.3 shows the results of our calculations [6.14] for TiO_6 and TiO_5 clusters. For simplicity we have presented only the relevant energy levels. (The entire complex of levels is presented in [6.14,15]. These results include the effects of the long-range Madelung potentials which would be present in a solid. In addition, for the TiO_5 "surface cluster", a 2 eV decrease in the repulsive Madelung potential at the Ti site is included in order to account for the change in the Madelung potential that occurs at a perovskite surface [6.16].

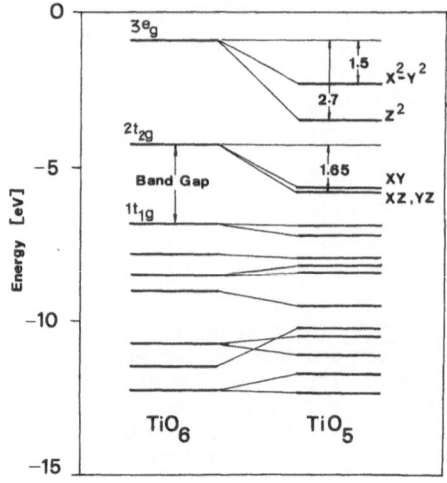

Fig.6.3. Comparison of some of the energy levels of the TiO_6 and TiO_5 clusters. The levels on the left are those of the TiO_6 cluster. Their relation to the levels of the TiO_5 cluster (right) is indicated by the lighter lines. The xy, xz, and yz states are in the gap between the $1t_{1g}$ and $2t_{2g}$ states of the TiO_6 cluster

From Fig.6.3 it is seen that the $3e_g$ level splits into the $x^2 - y^2$ and z^2 levels with the z^2 state dropping down about 2.7 eV in energy. The $2t_{2g}$ state is split into a doublet, xz, yz, and a singlet xy. These latter three surface cluster states fall into the forbidden band gap region of the solid. (That is, between the TiO_6 cluster levels which represented the top of the valence band and the bottom of the conduction band.)

In n-type $SrTiO_3$ or TiO_2 these "band gap" states can be occupied with electrons thereby producing "Ti^{3+}" ions in place of the bulk "Ti^{4+}" ions. (The actual situation, however, is more involved than suggested so far. The position of the band gap states depends upon their electronic occupation. Figure 6.3 indicates the level positions when they are unoccupied. When electrons are added to these levels their energy is increased because of the Coulomb repulsion among the additional electron occupying Ti d orbitals.)

Electron transitions between the xy, xz and yz and the z^2 cluster levels are possible when the former are occupied states. HENRICH et al. [6.17] have observed surface electronic transitions of about 1.5 eV by electron energy loss experiments for TiO_2 and $SrTiO_3$. We have suggested that these transitions are probably between the xz and z^2 cluster surface levels [6.14].

The xz and yz states of the TiO_5 cluster possess favorable symmetries for interaction with the antibonding states of many types of molecules. For example, the carbon-carbon π^* state, the antibonding states of H_2, O_2, CO and other diatomic species can hybridize with the xz or yz states of the TiO_5 cluster. In a similar fashion the z^2 state can interact with the bonding orbitals of these molecules. Some typical geometries are illustrated in Fig.6.4.

A number of important observations can be made at this point. First, we note that the xz and z^2 cluster surface states provide a "couple" for charge transfer processes. Consider n-type $SrTiO_3$ or TiO_2. The occupied xz surface state can

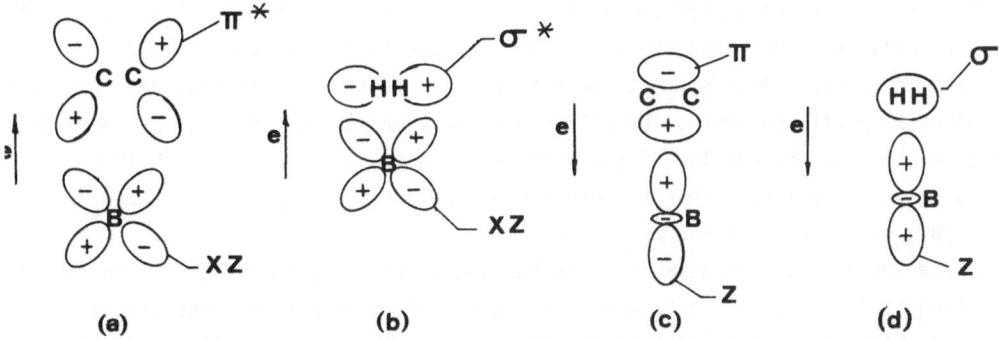

Fig.6.4a-d. A schematic of the orbitals and geometries of typical molecules that interact with the xz and z^2 orbitals of a transition metal ion. (a) and (b) illustrate the overlap of the xz state with antibonding molecular orbitals. (c) and (d) show the overlap between the z^2 state and bonding molecular orbitals. The arrows indicate the usual direction of charge transfer when these orbitals are hybridized to form a surface complex

hybridize with empty antibonding molecular orbitals. This hybridization provides a mechanism for transferring charge into the antibonding molecular orbital from the substrate. The effect is a weakening of the molecular bond and an increase of the molecular bond length. For example, for an organic molecule of the form $R_1-C = C-R_2$ (where R is some end group), filling the π^* antibonding level weakens the carbon-carbon double bond and increases the bond length [6.18]. This mechanism appears to be effective in the dissociation O_2 on $SrTiO_3$ and TiO_2 (see Sect.6.5). In a similar fashion the z^2 surface state can accept electrons from occupied molecular bonding states. The loss of electrons from a molecular bonding state also weakens the molecular bond.

The importance of the couple, however, is not just the fact that both processes can weaken the molecular bond but also that a greater redistribution of electronic charge can occur with a small *net* amount of charge being transferred. This is because the transfer of charge from the substrate to the molecule is limited by the Coulomb repulsion among the excess electrons on the molecule. That is, the formation of a polar molecule is energetically unfavorable. Similarly, the transfer of charge from the molecule to the z^2 level of the substrate is limited by Coulomb repulsion among the excess electrons on the surface cation. On the other hand, when both processes are simultaneously operative then the charge transfer of one process is partially cancelled by the charge transfer of the other process. This allows a greater redistribution of electrons for a given net charge transfer. Depending on the degree of electron redistribution, chemisorption or dissociative chemisorption may occur. The latter process is aided by the increase in molecular bond length that occurs when electrons are injected into the antibonding molecular state or withdrawn from the bonding state.

The energy separation between a molecular bonding level and the z^2 surface state will usually be larger than the separation between the molecular antibonding level and the xz surface state so that the charge transfer to the molecular antibonding level usually will dominate. The difference, however, is not as large as might be expected because the overlap of the molecular bonding state with the z^2 (Sigma type) tends to be larger than the overlap of the antibonding level with the xz (Pi type) by a factor of two or greater.

There are empirical models that can be used to estimate the molecular and chemisorption bond strengths and bond lengths as well as criteria for dissociative chemisorption from a knowledge of the bonding geometry. The so-called Bond Order models have proved very valuable in surface reactions on the transition metals [6.19] and they have also been applied to the oxides [6.12].

6.3 Bulk and Surface Electronic Properties of the Perovskites

6.3.1 Bulk Electronic Properties

In a solid perovskite the energy levels may be broadened into energy bands or, in some cases where electron correlation is very large, the d electrons may remain localized. The latter case is usually true for the magnetic oxides while for the nonmagnetic oxides conventional energy band theory applies [6.20]. Thus for TiO_2, Ti_2O_3 or any of the perovskites in columns one and three of Table 6.1 the electronic states are delocalized band states. We refer to the class of perovskites which possess energy bands as "d-band perovskites". In this section we discuss how some of the simple ideas developed for the cluster model carry over into the energy band picture. In order to facilitate the discussion it is necessary to first review briefly the nature of the bulk energy bands.

The perovskite solid is composed of a periodic array of ABO_3 unit cells. The positions of the unit cells are specified by a set of vectors, R_B, which locate the B ion in each unit cell. The oxygen ions are located at a distance a along the three coordinate axes from a cation site in the unit cell and the A ions are at the positions a(1,1,1) relative to the B ions. There are 14 energy bands derived from the 2p and nd orbitals of each unit cell; all of which are needed in describing the electronic properties of the perovskites. The bands derived from the A-ion orbitals, however, play only a very minor role [6.16] and will be neglected for the moment.

The energy band wave functions for the solid may be expressed in an LCAO form similar to that used for the cluster.

$$\psi_{\underline{k}\nu}(\underline{r}) = \frac{1}{\sqrt{N}} \sum_{\underline{R}_B} \left[a_\lambda^{\underline{k}\nu} \phi_\lambda(\underline{r} - \underline{R}_B) + \sum_i b_{i\lambda}^{\underline{k}\nu} \phi(\underline{r} - \underline{R}_B - \underline{r}_i) \right] \exp(i\underline{k} \cdot \underline{R}_B) \quad . \tag{6.2}$$

The wave functions, $\psi_{k\nu}$, are characterized by a wave vector $\underline{k}$ which lies in the first Brillouin zone and a band index, ν. The quantity, $\exp(i\underline{k}\cdot\underline{R}_B)$ is the Bloch-wave phase factor, the vectors $\underline{R}_B$ and $\underline{R}_B + \underline{r}_i$ locate the cations and anions, respectively, and the coefficients $a_\lambda^{k\nu}$ and $b_{i\lambda}^{k\nu}$ specify the amplitudes of the cation and anion orbitals for the νth band. (A factor of $1/\sqrt{N}$ is introduced for normalization purposes).

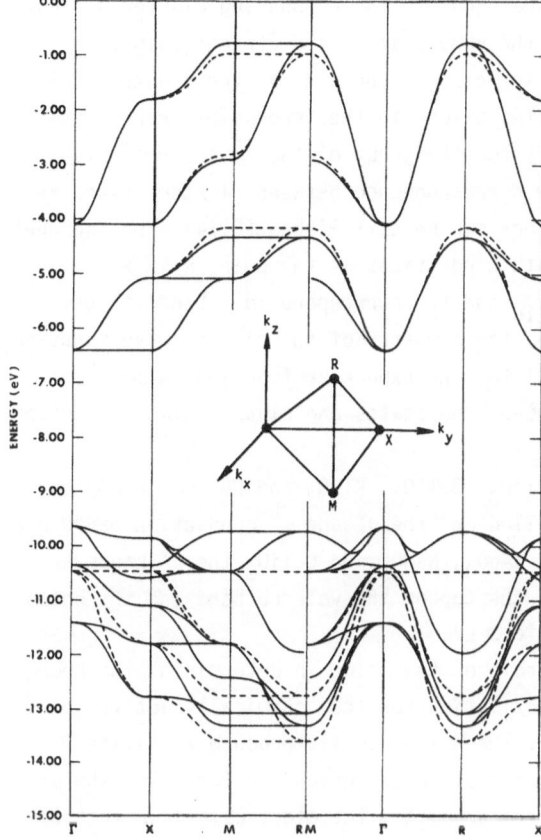

Fig.6.5. The energy bands of $\overline{\text{SrTiO}_3}$. σ^* and π^* are empty conduction bands, σ^o and π^o are occupied nonbonding oxygen bands and σ^* and π^* are occupied valence bands. The dashed lines are for a simple analytical LCAO model of the energy bands. For this model the π^o and σ^o are flat bands. The cubic Brillouin zone is shown in the inset

The energy of a band state, $E_{k\nu}$, depends on the band index, ν, and the wave vector, $\underline{k}$. The curves generated by $E_{k\nu}$ for a fixed ν as a function of $\underline{k}$ are called energy bands. Figure 6.5 shows the energy bands of $SrTiO_3$ derived from a semi-empirical LCAO model [6.16]. This band structure is typical of that found for all the d-band perovskites regardless of the calculation model employed.

The σ^* conduction band states have wave functions that are admixtures of the e_g symmetry (z^2 and $x^2 - y^2$) d orbitals with the 2p orbitals whose lobes are oriented parallel to the B-O axes. The σ^* band is the analog of the $3e_g$ cluster states. The "π^*" conduction band states have wave functions that are admixtures of the t_{2g} symmetry (xy, xz, yz) d orbitals with 2p orbitals oriented perpendicular to the B-O axes. These bands are the analogs of the $2t_{2g}$ cluster states. The energy bands labeled π° and σ° have wave functions that are primarily composed of combinations of oxygen 2p orbitals. These bands are analogous to the nonbonding cluster states.

Inspection of the wave functions for the band states at different points in the Brillouin zone shows that the local symmetry is the same as the cluster states at either Γ or R [6.14]. The cluster states belong to the irreducible representations of the O_h group. The point group symmetry for the group of the k-vector is also O_h at Γ and R. An approximate one-to-one correspondence between cluster states and band states at Γ or R in the Brillouin zone can be established [6.14]. The "gerade" states e_g, t_{2g}, t_{1g} and a_{1g} correlate with band states at the point R in the Brillouin zone. The "ungerade" states, t_{1u} and t_{2u} correspond to Γ-band states. The nature of this correspondence is that the symmetry of the cluster wave function is repeated throughout the entire crystal for the band wave function. A more complete discussion of the correlation between band states and cluster states is given in [6.14].

For insulating d-band perovskites, $SrTiO_3$, $BaTiO_3$, $KTaO_3$ and WO_3 are examples, the σ, σ°, π and π° valence bands are filled and the σ^* and π^* conduction bands are unoccupied. For the metals such as ReO_3, $KMoO_3$, $NaWO_3$ and $LaTiO_3$ the π^* bands are partially occupied. The band gap between the top of the valence band and the bottom of the π^* conduction band ranges from 1 to 3 eV.

Up to this point we have ignored the role of the A ion in determining the energy bands. This turns out to be a valid approximation for the "chemically active" bands displayed in Fig.6.5. The outer states of the A ion usually produce an s-like band which lies about 10 eV above the σ^* bands. Levels which lie far below the σ bands derived from the core states are also produced. The electrostatic potentials produced by the charged A ion must be included in the Madelung potentials acting on the B and O ions but the bands derived from the A ion orbitals may be neglected.

The dashed curves in Fig.6.5 are the results of a simple analytical LCAO model in which only nearest neighbor ions are assumed to have significant overlap of orbitals [6.16,21-24]. As can be seen, the energy bands are well represented by this simpler model. The major difference between the solid and dashed curves is that the nonbonding π° and σ° bands are collapsed into flat (i.e., dispersionless) bands for the analytical model. This is because the oxygen-oxygen overlap is neglected in the model. In the remainder of this paper the results discussed are derived from this analytical LCAO model. The model produces reasonably reliable results for the qualitative features of the electronic structure. For example,

Fig.6.6 shows a comparison of the density of states derived from the analytical model with the numerical results obtained by MATTHEISS [6.20] for $SrTiO_3$ and $KTaO_3$ from his APW (augmented plane wave) calculations. Equally good results are obtained for the metals ReO_3 and $NaWO_3$ [6.21] and for other perovskites. Good agreement between model calculations and experimental data on the reflectivity from $SrTiO_3$ by CARDONA [6.25] has also been demonstrated [6.22].

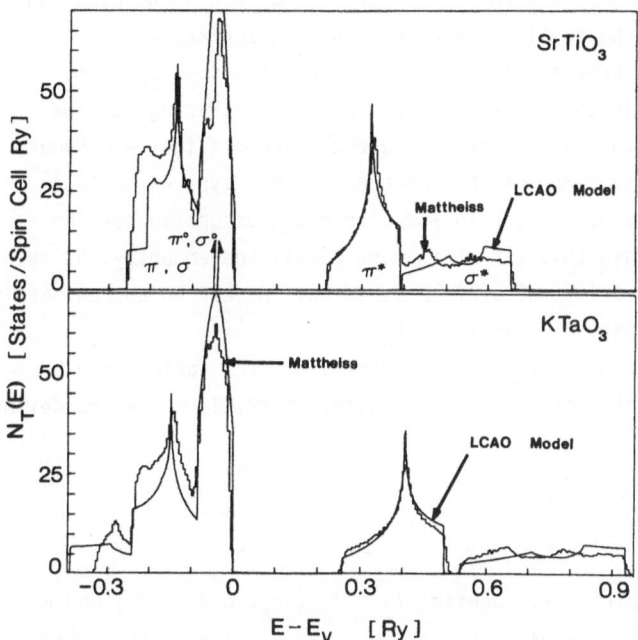

Fig.6.6. Comparison of the density of states calculated from the analytical LCAO model with numerical histograms based on APW calculations of [6.20]

6.3.2 Surface Electronic Properties

When a solid is terminated by a surface, new electronic states can appear associated with the surface and derived from the bulk band structure. Of particular interest for the perovskites are surface states which lie in the forbidden energy band gap region associated with coordinatively unsaturated transition metal bonds.

The (100) surface of a cubic perovskite such as $SrTiO_3$ has two possible configurations. In one case, type II, the oxygen and Sr ions are on the surface. The discussion here will be focused on the type I surface which contains the B ions and the O ions.

Detailed discussions of the surface energy bands based on the analytical LCAO model have been presented elsewhere [6.16,23,24,26]. In this chapter we shall only discuss some of the important results.

The surface energy bands depend upon the nature of the surface perturbations, which are not known a priori. For an ideal (100) type I surface of $SrTiO_3$ a major surface perturbation arises from the decrease in the magnitude of the repulsive Madelung potential acting at the Ti cation site. This decrease is about 2 eV for an ideal surface [6.16,27]. The Madelung potential at the surface oxygen site is nearly unchanged from the bulk value [6.16,27].

Other types of surface perturbations are also involved in determining the surface bands. One such perturbation is the change in the crystal field and additional crystal field splitting due to the axial field created by the surface and missing ligands. The axial field is not expected to be large [6.16]. A second type of perturbation is due to changes in the interlayer spacing and "puckering" of the surface in which, for example, the surface oxygens are displaced slightly without destroying the two-dimensional symmetry of the surface. Another type of surface perturbation comes about because of changes in the electronic occupation of the surface orbitals. We shall discuss this effect in more detail subsequently. At this point we simply comment that a variety of surface perturbations can be approximately represented by a surface perturbation parameter, ΔE_d.

Much of the information about the electronic properties of the surface can be derived from the local density of states (LDOS) functions, $N_\alpha(\underline{R},E)$, which are defined by

$$N_\alpha(\underline{R},E) = -\frac{1}{\pi} \, \text{Im}\left\{\sum_n \frac{|a_\alpha^n(\underline{R})|^2}{E-E_n+i0^+}\right\} \quad . \tag{6.3}$$

The subscript α denotes the particular orbital ($\alpha = d_{xz}$, p_x, d_{z^2}, etc.) and $\underline{R}$ is a cation or anion lattice site anywhere inside or on the surface of the solid. The sum is over all of the states of the finite solid including surface states and E_n is the energy of the nth state. The quantity $a_\alpha^n(\underline{R})$ is the normalized amplitude of the αth basis orbital located on the $\underline{R}^{th}$ site for the wave function corresponding to E_n. "Im" indicates the imaginary part of the quantity in brackets and 0^+ is a positive infinitesimal.

(The expression for the LDOS given by (6.3) is exact for the orthogonalized atomic orbitals, the LÖWDIN orbitals [6.28], but only approximate if the basis orbitals of the LCAO wave functions are true atomic orbitals because (6.3) neglects orbital overlap between neighboring ions. The Löwdin orbitals have symmetry properties that are identical to those of true atomic orbitals [6.29] and are very similar to atomic orbitals when overlap is small, as it is for the perovskites. For simplicity, henceforth we shall assume that the basis orbitals used in the energy band wave functions are Löwdin orbitals.)

The LDOS functions are normalized so that the integral over all energy is unity

$$\int_{-\infty}^{\infty} N_\alpha(\underline{R},E)dE = 1 \quad . \tag{6.4}$$

For a uniform, infinite system $N_\alpha(\underline{R},E)$ is independent of $\underline{R}$ and $\sum_\alpha N_\alpha(\underline{R},E)$ is just the usual density of states. For a finite system $N_\alpha(\underline{R},E)$ depends strongly on $\underline{R}$ when $\underline{R}$ corresponds to a site on or near the surface. When a surface band is formed $N_\alpha(\underline{R}_s,E)$ ($\underline{R}_s$ = surface site) will be large in the energy range of the surface band. Conversely, because of conservation of density (6.4), there must be a depletion of the density within the energy range of the bulk energy bands. This depletion signifies that the αth orbital on the surface participates only weakly in the bulk-like states of the solid.

In the following discussions we shall present results for $N_{xy}(E) \equiv N_{xy}(\underline{R}_s,E)$ ($\underline{R}_s$ is a surface cation site) and $N_{d\sigma}(E) \equiv N_{z2}(\underline{R}_s,E) + N_{x2-y2}(\underline{R}_s,E)$. These functions specify the π-type and σ-type d-orbital LDOS, respectively. The total d-orbital density of states for a surface cation is the sum of the various d-orbital contributions

$$N_d(E) = N_{xz}(E) + N_{yz}(E) + N_{xy}(E) + N_d(E) \quad . \tag{6.5}$$

The character of the surface electronic structure is illustrated in Figs.6.7 and 8, where the surface perturbation parameter ΔE_d is 0.0 and -2.0 eV, respectively Figure 6.7 shows the case of a surface characterized by $\Delta E_d = 0.0$. This case corresponds to a surface that is perfectly terminated. The left-hand panel shows the bulk continuum of π and π^* band states (cross-hatched areas) as a function of the k-vector along the Γ to X axis in the Brillouin zone. The adjacent panel shows the d-orbital contribution to the local density of states (LDOS) for a B cation located on a (100) type I surface (solid curve). The dashed curve is the LDOS for a cation infinitely far from the surface. The right-hand panels show similar results for the σ and σ^* bands and the e_g symmetry component of the LDOS for bulk and surface cations. The d orbitals are not involved in the nonbonding states and hence no LDOS is shown adjacent to the π^o and σ^o bands.

It is noted that there are no localized surface bands for $\Delta E_d = 0.0$. The surface LDOS is confined to the same energy region as the bulk LDOS. There are, however, substantial differences between the surface cation LDOS and the bulk LDOS. The surface cation LDOS is depleted at the top of the π^* and σ^* bands and enhanced at the bottom of these bands. This redistribution of density corresponds to the presence of virtual or resonance surface states.

When a nonvanishing surface perturbation ($\Delta E_d < 0$) exists then highly localized surface bands appear in the forbidden band gap. This can be seen in Fig.6.8 for $\Delta E_d = -2.0$ eV. Surface bands, indicated by the dotted curves, are formed below the π^* and σ^* bands. It is also important to note that the surface cation LDOS is completely altered from the $\Delta E_d = 0.0$ case. There is a large density below the corresponding conduction bands but almost no density within the bulk continuum of states. This means that the d electrons on surface cations do not participate in

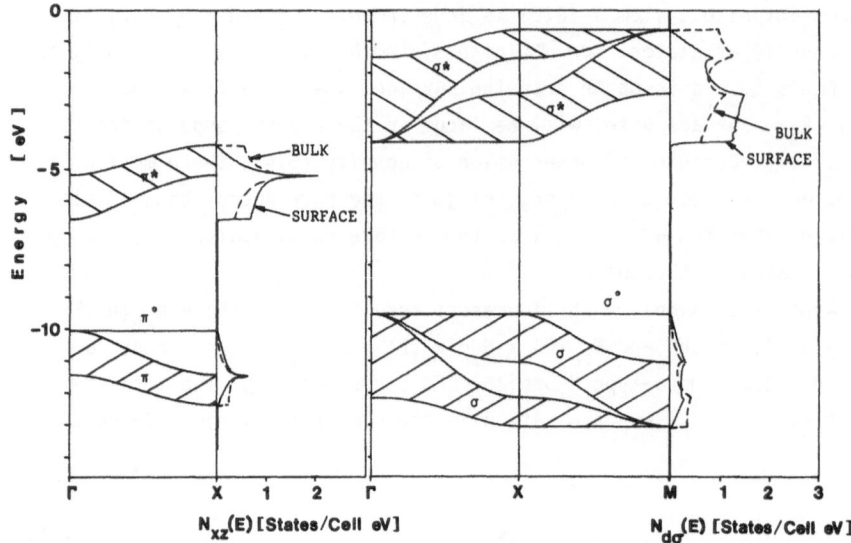

Fig.6.7. The local density of states of a cation for a (100)surface of SrTiO$_3$ with surface perturbation parameter $\Delta E_d = 0$. The cross-hatched areas indicate the energy-$k_\parallel$ vector space occupied by the bulk energy bands. $N_{xz}(E)$ is the xz component local density of states (LDOS) for a cation. In the second panel the solid curves are the LDOS for a cation on a type I, (100) surface of SrTiO$_3$. The dashed curves give the corresponding LDOS for a cation far from the surface (i.e., bulk cation). $N_{d\sigma}(E)$ is the LDOS associated with the e$_g$ symmetry d orbitals. The LDOS for both surface and bulk cations is given in the right-hand panel

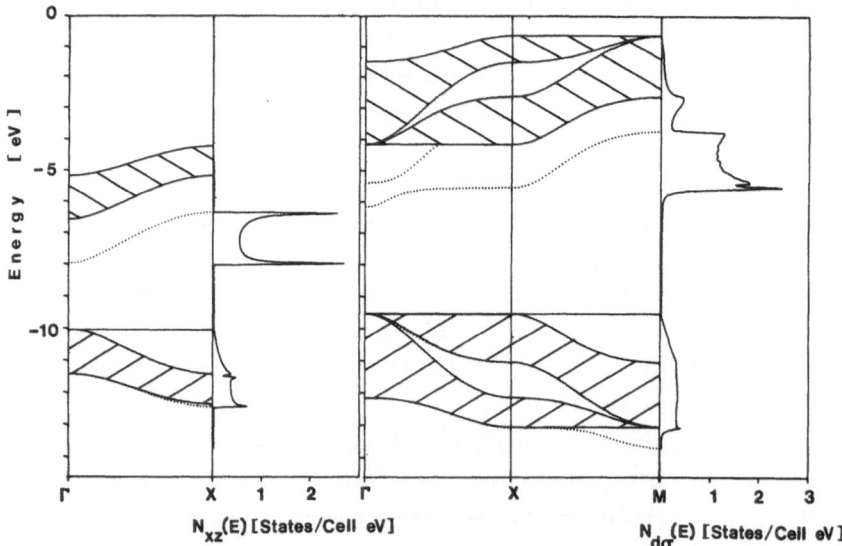

Fig.6.8. The local density of states for a (100) surface of SrTiO$_3$ with $\Delta E_d = -2$ eV. The dotted curves are surface energy bands. The cross-hatched areas indicate the bulk continuum of states. $N_{xz}(E)$ and $N_{d\sigma}(E)$ are local density of states functions as described in Fig.6.7

the bulk energy band states, but rather form their own band of states in the band gap region. The wave functions for these surface bands are highly localized on the first one or two atomic layers parallel to the surface. Figure 6.9 shows how the LDOS evolves from the surface to the bulk LDOS on going into the solid starting from the surface, (n = 0) for ΔE_d = -2 eV. The LDOS "heals" very rapidly. Over 90% of the charge density of a surface state is localized on the first layer.

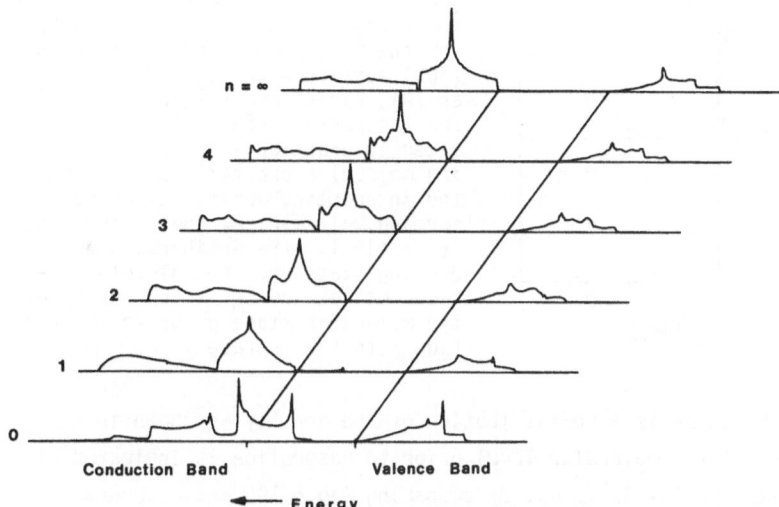

Fig.6.9. The total local density of states for cations as a function of distance from a (100) type I surface of SrTiO3. The surface is n = 0 and n = 1,2,3... are layers parallel to the surface containing the B cations. The interlayer spacing, 2a, is about 4 Å. The LDOS converges rapidly to the bulk function with increasing distance from the surface

6.3.3 Qualitative Features of Chemisorption

The importance of these results is that the interaction of molecules with the surface of a perovskite is dominated by the surface states and not by the bulk states. Thus there will be little or no *obvious* relation of the bulk electronic structure to the chemisorptive or catalytic properties of such solids. Because of the highly localized nature of the surface energy band states the formation of highly localized chemisorption bonds is possible.

To illustrate this feature we have calculated the LDOS for a molecule chemisorbed on a Ti ion on a (100) perovskite surface. For concreteness, we assume the molecule has an antibonding molecular state that hybridizes with the d_{xz} orbital of the Ti ion as shown in Fig.6.4a. The LDOS associated with the d_{xz} orbital for a clean surface with ΔE_d = -2 eV is shown in Fig.6.10b. For reference, the LDOS for ΔE_d = 0 is shown in Fig.6.10a. When the molecule chemisorbs to the surface

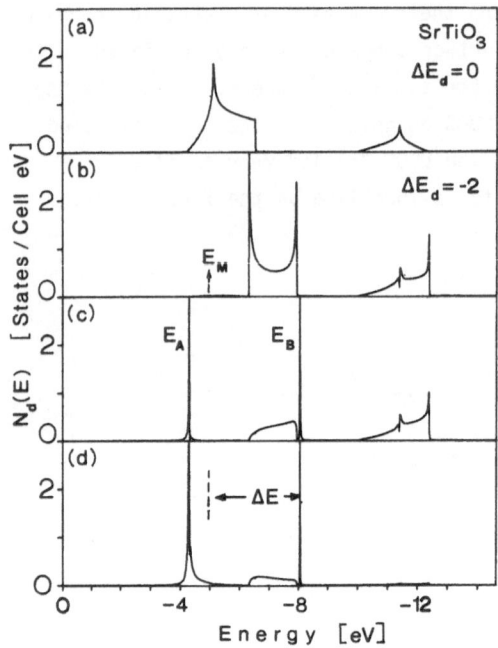

Fig.6.10a-c. Local densities of states for a chemisorption process. (a) LDOS for the xz component of a cation on a (100) surface with $\Delta E_d = 0$. (b) same as (2), except $\Delta E_d = -2$ eV. (c) LDOS for the surface cation ($\Delta E_d = -2$ eV) after interaction with an antibonding molecular orbital. (d) LDOS for the interacting orbital of an adsorbing molecule. E_A and E_B indicate, respectively, the antibonding and bonding states of the molecule-surface complex and E_M is the energy of the molecular state prior to interaction with the surface

which has $\Delta E_d = -2$ eV, there is a redistribution of the density as shown in Fig.6.10c. The energy of the molecular level prior to adsorption is indicated by E_M which for this illustration is -5 eV. By comparing Fig.6.10b and c several features are evident. The LDOS in the valence band is only slightly modified. The surface state LDOS is, however, significantly altered. The surface band states have interacted with the molecular state to form bonding and antibonding chemisorption states. The bonding state is just below the bottom of the surface energy band edge and the antibonding state lies above E_M in the energy region of the π^* band. The LDOS in Fig.6.10c indicates the Ti surface ion has hybridized with the molecule to form a chemisorption bond which consists of a localized antibonding state above E_M and a localized bonding state below the surface band. If the Fermi level lies above the surface band edge then the chemisorption bonding state will be occupied. Figure 6.10d shows the LDOS associated with the molecular orbitals. One notes that the molecule has a sizeable LDOS with the surface band as well as in the chemisorption states. This means that the adsorbed molecule participates in *delocalized* surface energy band states. This is significant because it provides a mechanism for an adsorbed molecule to "communicate" electronically over large distances with other adsorbed molecules or with substrate atoms. It is also important because the band width of the molecular state is increased from zero to about 1.5 eV. A second adsorbing species will also have a LDOS with density in this surface band region. Therefore, the substrate provides a mechanism for energy resonance between two species whose energy levels may be widely separated in the gas

phase. The delocalized states provide a mechanism for catalytic reactions on a surface between species that are well separated.

Another aspect to mention is that the participation of the adsorbed molecule
in the surface band states can effectively reduce the Coulomb barriers to charge
transfer. The coupling of the molecule to these band states allows the solid state
screening effect to reduce the bare Coulomb potential. Transfer of an electron
from the substrate to the molecule may proceed with relative ease because the wave
function of the adsorbed molecule has a very long ligand tail — that is the solid
itself becomes an extended ligand of the molecule. This feature permits different
charge states of the molecule to lie close in energy by means of the same mechanism
suggested by HALDANE and ANDERSON [6.30] for multiple charge states of impurities
in semiconductors.

The lowering of the energy from E_M for the free molecule to below the surface
band for a chemisorped molecule is directly related to the (exothermic) energy of
chemisorption. Chemisorption is clearly energetically favorable for the example
presented, provided the Fermi level lies below the chemisorption antibonding state.
If the Fermi level is raised above the antibonding state then chemisorption will
be less favorable or unfavorable depending on the details of the LDOS functions.

As a final point, it should be noted that the antibonding chemisorption state
has a very narrow band width even though it lies in the continuum of bulk band
states. This illustrates the importance of considering the surface LDOS rather
than the bulk density of states in understanding the mechanisms of chemisorption
and catalysis.

6.3.4 Effect of Electron-Electron Interactions

The cation LDOS in and below the valence bands determines the amount of d-orbital
admixture in the valence band states. It is clear from Figs.6.7 or 8 that the
participation of d orbitals in the valence band states is sizeable (approximately
30%). This means that there is significant covalent bonding between the cations
and the oxygen ions.

In Fig.6.7 one sees that the LDOS for the bulk and surface cations are quite
different. However, the integrated area of the LDOS in and below the valence bands
is approximately the same for a surface cations as it is for a bulk cation (for
the case $\Delta E_d = 0$). For an insulating material the Fermi level is at the middle of
the band gap and the valence bands are filled states. Therefore, the total charge
residing in d orbitals is the same for the bulk ions as it is for the surface ions.
Thus the model is charge self-consistent for this case. This circumstance do not
hold for $\Delta E_d \neq 0$.

An analysis of the LDOS of Fig.6.8 for $\Delta E_d = -2$ eV, reveals that the integrated
density in or below the valence bands (but excluding the LDOS in the band gap) is
much larger than for the $\Delta E_d = 0$. This means that the surface cations have a larger

admixture in the valence band states. We refer to this as "surface enhanced covalency" [6.24]. The effect of surface enhanced covalency is important. In Fig.6.8 the charge residing on a surface cation is larger than that for a bulk cation by about one electron. Therefore, a crude picture of this situation would be a solid having "Ti^{+4}" ions in the interior and "Ti^{+3}" ions on the surface. Because of the change in the charge of the surface cations the model calculation is no longer charge self-consistent. When the charge on the cation changes there is a corresponding change in the anion charge because of conservation of charge. Thus if we have "Ti^{+3}" ions on the surface the charge on the oxygen ions at or near the surface must also be reduced. We can picture the surface unit cell as corresponding to $Sr^{+2} Ti^{+3} (0^{-1.66})_3$. These modifications produce two effects. First the reduction of the oxygen charge will *reduce* the repulsive Madelung potential acting at the cation sites and this effect tends to lower the surface states in energy. On the other hand, the Coulomb repulsion among the extra electrons now occupying the surface d orbitals is greatly increased. For example, the ionization potential for $Ti^{+3} \rightarrow Ti^{+4}$ is 43 eV while for $Ti^{+3} \rightarrow Ti^{+2}$ it is only 24 eV [6.31]. Therefore, for a free ion the Coulomb repulsion due to an additional d electron is 19 eV. When the ion is in a solid the Coulomb repulsion is greatly reduced by polarization and screening. As examples, Fe^{+2}, Fe^{+3}, and Fe^{+5} in $SrTiO_3$ all have energy levels within a 3 eV range [6.32]. The exact reduction of the intra-atomic Coulomb repulsion for the Ti ions of $SrTiO_3$ is not known but some estimates can be made. From the ionization energies of Fe ions and their energy levels in $SrTiO_3$ it appears that the intra-atomic Coulomb repulsion is reduced to approximately 5% of its free ion value. This is probably a lower bound for a Ti surface ion. Another estimate is obtained by simply dividing the Coulomb and Madelung potentials by the dielectric constant, which is about 5. This implies that the Coulomb repulsion would be about 20% of the free ion value. In the calculations described here we assume the Coulomb integral to be 4 eV.

The surface perturbation parameter ΔE_d is not known a priori. It is necessary to assume a value of ΔE_d and calculate the LDOS. From the LDOS the charges on the cations and anions can be determined if the Fermi level is specified. Once the charges are known then the Madelung and intra-atomic potentials can be calculated and hence the surface perturbation parameter ΔE_d can be determined. This process must be repeated until self-consistency is achieved; that is the ΔE_d assumed equals the ΔE_d calculated from the LDOS.

It should be noted here that the value of ΔE_d depends on a delicate balance between the Madelung potential and the intra-atomic Coulomb repulsion. Estimates of these quantities for $SrTiO_3$ or TiO_2 suggest that the intra-atomic Coulomb repulsion dominates the Madelung term. The effect of this is to make the occupation of surface d orbitals energetically unfavorable. For example, for n-type $SrTiO_3$ the effect of charge self-consistency reduces the ΔE_d parameter from -2 eV to about 0 eV and the surface bands shown in Fig.6.8 are pushed very close to the conduction band edge so that the LDOS approaches that of the perfectly terminated surface (Fig.6.7).

Therefore, according to the model, it may not be possible to occupy the surface bands by simply raising the Fermi level (by reduction of the oxide, for example). Raising the Fermi level may force the surface band out of the band gap region. This conclusion depends upon the dominance of the intra-atomic Coulomb repulsion over the Madelung potential change at the surface. Changes in the surface condition may alter this situation. For example, the existence of surface oxygen vacancies may cause the Madelung potential to dominate the intra-atomic Coulomb repulsion. In Sect.6.5 we shall describe experimental results which indicate that surface states similar to those shown in Fig.6.8, for $\Delta E_d = -2$ eV, occur for $SrTiO_3$ samples containing surface oxygen vacancies.

In concluding the discussion of the surface energy bands, some comments on the relation of these bands to the results of the cluster model can be made. There is a correlation between the bulk energy bands and the BO_6 cluster levels. As we have seen the relationship is not as simple as assuming that the cluster levels correspond to the center of an energy band. The symmetry of the cluster states must also be considered. As was previously discussed, the BO_6 cluster states correspond to band states at either Γ or R in the Billouin zone. They belong to the irreducible representations of the cubic "O_h" group. The reason for the correspondence of the cluster levels to Γ or R is that the point group of the k-vector is also "O_h" at Γ and R. In a similar fashion a correlation between the states of the BO_5 surface-like cluster and the surface states of a perovskite with a (100) surface can also be established. The BO_5 cluster has C_{4v} symmetry and its levels can be related to the surface energy band states at M for the σ bands and at X for the π bands. In Sect.6.2 we indicated how the $x^2 - y^2$, z^2 and xz, xy, and yz cluster levels could bond to molecular orbitals in chemisorption. In the band model these cluster states are replaced by surface bands. The mechanisms of chemisorption involving the surface bands will be qualitatively similar to those for bonding to the discrete cluster levels. However, the feature of long-range interaction between surface adsorbed species via the delocalized surface bands is absent in the cluster model.

6.4 Surface Oxygen Vacancy States (SOVS)

The SOVS (surface oxygen vacancy states) may well play a vital role in the surface chemistry of the perovskites. Vacancy sites can provide geometrically and electronically ideal locations for the adsorption of oxygen, hydroxyls, H_2O and other molecular species. SOVS are, in fact, transition metal d-electron states since the wave functions for the vacancy states (F and F^+-like centers) are made up primarily from the d orbitals on cations adjacent to the vacancy site.

The problem of isolated surface vacancies on a (100) d-band perovskite can be solved analytically using our LCAO model. Neglecting the possibility of structural relaxation there will be again the two competing effects which perturb the two cations adjacent to the surface vacancy. First, the removal of the negatively charged "O^{-2}" ion reduces the magnitude of the repulsive Madelung potential at the cation sites. This increases the ionization energy of the d-electron states and allows the cations to bind electrons in the civinity of the vacancy. The addition of electrons to the cations, however, introduces additional Coulomb repulsion which decreases the effective d-electron ionization energy. In this instance the charge in the Madelung potential dominates the Coulomb repulsion so that electrons can be bound in the SOVS. In order to have a well-localized SOVS the energy must lie in the forbidden energy gap between the valence and conduction bands. The position of the SOVS depends upon the details of the electronic structure of the transition metal oxide.

We have obtained solutions for the SOVS using the LCAO model. The relevant information concerning their energy and band width may be obtained from the LDOS for a cation adjacent to the vacancy.

We can characterize the SOVS solutions by a parameter ΔE_0 which expresses the difference between the reduction in the repulsive Madelung potential (acting at the surface cation site) due to the loss of an "O^{-2}" ion and the increase in the intra-atomic Coulomb repulsion that occurs when the SOVS are occupied with electrons. For a singly or doubly occupied SOVS the Madelung term is expected to dominate so that $\Delta E_0 < 0$.

Figures 6.11,12 illustrate how the LDOS of the SOVS depends upon ΔE_0. Figure 6.14 shows the surface cation LDOS for $\Delta E_0 = 0$; (a hypothetical case for which the Madelung and intra-atomic potentials are equal). There are no localized SOVS for this example. However, when $\Delta E_0 < 0$ discrete localized SOVS appear in the band gap region as shown in Fig.6.12 for $\Delta E_0 = -2$ eV. There are three SOVS derived from the t_{2g} symmetry d orbitals. The wave functions belong to the C_{2v} group and are localized principally on the two cations adjacent to the vacancy. There is also some 2p-orbital amplitude on neighboring anions. The density of states for a localized SOVS is a δ function but the spectral weight is not unity; that is, an integral number of electrons are not necessarily bound in the d orbitals of the SOVS. The heights of the vertical lines in Fig.6.12 indicate schematically the relative spectral weight of a SOVS. The total integrated density over all energy is three in all cases shown in Figs.6.11,12.

As the parameter ΔE_0 becomes more negative the SOVS move further into the band gap region. As they do, a second set of vacancy states appears below the valence

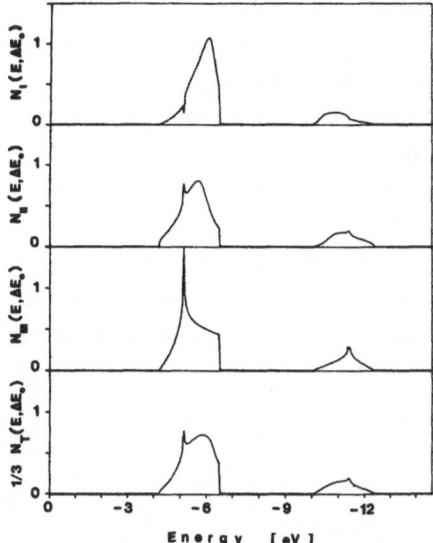 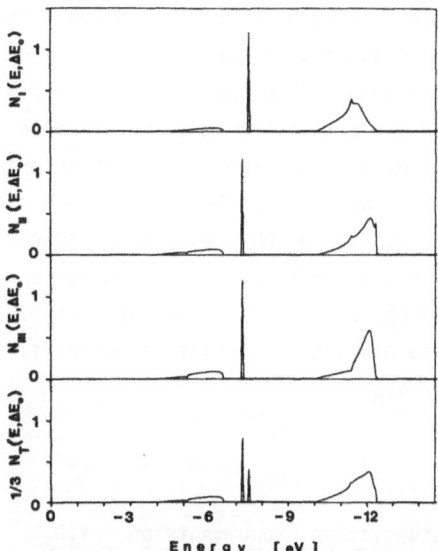

Fig.6.11. Local densities of states for a surface cation adjacent to an oxygen vacancy. N_I, N_{II}, and N_{III} are LDOS functions for the xz, yz and xy orbitals, N_T is the sum of the three densities. The vacancy perturbation parameter is $\Delta E_0 = 0$ for the example and no localized vacancies occur in this case. (Structure associated with the σ bands is not shown)

Fig.6.12. The LDOS for a surface cation adjacent to a surface oxygen vacancy for $\Delta E_0 = -2$ eV. The various functions are the same as in Fig.6.11. The sharp peaks are localized surface oxygen vacancy states which occur in the band gap for $\Delta E_0 < -2$ eV. (Structure associated with the σ bands is not shown)

band and the spectral weights of band gap states decrease. These features can be seen in Fig.6.12.

The SOVS can be active surface sites for processes of chemisorption and catalysis. They may act as either a source or sink for electron transfer depending upon the surface environment and their electronic occupation. The precise nature of the SOVS clearly depends on ΔE_0, which in turn depends on a knowledge of the Madelung and intra-atomic Coulomb potentials. For the results presented in our discussion here the character of the SOVS was studied as a function of ΔE_0, treating ΔE_0 as an adjustable parameter. The actual value of ΔE_0 is not known with any certainty at this time. However, it appears from UPS (ultraviolet photoemission spectroscopy) studies [6.33-35] of Ar-ion bombarded $SrTiO_3$ and TiO_2 that SOVS can occur in the band gap region. From such information it may be inferred that $\Delta E_0 < 0$. This means that the reduction of the Madelung potential is greater than the intra-atomic Coulomb repulsion, at least for a singly occupied SOVS. From the position of the vacancy state band in the UPS data it appears that ΔE_0 lies in the range of -1 to -3 eV.

The distance between cations adjacent to the vacancy is about 4 Å. This separation is so large that there is only a weak *direct* interaction between the d orbitals involved in the SOVS. Consequently the occupation of SOVS may lead to localized surface states which approximate "Ti^{+3}" ions adjacent to the vacancies.

On the other hand, the d orbitals interact indirectly through the oxygen ions to form band states and therefore when a sizeable fraction of the surface oxygen sites are vacant the SOVS will be broadened into a band. In such a case the distinction between vacancy states and surface energy bands becomes unclear. A surface band approach is described in Sect.6.5 where the surface oxygen vacancies are assumed to provide an attractive potential that contributes to the surface band parameter ΔE_d.

6.5 Photoemission Experiments on $SrTiO_3$

6.5.1 Surface State Emission

The first attempt to observe the theoretically predicted surface energy bands of $SrTiO_3$ was reported by POWELL and SPICER [6.36]. They performed UPS photoemission experiments on n-type reduced and doped $SrTiO_3$ samples which were fractured in vacuum. No significant density of surface states in the band gap region was detected.

Recently, a more extensive study of n-type $SrTiO_3$ (and TiO_2) surfaces has been taken up by HENRICH et al. [6.33-35]. They investigated vacuum fractured, bombarded and annealed samples. Surface states were found in the band gap region for both bombarded and annealed samples but for vacuum fractured surfaces only very weak band gap emission was observed. Figure 6.13 shows their UPS data for a vacuum fractured n-type $SrTiO_3$ sample. The LEED pattern indicated a well-ordered, unreconstructed, surface. The empty conduction bands are to the right of the Fermi energy, E_F. The point labeled E_v marks the top of the valence band. The large minimum is approximately at the bottom of the σ band. The peak nearest the band gap is due to emission from the π^o and σ^o nonbonding oxygen bands. The rapid rise at the far left is background emission arising from secondary electrons and inelastically scattered electrons.

The inset shows the band gap emission after correction of the data for the effects of an extraneous line in the optical source. The lower part of the inset shows a surface band in the band gap with a small electronic occupation (estimated by HENRICH et al. to be less than 0.1 electrons per surface unit cell). The origin of this emission is not certain but is probably arises from surface impurities.

The absence of a strong surface state emission in the band gap region for the vacuum fractured $SrTiO_3$ samples is consistent with the results of POWELL and SPICER

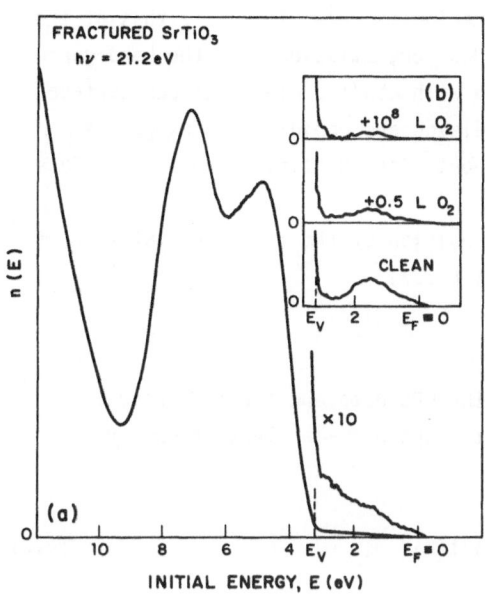

(a)

INITIAL ENERGY, E (eV)

Fig.6.13a,b. The photoelectron energy
distribution curve for a (100) surface of vacuum fractured n-type SrTiO₃.
E_V is the top of the valence band and
E_F is the Fermi level. The empty conduction bands are to the right of E_F.
(a) The first peak to the left of E_V
is due to emission from the oxygen
nonbonding bands, the second peak is
due to the   and   valence bands. The
deep minimum occurs at the bottom of
the valence band. The rapidly rising
portion below the valence band is background emission due to secondary and
inelastically scattered electrons.
(b) The inset shows the emission in the
band gap region after correction of
the data for an extraneous source line.
The emission in the band gap is quenched
by exposure to oxygen

and suggests that the effect of intra-atomic Coulomb repulsion has expelled the
surface states from the band gap in n-type material as suggested previously and
as represented by Fig.6.7.

The results of another study of n-type(100) SrTiO₃ by HENRICH et al. are shown
in Fig.6.14. The sample was subjected to Ar-ion bombardment which produced Sr and O
vacancies and surface disorder. The sample was then vacuum annealed at 1100 K.
After annealing, the LEED pattern was sharp and indicated that the surface was
ordered in an ideal (100) type I structure. Auger measurements were found to be
consistent with an ideal (100) type I structure with a concentration of about 20%
surface oxygen vacancies.

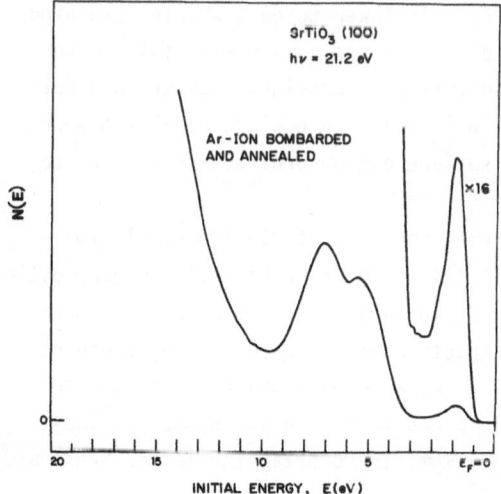

INITIAL ENERGY, E(eV)

Fig.6.14. Photoelectron energy distribution curve for an Ar-ion bombarded
high temperature annealed (100) SrTiO₃
surface. The emission from the bandgap
region, between E_F and E_V, is due to
surface states

The photoemission from the valence band region is very similar to that of the vacuum fractured sample shown in Fig.6.13. However, emission from the band gap region corresponding to a surface state band with about 2 electrons per surface unit cell has appeared. This emission band disappears rapidly when oxygen is allowed into the vacuum system and this supports its identification as a surface band.

Studies of the surface structure and composition of the (111) crystal face of $SrTiO_3$ have been reported by LO and SOMORJAI [6.37].

6.5.2 Comparisons with Theory

In order to explore the comparison between the UPS data and the theoretical model previously discussed, we have calculated approximate electron energy distribution curves N(E) as follows

$$N(E) \propto \sum_{n=0}^{\infty} \exp[-2n\lambda(E)a][N_d(n,E) + N_p(n,E)]f(E) \quad , \tag{6.6}$$

where $\lambda(E)$ is the energy-dependent escape depth [6.38] (11 to 18 Å), n is the number of unit cells from the surface and 2a is the unit-cell layer spacing. N_d and N_p are the LDOS functions for the d and p orbitals as functions of the layer number and energy. The function f(E) is the electron occupation probability; the Fermi distribution function.

To compare directly with experimental data we convolute the function N(E) with a gaussian function

$$<N(E)> \propto \int_{-\infty}^{\infty} N(E') \exp\{-[(E - E')/R]^2\}dE' \quad , \tag{6.7}$$

where the resolution parameter R, which accounts for spectrometer resolution, secondary and inelastically scattered electrons is taken to be 1 eV. In obtaining N(E) for $SrTiO_3$, the analytical LCAO model discussed previously was utilized to calculate the LDOS functions. The LCAO parameters were obtained from the APW calculations of MATTHEISS [6.20] as described in [6.21]. The Coulomb repulsion among the excess (over bulk) electrons occupying surface cation orbitals is assumed to be 4 eV.

A comparison of the theoretical results with the data of HENRICH et al. for the vacuum fractured sample of $SrTiO_3$ (Fig.6.14) is shown in Fig.6.15. Figure 6.15a shows the function N(E) of (6.6) with $\Delta E_d = 0$ (for which no surface states appear in the band gap region). The valence band structure is characterized by features (1)-(3). Feature (1) is due to the nonbonding oxygen bands, while (2) is due to the valence bands involving admixtures of Ti d orbitals with t_{2g} symmetry. The valence bands involving Ti d orbitals with e_g symmetry overlap the latter bands and

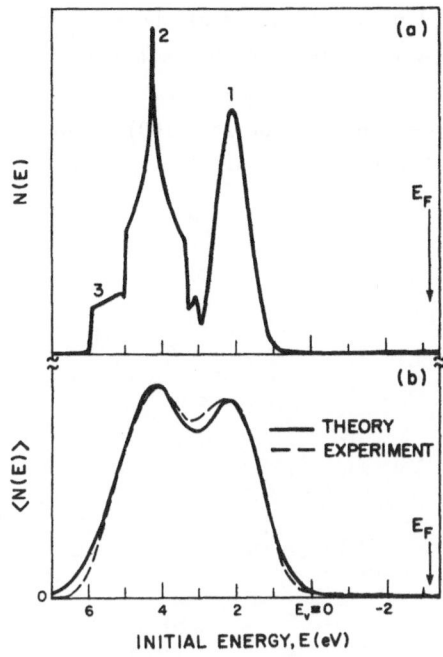

Fig.6.15. (a) N(E) for a vacuum fractured (100) SrTiO3 surface with $\Delta E_d = 0$. (b) Comparison of <N(E)> (solid curve) with UPS data (dashed curve)

and extend below them to produce the edge labeled (3). Figure 6.15b shows the broadened density of states <N(E)> (solid curve) obtained from N(E) in Fig.6.15a, together with the UPS spectrum for the vacuum-fractured $SrTiO_3(100)$ surface (dashed curve), after subtraction of a smooth background due to inelastically scattered electrons. The agreement between theory and experiment is seen to be excellent.

In general, UPS spectra cannot be compared directly with initial-state densities because of the energy dependence of the transition matrix elements and of differences in final densities of states. However, the $SrTiO_3$ UPS spectra of POWELL and SPICER [6.36], and of DERBENWICK [6.39] indicate that the shape of the valence band emission changes slowly with photon energy, $h\nu$, for $11.7 \leq h\nu \leq 21.2$ eV. This is because the final states for the transitions are the Sr(4s) and Ti(4s) and (4p) bands, which are presumably nearly plane-wave-like and have large bandwidths; as a result, the final-state density is roughly constant. The $SrTiO_3$ valence band shape seen in XPS spectra [6.40], where final-state effects are less important, is also very similar to that seen in UPS spectra taken at 21.2 eV.

In the case of the bombarded-annealed $SrTiO_3$ sample (Fig.6.14), we assume that the surface oxygen vacancies produce an attractive surface potential which contributes to ΔE_d and allows occupied surface states to exist in the band gap region. The average surface oxygen vacancy potential at a cation site may be estimated by representing the vacancies as a planar lattice of +2 point charges compensated by a uniform background of negative charge [6.27]. A surface vacancy concentration

of 20% produces an average (attractive) potential of about -10 eV at the surface
cation sites. The self-consistent surface state calculation for this situation
corresponds to that of Fig.6.8 with ΔE_d = -2 eV.

A comparison of the theoretical and experimental results for the (100) SrTiO$_3$
bombarded-annealed sample is shown in Fig.6.16. Figure 6.16a shows the function
N(E) calculated from (6.6) with ΔE_d = -2 eV.

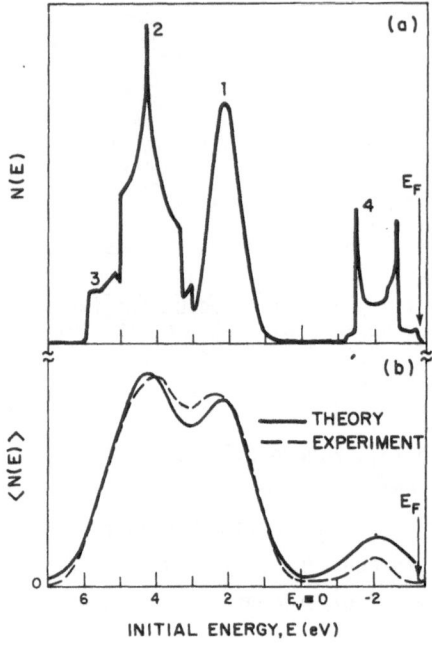

Fig.6.16. (a) N(E) for a Ar-ion bombarded
and annealed (100) SrTiO$_3$ surface with
ΔE_d = -2 eV. (b) Comparison of <N(E)>
(solid curve) with UPS data (dashed curve)

The valence band region [features (1)-(3)] is almost identical to that for the
vacuum fractured surface in Fig.6.15a, but the surface-state band [feature (4)]
has appeared in the bulk band gap. The solid curve in Fig.6.16b shows the resultant
<N(E)> curve, and the dashed curve is the UPS spectrum [6.35] for an Ar-ion bom-
barded and annealed SrTiO$_3$(100) surface.

It is seen in Fig.6.16b that the position and shape of the theoretical <N(E)>
curve agrees well with the UPS spectrum. However, the intensity of the surface
state band is too large by approximately a factor of two. There are a number of
possible explanations for this discrepancy. First, it is possible that the emission
from the surface states has a strong angular dependence. The method of electron
collection used in the UPS experiment may be less efficient for the surface state
electrons.

The assumption of a much longer escape depth, about 40 Å, would reduce the surface state emission relative to the valence band emission by a factor of two. However, such a large escape depth seems unlikely. It is also possible that some type

of band splitting occurs. For example, the d_{xy} surface band could be lower than the d_{yz} and d_{xz} bands with the latter bands lying above E_F. Finally, we note that the surface band could be spin polarized. The assumption of a spin correlation energy of about 2 eV leads to the theoretical prediction of ferromagnetic surface energy bands corresponding to about 2.3 Bohr magnetons per surface unit cell. The intensity for the spin-polarized surface state band is in excellent agreement with the data of Fig.6.16. A definitive interpretation of the UPS surface state emission will be greatly facilitated when angle-resolved and spin-polarized photoemission data become available.

6.5.3 Oxygen Adsorption on (100) SrTiO₃

HENRICH et al. [6.33] have performed a series of UPS experiments to study the nature of O_2 adsorption on vacuum fractured, bombarded, and bombarded and annealed samples of $SrTiO_3$. They recorded the UPS spectra as a function of O_2 exposure, for exposures ranging from 0.5 to 10^8 L(1 L $\equiv 10^{-6}$ Torr s). The effect of O_2 adsorption is displayed in a UPS "difference spectrum" which is obtained by subtracting spectra for different O_2 exposures.

HENRICH et al. found that O_2 adsorption occurs in two phases. Phase I occurs for O_2 exposures $\lesssim 30$ L and is characterized by a sticking coefficient between 0.2 and 1. Phase II proceeds for exposures $\lesssim 30$ L up to 10^8 L. The sticking coefficient for phase II is 10^{-3} or less.

For phase I adsorption, on bombarded or bombarded and annealed samples, the (band gap) surface state emission is dramatically reduced as the O_2 exposure is increased. In addition, for all types of surfaces, there is observed an enhancement of the emission from the valence band region. These two effects are illustrated in Fig.6.17a, which shows the difference spectra, $\Delta n(E)$, for O_2 adsorption on a bombarded $SrTiO_3$ surface.

Figure 6.17a shows $\Delta n(E)$, the difference of the UPS spectra for 30 L and 0 L, for phase I. The negative region of $\Delta n(E)$, from -4 to 0 eV, represents the loss of the surface state emission band that occurs when O_2 is adsorbed. The positive part of $\Delta n(E)$, from 0 to about 5 eV, represents an enhancement of emission from the valence band region. The rise in $\Delta n(E)$ beyond 5 eV is due to changes in the inelastic background and is of no importance in this discussion. The difference spectrum shown in Fig.6.17b is for phase II. It is characterized by a featureless enhancement of the valence band emission. The peak near 8 eV in $\Delta n(E)$ is presumed to be due to inelastic scattering and has no direct relation to the O_2 adsorption process.

HENRICH et al. [6.33] have suggested that the results for phase I imply that O_2 is dissociated and adsorbed as O^{2-} ions on bombarded and bombarded-annealed $SrTiO_3$. Molecular oxygen is probably dissociated at a surface cation site by means of the catalytic mechanisms described in Sect.6.2. The electrons associated with

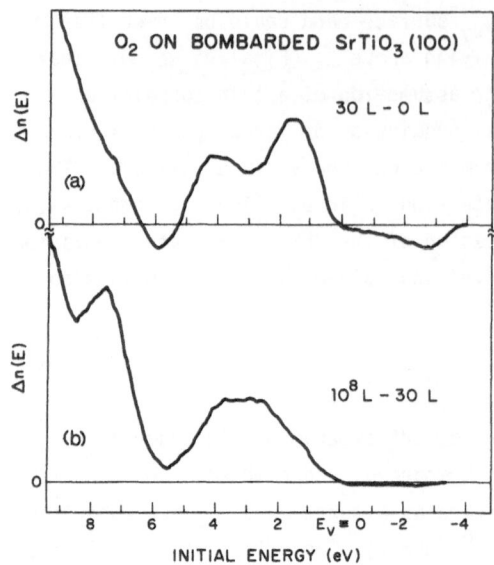

Fig.6.17a,b. UPS difference spectra for O_2 adsorption on bombarded $SrTiO_3$. (a) phase I, (b) phase II

the surface states, which occupy the surface t_{2g} d orbitals, are transferred to the antibonding states of the O_2 molecules causing dissociation. This charge transfer process is evident from the depletion of the surface state emission (negative region of $\Delta n(E)$ in Fig.6.17a). The oxygen ions produced by the dissociation of the O_2 molecules can become bonded to a surface cation or migrate to an oxygen vacancy site where they are chemisorbed as lattice O^{2-} ions. In either case, these adsorbed oxygen ions will have a LDOS that reflects the valence energy band structure. Evidence supporting this interpretation is found in the enhancement of the valence band emission in the region from 0 to 5 eV in Fig.6.17a. In fact, $\Delta n(E)$ in that region is very similar to the total valence band emission shown in Fig.6.15b. A similar enhancement of the valence band emission is also found for phase I adsorption on vacuum fractured $SrTiO_3$ samples.

The interpretation of phase II data is less clear and the data are different for different types of surfaces. It seems probable, considering the small sticking coefficient, that phase II involves weak chemisorption or physical adsorption of some form of molecular oxygen.

6.6 Summary

In the preceding sections we have described some of the features of the surface electronic properties of the d-band perovskites. Attention was focused on the d-electron surface states than can occur at energies within the forbidden band gap of the bulk material. These states are believed to be important in surface reactions and chemisorption on the perovskites.

Coordinatively unsaturated surface cations have been discussed in the past as a source of active surface sites. In this chapter we have investigated the details of the electronic structure of such active sites. The BO_5 cluster surface model (coordinatively unsaturated B ion) was discussed and it was shown that cation surface states of e_g and t_{2g} symmetry occur at energies within the gap between $2t_{2g}$ and $1t_{1g}$ levels of the BO_6 (fully coordinated) cluster. These surface states can act in concert with an interacting molecule to form hybridized molecule/substrate covalent bonds. It was shown schematically how the particular symmetry of the surface orbitals could reduce the net charge transfer required to effect a surface reaction.

In Sect.6.3 the bulk and surface electronic structure of the d-band perovskite was reviewed. It was shown how the electronic levels of the BO_6 cluster could be related to the conduction and valence bands of a typical d-band perovskite. The cluster surface states were then related to the surface energy bands.

The dependence of the surface energy bands on the surface perturbations and electronic occupation was described. It was shown that there is a delicate balance between the interatomic Coulomb (Madelung) potential and the intra-atomic Coulomb repulsion among excess electrons occupying surface d orbitals. In the absence of surface oxygen vacancies it was suggested that the intra-atomic repulsion would dominate the change in the Madelung potential and that surface bands on n-type perovskites would be forced out of the band gap region. On the other hand, it was found that a concentration of surface oxygen vacancies of the order of 10 to 20% would be sufficient to stabilize occupied surface bands in the band gap region.

We also discussed the nature of the electronic structure of surface oxygen vacancy states. Initial results on the SOVS derived from the t_{2g} symmetry d orbitals were presented and it was concluded that the reduction in the Madelung potential due to the removal of an oxygen ion would dominate the intra-atomic Coulomb repulsion among electrons occupying the states for single and perhaps double occupation of the SOVS. Bonding of a molecular species to a SOVS could result in a charge transfer of electrons from occupied SOVS to the interacting molecule. Conversely, empty available SOVS could accept electrons from electrophobic species. Therefore the SOVS may act as either a source or sink for electron transfer.

In Sect.6.5 a review of some of the recent photoemission experiments on (100) $SrTiO_3$ surfaces was presented. Theoretical calculations of the electron energy distribution curves were compared with the experimental data. Model predictions of the electron energy distribution curves were found to be in excellent agreement with experiment for vacuum fractured $SrTiO_3$ (100) surfaces. Comparison of theory with data for the annealed surface suggested that a band splitting mechanism is involved. These studies suggest that the surface oxygen vacancies created by Ar bombardment and remaining after high temperature annealing are sufficient to alter the surface conditions in such a way as to allow occupied surface bands to penetrate deep into the band gap region.

The precise condition of the surfaces of the n-type materials used in UPS studies is not known and the ideas described must remain tentative until such information is known with certainty.

Section 6.5.3 gave a brief discussion of the experiments on oxygen adsorption and presented some tentative explanations.

It should be clear from the discussions in this chapter that there is still great uncertainty about the surface electronic properties of the d-band perovskites. These materials require and deserve much more experimental and theoretical attention.

Acknowledgements. The authors wish to express their appreciation to Dr. V. Henrich for valuable discussions and for his permission to use Figs.6.13-17.

References

6.1   J.B. Goodenough: *Metallic Oxides*, in Prog. Solid State Chem. *5*, 145 (1971)
6.2   L.A. Pedersen, W.F. Libby: Science *176*, 1335 (1972)
6.3   D.B. Meadowcraft: Nature *226*, 847 (1970)
6.4   R.J.H. Voorhoeve, D.W. Johnson, Jr., J.P. Remeika, P.K. Gallagher: Science *195*, 827 (1977)
      R.J.H. Voorhoeve, J.P. Remeika, D.W. Johnson, Jr.: Science *197*, 353 (1972)
6.5   W.F. Libby, S. Aegerter: UCLA Tech. Rpt. (unpublished)
6.6   A.C.C. Tseung, H.L. Beven: J. Electroanal. Chem. Interfacial Electrochem. *45*, 429 (1973)
6.7   J.G. Mavroides, D.I. Tchernev, J.A. Kafalas, D.F. Kolesar: Mater. Res. Bull. *10*, 1023 (1975)
6.8   M.S. Wrighton, A.B. Ellis, P.T. Wolczanski, D.L. Morse, H.B. Abrahanson, D.S. Ginley: J. Amer. Chem. Soc. *98*, 2774 (1976)
6.9   R.D. Nasby, R.K.-Quinn: Mater. Res. Bull. *11*, 985 (1976)
6.10 H. Kung, H.A. Jarrett, A.W. Sleight, A. Ferretti: J. Appl. Phys. *48*, 2463 (1977)
6.11 S.N. Frank, D. Laser, K.L. Hardee, A.J. Bard: NBS Special Publication 455, 149 (1976)
6.12 T. Wolfram, F.J. Morin, R. Hurst: NBS Special Publication 455, *Electrocatalysis on Non-Metallic Surfaces*, Proc. Workshop held at NBS, Gaithersburg, Md., Dec. 9-12, 1975 (Issued Nov. 1976)
6.13 R.L. Burwell, Jr.: NBS Special Publication 455, *Electrocatalysis on Non-Metallic Surfaces*, Proc. Workshop held at NBS, Gaithersburg, Md., Dec. 9-12, 1975 (Issued Nov. 1976)
6.14 T. Wolfram, R.R. Hurst, F.J. Morin: Phys. Rev. B*15*, 1151 (1977)
6.15 J.A. Tossel, D.H. Vaughan, K.H. Johnson: Am. Mineral. *59*, 319 (1974)
6.16 T. Wolfram, E.A. Kraut, F.J. Morin: Phys. Rev. B*7*, 1677 (1973)
6.17 V.E. Henrich, H. Zeiger, G. Dresselhaus: Phys. Rev. Lett. *36*, 1335 (1976)
6.18 A more detailed discussion of this concept may be found in T. Wolfram, F.J. Morin: Appl. Phys. *8*, 125 (1975)
6.19 H. Weinberg, R.P. Merrill: Surf. Sci. *33*, 493 (1972)
      H. Weinberg, M. Lambert, C.M. Comrie, J.W. Linnett: Surf. Sci. *30*, 299 (1972)

6.20 L.F. Mattheiss: Phys. Rev. B*6*, 4718 (1972)
6.21 Ş. Ellialtıoğlu, T. Wolfram: Phys. Rev. B*15*, 5909 (1977)
6.22 T. Wolfram: Phys. Rev. Lett. *29*, 1383 (1972)
6.23 Ş. Ellialtıoğlu, T. Wolfram: "Surface Electronic Structure of d-Band
      Perovskites: Study of the ·π-Bands", Phys. Rev. B *18*, 4509 (1978)
6.24 T. Wolfram, Ş. Ellialtıoğlu: Appl. Phys. *13*, 21 (1977)
6.25 M. Cardona: Phys. Rev. A *140*, 651 (1965)
6.26 F.J. Morin, T. Wolfram: Phys. Rev. Lett. *30*, 1214 (1973)
6.27 E.A. Kraut, T. Wolfram, W.F. Hall: Phys. Rev. B*6*, 1499 (1972)
6.28 P.-O. Löwdin: J. Chem. Phys. *18*, 365 (1950)
6.29 J.C. Slater, G.F. Koster: Phys. Rev. *94*, 1498 (1954)
6.30 F.D.M. Haldane, P.W. Anderson: Phys. Rev. B*13*, 2553 (1976)
6.31 C.E. Moore: *Atomic Energy Levels*, U.S. Dept. of Commerce, NBS, Vol.III,
      Circular 467 (1958)
6.32 F. Morin, J. Olivers: Phys. Rev. B *8*, 584 (1973)
6.33 V.E. Henrich, G. Dresselhaus, H.J. Zeiger: In *Elastic Electron Tunneling
      Spectroscopy*, ed. by T. Wolfram, Springer Series in Solid-State Sciences,
      Vol.4 (Springer, Berlin, Heidelberg, New York 1978)
6.34 V.E. Henrich, H.J. Zeiger, G. Dresselhaus: Solid State Res. Rpt., Lincoln
      Laboratory, M.I.T. (1976:2) p.39
6.35 V.E. Henrich, G. Dresselhaus, H.J. Zeiger: Phys. Rev. B*17*, 4908 (1978)
6.36 R.A. Powell, W.E. Spicer: Phys. Rev. B*13*, 2601 (1976)
6.37 W.J. Lo, G. Somorjai: Phys. Rev. B*17*, 4942 (1978)
6.38 D.A. Shirley: J. Vac. Sci. Technol. *12*, 280 (1975)
6.39 C.F. Derbenwick: Ph.D. Thesis, Stanford University (1970) unpublished
6.40 F. Battye, H. Höchst, A. Goldman: Solid State Commun. *19*, 269 (1976)
      S.P. Kowalezyk, F.R. McFeely, L. Ley, V.T. Gritsyna, D.A. Shirley: Solid
      State Commun. *23*, 161 (1977)

7. Theoretical Issues in Chemisorption

T. L. Einstein, J. A. Hertz, and J. R. Schrieffer

With 19 Figures

This chapter has two principal aims, one tutorial and the other critical. The first is to illuminate as much as possible of the physics of chemisorption through simple models which allow transparent calculations. The second goal, which is somewhat more elusive, is to examine just how valid and relevant various models and theoretical methods are — what questions they can answer, how accurately they do so, and what insight they give us. Although in a sense, the tutorial part of the chapter belongs at the beginning of a book like this and the critical part at the end, combining the material in a single chapter allows a coherent review of the essential theoretical aspects of the problem of chemisorption. These aims also give this chapter a flavor somewhat different from that of the rest of the book. We are more concerned with qualitative than quantitative answers (and questions).

One reason for this flavor is also the reason chemisorption is a fascinating problem to theoretical physicists: In many real systems it is what we call an "intermediate coupling" problem, where there is no obvious small parameter in powers of which to expand a solution around some simple limiting case. Thus the initial problem is always the qualitative one of identifying the "right physics" to generate a sensible approximate starting point. It is in this spirit that we approach the theory of chemisorption here. We hope that our account will be useful both in conveying the physics of the problem to theoreticians who work in other areas and in communicating the interesting theoretical issues to experimentalists.

We begin with a discussion of local density approximations. As previous chapters have shown, these methods have been impressively successful in calculating certain kinds of quantities. However, they do have serious short-comings, particularly insofar as nonlocal and energy-dependent effects are concerned. We discuss these deficiencies critically, thereby setting the stage for the Anderson model introduced in Sect.7.3. Although this model and its generalization take us out of the realm of ab initio calculations and force us to fit its parameters experimentally, it permits formal theoretical pursuit of effects treated inadequately in local density approximation schemes. An important example is the effect of correlation on the excitation spectrum, as measured by a photoemission experiment. Section 7.3 focuses on Hartree-Fock theory for this model, while in Sect. 7.3 we describe the various attempts to go beyond the Hartree-Fock level and confront the large correlation effects. Section 7.5 deals with interactions between adsorbed atoms, and Sect.7.6 relates the Green's functions of the theory to aspects of experimentally measured photoemission spectra. Section 7.7 contains a few concluding remarks.

7.1 Local Density Approximations and Generalizations

Much of the calculational work described in the preceding chapters has been based on ab initio techniques. One such technique is the Hartree-Fock (HF) scheme. However, the HF scheme is difficult to implement in surface calculations and neglects correlation effects. In an attempt to improve the accuracy and reduce the computational complexities of the HF approximation, SLATER [7.1], and KOHN and SHAM [7.2] introduced effective one-electron potential schemes which include exchange and correlation in a simplified form. In these schemes, the exchange and correlation contributions to the effective potential at a point $\underline{r}$ depend only on the density of electrons $n(\underline{r})$ at that point, rather than involving the one-electron density matrix $n_1(\underline{r},\underline{r}')$ which enters the HF approximation. Considering their simplicity, these local density functional (LDF) methods and subsequent generalizations have been remarkably successful in treating the ground state properties and excitations of atoms, molecules and solids. Nevertheless, such methods lead to quantitative and in some cases qualitative errors, particularly in problems involving image potential effects, partly filled narrow resonance levels of adsorbates, etc., as discussed below.

The density functional theory is discussed most thoroughly in Chap.3. There the exchange correlation potential is defined by

$$V_{xc}(\underline{r}) = \frac{d}{dn(r)} \{n(r)E_{xc}[n(\underline{r})]\} \quad . \tag{7.1}$$

Thus, within the LDF approximation, the effective electrons move in the classical electrostatic potential and the exchange-correlation potential is given by that of a uniform electron gas whose density is equal to that of the actual system at the point in question, $n(\underline{r})$. Were $n(\underline{r})$ to vary slowly over the size r_{xc} of the locally defined exchange-correlation hole, this approximation would be valid. Unfortunately, for many systems of interest $n(\underline{r})$ varies rapidly, say by a factor of two or more, over the distance $r_{xc}(\underline{r}_0)$ about a given point $\underline{r}_0$. In spite of this fact, cancellations occur in calculating the total energy, so that the local density assumption works considerably better than one might expect.

To see this, we note that the exact exchange-correlation energy is given by

$$E_{xc} = \frac{e^2}{2} \int \frac{n(\underline{r})[g(\underline{r},\underline{r}')-1]n(\underline{r}')}{|\underline{r}-\underline{r}'|} \, d\underline{r} d\underline{r}' \quad , \tag{7.2}$$

where $g(\underline{r},\underline{r}')$ is the pair correlation function which expresses how the electron charge density at $\underline{r}'$ is suppressed by exchange and correlation if an electron is located at $\underline{r}$. It is convenient to define an exchange-correlation charge density which is a function of $\underline{r}'$ for an electron located at $\underline{r}$ as

$$n_{xc}(\underline{r},\underline{r}') = [g(\underline{r},\underline{r}') - 1]n(\underline{r}') \quad . \tag{7.3}$$

A sum rule on $\underline{r}'$ follows from conservation of the total number of electrons,

$$\int n_{xc}(r,r')dr' = -1 \quad . \tag{7.4}$$

Since in most systems $n(\underline{r},\underline{r}')$ drops off rapidly as $|\underline{r} - \underline{r}'|$ increases, (7.4) shows that exactly one electron is missing from the vicinity of a given electron relative to the idealized situation in which exchange and correlation were absent. The LDF scheme automatically satisfies this sum rule.

The question then is how strongly the shape of the exchange-correlation hole in inhomogeneous systems is distorted from the spherical symmetry assumed in the LDF scheme. GUNNARSSON et al. [7.3] studied the hydrogen atom as an exactly soluble model. In this case $g(\underline{r},\underline{r}') = 0$ and $n_{xc}(\underline{r},\underline{r}') = -n(\underline{r}')$, i.e., it is centered around the nucleus and independent of $\underline{r}$. Now $r_{xc}(n)$ decreases as n increases. Near the nucleus, n is large but falls rapidly as one moves to large r. Therefore, one underestimates r_{xc} near the nucleus and obtains too large a contribiton to E_{xc} [see (7.2)] from this region. For large r, n(r) falls exponentially and r_{xc} grows exponentially, rather than proportional to r, giving too small a contribution to E_{xc} from this region. In addition, the hole has the wrong shape, being centered on the electron rather than the nucleus. In Fig.7.1 the contribution to E_{xc} coming from the electron being a distance r from the nucleus is plotted for the LDF, LSDF (local spin density functional) and exact calculations.

When unpaired spins are present, the LDF scheme can be improved by allowing V_{xc} to become spin dependent (LSDF) as in the spin unrestricted Hartree-Fock scheme. This extension was developed by a number of authors [7.4] and corresponds to replacing (7.2) by writing

$$E_{xc} = \int n(\underline{r})\varepsilon_{xc}\Big(n(\underline{r}), \zeta(\underline{r})\Big)d\underline{r} \tag{7.5}$$

where ζ is the degree of spin polarization

$$\zeta(\underline{r}) = \frac{n_+(\underline{r}) - n_-(\underline{r})}{n(\underline{r})} \tag{7.6}$$

$$n(\underline{r}) = n_+(\underline{r}) + n_-(\underline{r}) \tag{7.7}$$

$\varepsilon_{xc}(n,\zeta)$ is the exchange-correlation energy of a uniform electron gas of density n and spin polarization ζ. Solving the LDF and LSDF equations self-consistently leads to the total energy of the H atoms of -12.25 eV and -13.38 eV as compared to the exact result -13.60 eV.

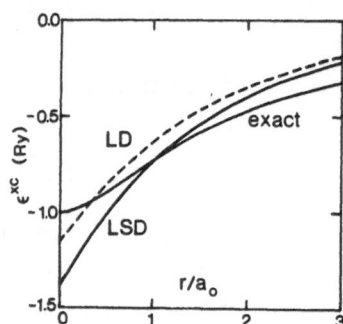

Fig.7.1. The contribution to the exchange-correlation energy of the hydrogen atom for the electron at distance r from the nucleus. The LDF, LSDF, and exact results are indicated [7.3]

To see how the local density errors tend to cancel, consider the contribution to E_{xc} for a hydrogen atom from a spherical shell a distance r_0 away from the electron at $\underline{r}$,

$$\varepsilon_{xc}(\underline{r},r_0) = \frac{e^2}{2r_0} \int n_{xc}(\underline{r},\underline{r}')\delta(|\underline{r} - \underline{r}'| - r_0)d\underline{r}' \tag{7.8}$$

where

$$E_{xc} = \int n(\underline{r})\varepsilon_{xc}(\underline{r},r_0)dr_0d\underline{r} \quad . \tag{7.9}$$

From (7.5) it is clear that for each $\underline{r}$ only the spherical average with respect to $\underline{r}$ of $\varepsilon_{xc}(\underline{r},\underline{r}')$ enters in determining E_{xc} for any system. Since the sum rule (7.4) fixes the integral over all $\underline{r}'$, some cancellation of errors in the shape of the hole is to be expected. To see this for the H atom [7.3] $\varepsilon_{xc}(\underline{r},\underline{r}_0)$ is plotted in Fig.7.2 for the LSDF and exact theory as a function of r_0 for three locations of the electron, $r = 0$, a_0 and $2a_0$. The area under the curve gives the contribution to ε_{xc} for the electron at r. While the LSDF results are poor in the $r = 0$ and $r = 2a_0$ cases, they are good for $r = a_0$. Due to the large weight of the region

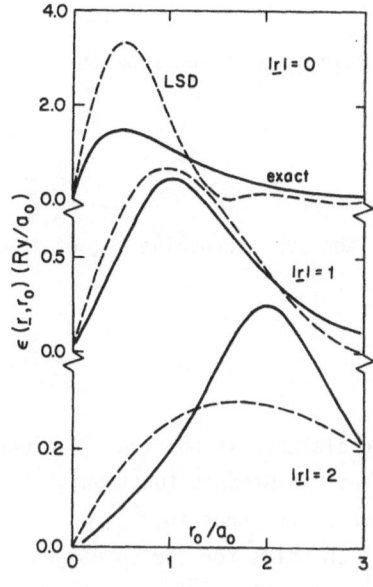

Fig.7.2. The contribution to the exchange-correlation energy of the hydrogen atom from a spherical shell of radius r_0 centered around the electron a distance r from the nucleus. The dashed curves are the results of the LSDF scheme and the solid curves are the exact results [7.3]

$r \sim a_0$ and an approximate cancellation of errors for $r < a_0$ and $r > a_0$, one obtains a reasonable answer for the total energy, despite sizeable errors in $\varepsilon_{xc}(r,r')$.

A number of attempts have been made to improve the local density schemes by including terms involving low orders in ∇n [7.5], with limited success, due to the poor convergence of the gradient expansion. Recently, GUNNARSSON et al. [7.6] have proposed replacing $\varepsilon_{xc}[n(r)]$ by $\varepsilon_{xc}[\overline{n(r)}]$ in calculating ε_{xc}, where $\overline{n(r)}$ is a weighted density

$$\overline{n(r)} = \int \omega[r - r', n(r)] n(r') dr' \qquad (7.10)$$

with ω determined so that the energy properly reduces to the leading gradient correction for slowly varying n(r). When applied to the He atom the error of 0.28 Ry of the LDF scheme is reduced to -0.01 Ry. LANGRETH and PERDEW [7.7], considering the surface energy problem in terms of electronic fluctuations of various wavelengths, have interpolated between the LDF approximation, which is presumably accurate for short wavelength fluctuations (i.e., short distances, over which the density varies fractionally by only a small amount), and the random phase approximation (RPA) for long wavelengths, where that scheme becomes exact. They conclude that the LDF scheme is accurate within 10% for the exchange-correlation energy of a jellium surface with considerable improvement for this hybrid LDF-RPA scheme. To date, this improved scheme has not been applied to chemisorption problems.

7.1.1 Excitation Spectrum

For a uniform electron gas, the energy required to remove an electron from the
Fermi surface is the Fermi energy

$$E_F = \frac{\partial E}{\partial N} \quad . \tag{7.11}$$

Within the LDF scheme this quantity is identical with the LDF eigenvalue E_{k_F} of the
highest filled level

$$E_{k_F} = \frac{\hbar^2 k_F^2}{2m} + V_{xc}(n) \quad . \tag{7.12}$$

However, for $|k| < k_F$, the correct energy to remove an electron is not the LDF eigen-
value E_k but rather is given by the pole of the one-electron Green's function,
leading to an effective mass $m^* \neq m$. The LDF eigenvalue corresponds to $m^* = m$,
since $V_{xc}(n)$ is velocity independent. This same situation holds for the inhomogeneous gas. Therefore, the LDF eigenvalues *cannot* be used directly in interpreting
spectroscopy. This situation is acute for an atom, where the LDF eigenvalues are
closer to the average of the ionization and affinity energies for a partly filled
level, rather than the ionization energy itself, a shift of a 3-6 eV in typical
cases. The reason for this is the localized nature of the atomic orbitals, which
leads to a large potential change when one electron is removed. Actually, if one
removes an electron from state i, one wants $E_{tot}(n_i)-E_{tot}(n_i - 1)$ rather than
$(\partial E_{tot}/\partial n_i)|_{n_i}$; these differ significantly for localized states. To handle this
problem, SLATER suggested expanding $E_{tot}(n_i)$ about the transition state $n_i - 1/2$
using only the first order term $(\partial E_{tot}/\partial n_i)|_{n_i-1/2}$ in calculating the difference
of the E_{tot}. This idea works reasonably well for localized states. However, for
the partly-filled resonance levels broadened by chemisorption the LDF states are
delocalized, so that the transition state does not shift the eigenvalue (the analog
of Koopmans' theorem of HF theory). An example of this situation is shown in
Fig.7.3, taken from the work of LANG and WILLIAMS [7.8] concerning the chemisorption
of Cl and Si on jellium. Here, the resonance levels of Cl(3p) and Si(3s, 3p) are
plotted as a function of d, the distance between the adsorbate nucleus and the
positive background edge. Since the 3p orbitals are only partly filled in the free
atoms, these LDF resonances approach the Fermi level as d → ∞. This result is
proper for calculating the total energy of the system, using (7.2) plus the associated expressions for kinetic energy and classical Coulomb interactions. The LDF
eigenvalues must not be used in interpreting one-electron spectroscopies in general,

---

$\hbar$ = h/2π (normalized Planck's constant)

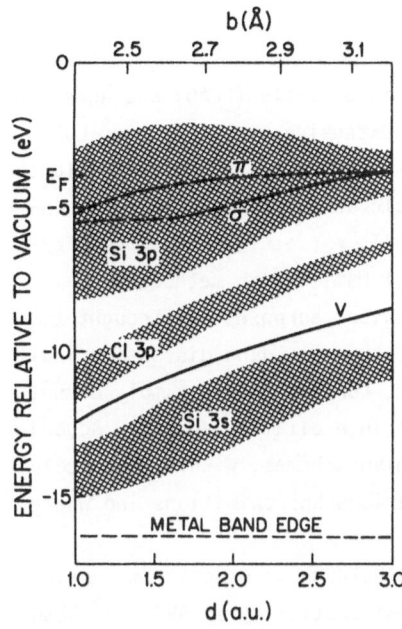

Fig.7.3. The LDF eigenvalue resonance levels of Si and Cl plotted as a function of metal-adatom separation d. As d → ∞ the eigenvalues of the partly filled 3p shells approach $E_F (\cong 4$ eV) rather than the atomic ionization energies (8.15 and 13.01 for Si and Cl, resp.). This illustrates that the LDF eigenvalues cannot be used in interpreting one-electron spectroscopies in general. [7.8]

as is clear since the actual ionization energies of free Si and Cl are 8.15 eV and 13.01 eV, respectively, rather than $E_f \sim 4$ eV as shown in the figure. For very broad levels, the atomic character of the states is reduced. In this limit the LDF eigenvalues may have relevance to spectroscopy.

The physical origin of the above complications is clear. For an open shell atom, above a metal surface, large fluctuations of charge occur on the atom within a one-electron theory. The size of the exchange-correlation hole which suppresses such fluctuations is not given by $\varepsilon_{xc}[n(\underline{r})]$ for $\underline{r}$ in the vicinity of the atom (as enters in LDF theory), but rather is related to the distance between the adatom and the metal. In addition, the hole is highly nonspherical. Finally, in spectroscopy, one is removing a full electron rather than an infinitesimal fraction of an electron. In situations where E_{tot} varies in a strongly nonlinear manner with an effective occupation number entering the measurement, (e.g., the population in a narrow resonance level), a Slater-like transition state should be used if one hopes to connect the LDF eigenvalues with spectroscopy. Unfortunately, for the continuum states entering a resonance, this is an ad hoc procedure which works properly only for the extreme atomic limit. For strong bonding, such a procedure overcorrects for the exchange-correlation hole effects in spectroscopy, since the LDF already gives a reasonable one-electron spectrum for a nearly uniform electron gas. A proper treatment of the case of intermediate bond strength remains to be developed.

7.2 The Anderson Model Picture

Historically, the use of linear combination of atomic orbitals (LCAO) and Anderson model Hamiltonians for chemisorption antedates the extensive application of LDF techniques described in other chapters of this book. Interest shifted to the latter methods because of the possibility of making credible ab initio calculations, unimpaired by worries about the parameters of the model, for at least some real experimentally accessible systems. In some eyes, model Hamiltonian methods were relegated to museum piece status, useful for pedagogical purposes but thought too oversimplified to apply seriously to data. However, the preceding discussion indicates that at their present level of sophistication, LDF techniques simply cannot address problems such as strong correlation effects in excitation spectra. Accordingly, in this section we retreat to model Hamiltonian schemes, within which calculations of these effects can be carried out in various approximations and where the physics remains at least partially transparent.

We begin, therefore, with the model proposed by ANDERSON to describe magnetic impurities in metals [7.9] and first applied to chemisorption by EDWARDS and NEWNS [7.10]. The Hamiltonian is

$$H = \sum_{k\sigma} \varepsilon_k c_{k\sigma}^+ c_{k\sigma} + \sum_{\sigma} \varepsilon_a c_{a\sigma}^+ c_{a\sigma} + U n_{a\uparrow} n_{a\downarrow}$$
$$+ \sum_{k\sigma} (V_{ka} c_{k\sigma}^+ c_{a\sigma} + V_{ka}^* c_{a\sigma}^+ c_{k\sigma}) \quad . \tag{7.13}$$

The first term describes the solid without any adatom in a one-electron approximation. The eigenstates (which are not Bloch states because the solid has a surface) are labeled by an index k; the corresponding eigenvalues are ε_k. In simple applications, we think of k as restricted to a single, nondegenerate energy band. The second and third terms described a highly idealized adatom—a single nondegenerate state at ε_a with a correlation energy U. That is, a single electron on the adatom has energy ε_a and a second one has energy ε_a + U. The U term originates in the mutual Coulomb repulsion (taking into account screening by all other electrons) between the two electrons on the adatom. There is no term like $(n_{a\uparrow})^2$ or $(n_{a\downarrow})^2$ because the Pauli principle prohibits two electrons of the same spin from occupying the same state. The final term describes a mixing of the orbitals of the substrate with that of the adatom and is responsible for the bonding. In the absence of correlation (U = 0), we would have a Hückel theory of the sort first given for chemisorption by GRIMLEY [7.11]. However, Coulomb interactions are seldom very small, especially for small molecules. An extreme situation occurs in hydrogen chemisorption. Let us consider a hydrogen atom on a transition metal surface with a d-band width of, say, 5 eV and a work function of roughly the same size (Fig.7.4). The atomic level

Fig.7.4. Schematic picture of an adatom near a metal surface

ε_a is just the negative of the H binding energy, 13.6 eV , relative to the vacuum. Thus in this case the ε_a level lies well below the bottom of the substrate band, and neglecting correlation we would expect that almost all of the adatom spectral weight lies in a bonding level near ε_a, with only a tiny tail above the Fermi level. That is, the adatom should have almost unit mean occupancy per spin and thus be negatively charged. This is not observed; adsorbed H is nearly neutral. The reason the one-body picture gives nonsense is the neglect of electron-electron interactions. If we consider the limit of small V for the moment, we note that if we try to doubly occupy the adatom, the second electron will have an energy $\varepsilon_a + U$. In the case of hydrogen, $\varepsilon_a + U$ is just the negative of the electron affinity, which is about 0.7 volts. This level lies well above E_F, so the second electron would rather sit in the substrate than on the adatom.

This example points up the importance of the correlation energy in chemisorption—U can be the largest characteristic energy in the problem. Accordingly, the remainder of this section is devoted to approximate solutions of the Anderson model (and generalizations of it) in which U and V are both finite. The simplest approaches one can try are perturbation theory in U or V around the exactly soluble limits of zero U or zero V. Lowest order self-consistent perturbation theory in U gives a Hartree-Fock or molecular field description. Perturbation theory in V leads to a valence-bond picture analogous to the Heitler-London theory for molecules. One can also try to include terms beyond the lowest orders, starting from either limit, in such a way as to get both limits right and interpolate reasonably in the intermediate region. We now survey these descriptions.

7.2.1 Hartree-Fock Calculations for Simple Tight-Binding Model Substrates: Surface Densities of States [7.12]

In this subsection, we review calculations carried out for Anderson models in the spin-independent Hartree-Fock approximation, with particular attention to the changes in the density of states, both on the adatom and in the substrate. In the Hartree-Fock approximation, an electron on the adatom is repelled by the potential due to the average occupancy of opposite-spin electrons. The one-electron level ε_a is raised, by a fraction of U equal to the mean occupation, to $\varepsilon_a = \varepsilon_a + U\langle n_{-\sigma}\rangle$. NEWNS [7.10] found that the Hartree-Fock calculation rather badly overestimated the degree of charge transfer for hydrogen on Ni, Cu, Ti and Cr. A similar conclusion pertains to the more sophisticated (but still Hartree-Fock) treatment of hydrogen on tungsten by LYO and GOMER [7.13].

Nevertheless, one can obtain useful information in the Hartree-Fock approximation about trends in surface densities of states as a function of adatom substrate hopping strengths and other parameters.

We begin with the model (7.13), writing the transfer term as

$$\hat{V} = V \sum_{\sigma} (c_{a\sigma}^{+}c_{0\sigma} + c_{0\sigma}^{+}c_{a\sigma}) \tag{7.14}$$

for the case where the adatom couples to only one atomic orbital (labeled 0) of a substrate which we shall describe by a single-band tight-binding model. We let G denote the Green's function of the adsorbed system and G^0 that of the clean substrate plus isolated adatom. Here and in what follows we shall be using a (generalized) Wannier basis in describing the substrate. Thus, for example, $\pi^{-1}|\text{Im}\{G_{ii}(E)\}| = \sum_{k}|<i|k>|^2\delta(E - \epsilon_k)$ measures the projection of the metal density of states onto the orbital centered at site i, and so forth.

We assume that all the matrix elements of G^0 are known; they can be obtained from those of the Green's function of the periodic infinite substrate by the method of KALKSTEIN and SOVEN [7.14]. We can find G quite simply: For the adatom the self energy is simply

$$\Sigma_v(E) = V^2 G_{00}^0(E) \tag{7.15}$$

so that

$$G_a(E) = [G_a^{0^{-1}}(E) - \Sigma_v(E)]^{-1} = \frac{G_a^0(E)}{1-V^2 G_a^0(E)G_{00}^0(E)} \tag{7.16}$$

and for the substrate

$$G_{ij}(E) - G_{ij}^0(E) = V^2 G_a(E)G_{i0}^0 G_{0j}^0 \quad . \tag{7.17}$$

These expressions are exact for a strict Hückel (U = 0) model; in the Hartree-Fock approximation U only enters through the self-consistent adjustment of the adatom energy level. We remark that (in the present model) the problem is solved once G_a is found, and that the influence of the substrate on the adatom comes entirely from atom 0.

To amplify the physical meaning of (7.16,17), we substitute $G_a^0(E) = (E - \epsilon_a)^{-1}$. Then we see from (7.16) that the adatom density of states becomes

$$\rho_a(E) = \pi^{-1}|\text{Im}\{G_a(E)\}| = \frac{\pi^{-1}V|\text{Im}\{G_{00}^0(E)\}|}{[E-\epsilon_a-V^2\text{Re}\{G_{00}^0(E)\}]^2+[V^2\text{Im}\{G_{00}^0(E)\}]^2} \quad . \tag{7.18}$$

(The argument E of the Green's functions in (7.18) has an positive infinitesimal imaginary part. We shall generally not write it explicitly.)

Thus the spectral weight formerly concentrated in a δ functional at ε_a is spread out into a continuum which is nonzero everywhere within the substrate band limits, plus possible new δ function peaks indicating bonding or antibonding at real roots of $(G_a^0)^{-1}(E) = \Sigma_v(E)$. The total change in density of states due to adsorption is

$$\Delta\rho(E) = -\pi^{-1} \text{Im}\left\{ G_a - G_a^0 + \sum_i (G_{ii} - G_{ii}^0) \right\}$$

$$= -\pi^{-1} \text{Im} \left\{ \frac{V^2 (G_a^0)^2 G_{00}^0}{1-V^2 G_{00}^0 G_a^0} + \frac{V^2 G_a^0}{1-V^2 G_{00}^0 G_a^0} \sum_i (G_{i0}^0)^2 \right\}$$

$$= -\pi^{-1} \text{Im} \left\{ \frac{V^2}{1-V^2 G_{00}^0 G_a^0} \left(-G_{00}^0 \frac{\partial}{\partial E} G_a^0 - G_a^0 \frac{\partial}{\partial E} G_{00}^0 \right) \right\}$$

$$= -\pi^{-1} \text{Im} \left\{ \frac{\partial}{\partial E} \ln(1 - V^2 G_a^0 G_{00}^0) \right\} \quad . \tag{7.19}$$

Associated with this change in density of one-electron state is a change in one-electron energies

$$\Delta E = 2 \int_{-\infty}^{E_F} (E - E_F)\Delta\rho(E)dE \tag{7.20}$$

where the factor of 2 comes from spin degeneracy and the E_F from the fact that the total number of electrons, rather than the Fermi level, is fixed [7.11,15].

If we add $\delta(E - \varepsilon_a)$ to $\Delta\rho$ we get the difference between the final density of states and that for the clean substrate, $\tilde{\Delta\rho}(E)$, which is the experimentally relevant quantity. This $\tilde{\Delta\rho}(E)$ is plotted in Fig.7.5 for a range of V. We see a simple broadening of the adatom level for weak V, and bonding and antibonding resonances, as mentioned in the preceding subsection, for intermediate V. As V increases further, these resonances split off from the band to become sharply defined energy levels formed by the mixing of the orbitals on the adatom and site 0; correspondingly, the negative change in the density of states in the middle of the band is, in the large V limit, simply the original density of states of site 0: in order to participate in this very strong chemical bond, the surface orbital must be effectively severed from the substrate. The surface molecule can then re-bond perturbatively to the indented solid.

Using (7.16 and 17), it is easy to decompose $\tilde{\Delta\rho}$ into its site-by-site components. Both sites a and 0 markedly exhibit the bonding and antibonding resonances of the surface molecule. For an adatom level (i.e., the energy of the orbital at 0 isolated from the substrate) near the middle of the substrate band these peaks are

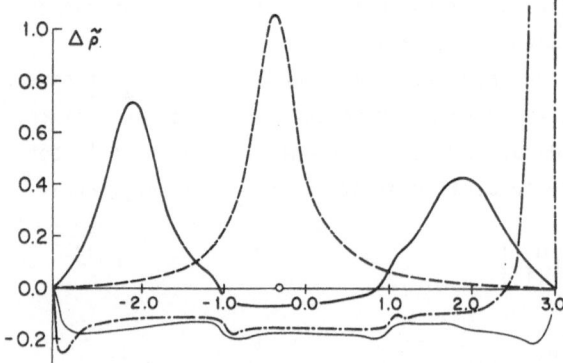

Fig.7.5. Total change in density of states ($\Delta\rho$) vs energy for four adatom-substrate hopping strenghts: V = 0.5 (dashed), 1.5 (heavy solid), 2.504 (dot-dashed), and 3.5 (light solid). The energy unit is one-sixth of the bandwidth. The adatom-level parameter ε_a, indicated by the small circle on the abscissa, is -0.3 relative to the energy zero at the band center. [7.12]

equally weighted on both sites. Were the adatom energy much lower than the center of the band, the bonding peak would be mostly on the adatom and the antibonding peak on O, and vice versa. The striking feature is that in spite of the appearance of (7.17), there is no component of the surface molecule resonances on the substrate neighbors of site O. Formally this absence comes from $(G_{i0}^0)^2$ contributing a substantial imaginary part near the resonance of G_a^0, wiping it out. Physically, this corroborates the surface molecule picture. In reality, the adatom-substrate hopping will most likely be of longer range than nearest neighbor, so that the molecule is not quite so well confined spatially.

Except for the site-wise decomposition, most of the preceding analysis has been carried well beyond this simple level. The Hartree-Fock approach can elucidate many features usually ignored in more sophisticated approaches. In the foregoing, we have assumed the adatom binds to a single surface atom, and so sits in the "atop" (A) position. In the case of "bridge" (B) bonding, (7.14) is replaced by

$$\hat{V} = V \sum_\sigma (c_{a\sigma}^+ c_{0\sigma} + c_{a\sigma}^+ c_{1\sigma} + h.c.)$$

$$= \sqrt{2}V \sum_\sigma \left[c_{a\sigma}^+ \left(\frac{c_{0\sigma} + c_{1\sigma}}{\sqrt{2}} \right) + h.c. \right]$$

(7.21)

where sites 0 and 1 are nearest neighbors on the surface. The associated orbital into which the adatom electron hops is $\phi_B(\underline{r}) = [\phi(\underline{r} - \underline{R}_0) + \phi(\underline{r} - \underline{R}_1)]/\sqrt{2}$. This symmetric bonding combination has lower energy than the individual orbitals. Correspondingly, in place of G_{00}^0 in (7,19), one must use the diagonal Green's function element G_{BB}^0 for this orbital. Like G_{00}^0, this quantity can be expressed as a sum of surface Green's functions $G_s^0(\underline{k}_{||}, E)$ over the surface Brillouin zone of allowed values of $\underline{k}_{||}$, but with an extra weighting factor which enhances the contribution of states near $k_x = 0$ and suppresses those with k_x near the zone boundary:

$$G^0_{BB}(E) = \frac{\sqrt{2}}{N_{\shortparallel}} \sum_{k_{\shortparallel}} \cos^2\left(\frac{k_x a}{2}\right) G^0_s(\underline{k}_{\shortparallel}, E) \tag{7.22}$$

where $N_{\shortparallel}$ is the number of sites in a layer and we have taken the lattice vector connecting two sites 0 and 1 in the x direction [7.12,15]. In general, this factor will then lower the mean energy and narrow the width of the density of states associated with G^0_{BB}.

For adatoms in a "centered" position these effects are further enhanced by bonding to the symmetrical combination θ_c of four orbitals. On a square lattice face, for instance, the new weighting factor becomes $2 \cos^2(k_x a/2) \cos^2(k_y a/2)$. Figure 7.6 shows these effects on the local density of states at the surface site viewed by the adatom and also the corresponding $\Delta\tilde{\rho}$ produced by single atom adsorption.

These group orbitals θ_B or θ_c are now the active participants in the chemisorption band. The depletion of states near the center of the band in $\Delta\tilde{\rho}$ corresponds to these states being hybridized into bonding and antibonding states. In multiband situations, one can still often focus on just a single subband which gets most affected by the chemisorption process. Likewise, the p or d symmetry of these active orbitals can be included by straightforward generalizations of the above formalism. For example, GADZUK has considered a variety of combinations in the surface molecular limit, in which the substrate is replaced by just the near neighbors on the surface (i.e., with broadening due to the rest of the solid neglected) [7.16].

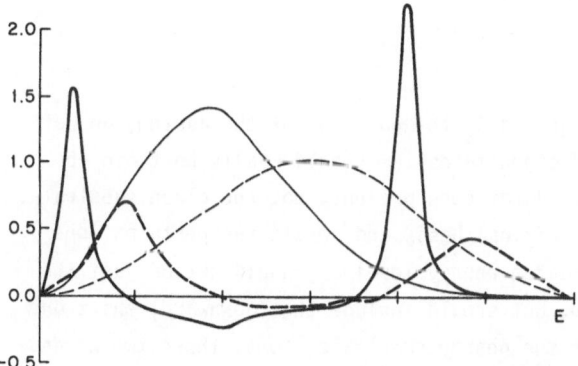

Fig.7.6. Illustration of the effect of an adatom sitting at the fourfold centered (C) position (solid curves) rather than the atop (A) position directly over a substrate site. The light curves give π times the (unperturbed surface) density of states that the adatom electron encounters in hopping to the bulk, and shows that the C curve is shifted down and narrowed. The heavy curves give the resultant $\Delta\tilde{\rho}$ for V = 1.5 and ε_a = -0.3; i.e., the heavy dashed curve gives the same quantity as the heavy solid curve of Fig.7.5. The grid of the abscissa is the same as in Fig.7.5. [7.12]

7.2.2 Improvements of the Anderson Model

Deficiencies in the foregoing theoretical description can be divided into two classes: those which are due to the inadequacies of the Hartree-Fock solution of the model, and those which can be traced to shortcomings of the model itself. In this subsection we discuss the latter sort and review briefly various attempts to improve or generalize the Anderson model, still solving it in the Hartree-Fock approximation.

Improvements in the model can be further classified into two groups: 1) The addition to the Hamiltonian of various kinds of interaction, either one-body or two-body, which were ignored in the simple model (7.13), and 2) improvements in the set of basis states which span the space within which the problem is to be solved, including nonorthogonality effects and extensions to larger bases.

We begin the discussion of the effects of additional interactions by restricting ourselves to systems with a single-band substrate and a single-orbital adatom. For this kind of system we may take the Anderson model Hamiltonian as a reference point, noting the microscopic origin of the terms in it and adding extra terms to describe interactions omitted from it [7.17]. We start with the one-body terms. The substrate part of (7.13) is generally taken to be the Hamiltonian of the relevant band of the clean substrate, written formally in the basis where it is diagonal. However, the presence of the ion core of the adatom alters the value of the one-electron energy levels on nearby substrate sites and the transfer matrix elements between them as well. That is, in the Wannier representation, matrix elements like

$$t_{ij} = \left\langle i \left| \frac{Ze^2}{|\underline{r} - \underline{R}_a|} \right| j \right\rangle \tag{7.23}$$

where Z is the substrate ionic charge and $\underline{R}_a$ the position of the adatom, do not vanish. Thus the eigenvalues ε_k and eigenstates $|k\rangle$ should really be those appropriate to the substrate with the adatom core present, not the clean substrate. Alternatively, one can stay in the original basis and retain the perturbations (7.23) explicitly. Similarly, the adatom energy level ε_a should not be just the bare ionization level of the adatom, but should include the (downward) shift due to the presence of the ion cores of the nearby substrate atoms. These two kinds of effects are the only purely one-body corrections to the naive Anderson model picture in the sort of model we are considering here. Turning now to two-body terms, there are first of all correlations in the substrate, arising from both intrasite and intersite Coulomb interactions. In most transition metals, the intersite interactions are fairly small, but intrasite ones can be quite relevant; they are responsible for the strong magnetic properties, for example. Thus one might try a Hubbard model description of the substrate, adding a term

$$H'_{Hubbard} = \sum_i I_i n_{i\uparrow} n_{i\downarrow} \qquad (7.24)$$

to the Hamiltonian. The correlation energies I_i are typically 3-5 volts, which is comparable to the bandwidth for the 3d series metals. (I_i can depend on the layer in which i lies, because translational invariance is broken by the presence of the surface.)

The corresponding correlation energy on the adatom is a basic ingredient of the Anderson Hamiltonian, but there are also Coulomb forces coupling substrate and adsorbate electrons. There are in principle many nonvanishing matrix elements, leading to terms of the general form

$$\sum_{\alpha\beta\gamma\delta,\sigma\sigma'} U_{\alpha\beta\gamma\delta} c^+_{\alpha\sigma} c^+_{\beta\sigma'} c_{\delta\sigma'} c_{\gamma\sigma} \qquad (7.25)$$

where α,β,γ, and δ label either substrate or adsorbate orbitals. The largest ones,

$$\sum_{i\sigma\sigma'} U_{iaia} n_{i\sigma} n_{a\sigma'} \qquad (7.26)$$

describe a simple Coulomb repulsion between substrate and adatom electrons.

Let us consider the effects of the dominant interactions, (7.24) and (7.26), in a Hartree-Fock approximation. The substrate correlation energy (7.24) is then replaced by

$$\sum_i I_i (\langle n_{i\uparrow}\rangle n_{i\downarrow} + \langle n_{i\downarrow}\rangle n_{i\uparrow} - \langle n_{i\uparrow}\rangle\langle n_{i\downarrow}\rangle) \quad . \qquad (7.27)$$

In the bulk, this just shifts the band by a constant amount and is not very interesting unless one finds solutions with $\langle n_{i\uparrow}\rangle \neq \langle n_{i\downarrow}\rangle$, indicative of a magnetic state. Near the surface, however, $\langle n_{i\sigma}\rangle$ can depend on i, shifting the effective Wannier energy levels, and these shifts may be modified further by the presence of the adatom. If the part of (7.27) present in the clean substrate is absorbed into the free substrate Hamiltonian, we are left with a shift

$$\delta\varepsilon_i = I_i (\langle n_i\rangle - \langle n_i\rangle^0) \qquad (7.28)$$

showing explicitly the dependence on the charge redistribution.

The Hartree-Fock decoupling of the substrate-adsorbate Coulomb term (7.26) leads to several kinds of terms. First, the sort

$$\sum_{i\sigma\sigma'} U_{iaia} \langle n_{i\sigma}\rangle n_{a\sigma'} \qquad (7.29)$$

shifts the adatom level ε_a upward, and

$$\sum_{i\sigma\sigma'} U_{iaia} <n_{a\sigma'}> n_{i\sigma} \tag{7.30}$$

make a corresponding upward shift of the levels of the orbitals on nearby substrate atoms. If the adatom is neutral, the terms (7.29,30) will nearly cancel the one-body shifts from the ion potentials discussed above. Thus when there is very little charge transfer one can safely ignore both effects, as in the naive Anderson model. However, the degree of charge transfer is something to be solved for self-consistently; a solution of the Anderson model which predicts a large charge transfer indicates that one really should have started with an improved model which included these interactions. The other sorts of term obtained in a Hartree-Fock approximation on (7.26) are

$$\sum_i U_{iaai} (<c_{i\sigma}^+ c_{a\sigma}> c_{a\sigma}^+ c_{i\sigma} + <c_{a\sigma}^+ c_{i\sigma}> c_{i\sigma}^+ c_{a\sigma}) \quad . \tag{7.31}$$

These act as corrections to the V_{ak} term in the Anderson Hamiltonian; the transfer matrix elements are changed by an amount proportional to the "bond charges" $<c_{i\sigma}^+ c_{a\sigma}>$. These terms are naturally important if the coupling U is strong.

One can go on in this vein. For example, terms like

$$\sum_{i\sigma} U_{iaaa} c_{i,-\sigma}^+ c_{a\sigma}^+ c_{a\sigma} c_{a,-\sigma} \rightarrow \sum_{i\sigma} U_{iaaa} <c_{i,-\sigma}^+ c_{a,-\sigma}> c_{a\sigma}^+ c_{a\sigma} + \cdots \tag{7.32}$$

lead to perturbations of the adatom level proportional to the bond charge, and so on.

The lesson of all of these points is that one can subsume many of the corrections to the Anderson model into changes in its coupling parameters, but that this makes these parameters subject to self-consistency conditions [7.17]. These problems are most relevant when the degree of charge transfer is large. In calculating such effects, it is naturally advantageous to organize correction terms into groups, such as the downward shift of ε_a due to the substrate ions and its nearly equal upward shift (7.29) due to the mean substrate electron density, which nearly cancel.

These additions to the Anderson model permit the examination of interesting properties of the adsorbed system. For example, the Hubbard term (7.24) can give rise to a ferromagnetic or antiferromagnetic instability and thus to large spin fluctuations near the Curie or Neél temperature of the substrate. Terms like (7.25) allow the coupling of these large spin fluctuations to the adsorbate electron in second order in V. This phenomenon has considerable importance because of the well-known anomalies near the Curie or Neél temperature of the substrate [7.18].

Nevertheless, such self-consistency is much easier to talk about than to calculate. A start at realizing such a scheme has been made by GRIMLEY and PISANI, [7.19] who enforced mutual self-consistency of the energy levels on the adatom and the nearest substrate atom. This two-atom level of self-consistency can be extended to larger clusters, but the convergence is rather slow [7.20].

We turn our attention now to interactions involving electrons in states not included in the original Anderson model. These may include excited levels of the adsorbed atom or molecule and degenerate bands in the substrate. More important in terms of relevant physics is the addition of the s-like conduction band electrons in the substrate. As a first approximation, we may regard them as uncoupled from the d electrons (except insofar as they screen the d-d Coulomb interactions so that intrasite repulsions are small). They will, however, have a long-range Coulomb interaction with each other and with the adsorbate ion and electrons. These give rise to important screening and image effects, of which we give a brief qualitative account here [7.13,21,22]. The Coulomb binding energy between a positively charged ion and its image charge, in the substrate for large distance x from the surface is

$$V_{im} = e^2/4x \quad . \tag{7.33}$$

Closer to the surface, quantum effects modify the image potential; it becomes nonlocal and energy dependent. A decent static local approximation for qualitative purposes is

$$V_{im} = \frac{e^2}{4(x+k_s^{-1})} \tag{7.34}$$

where k_s^{-1} is the Fermi-Thomas screening length.

The energy ϵ_a of the adsorbate ionization level is thus raised by V_{im}, which can amount to a few eV. Correspondingly, the energy of the affinity level $\epsilon_a + U$ is *lowered* by the same amount, i.e., U is lowered to

$$U_{eff} = U - 2V_{im} \quad . \tag{7.35}$$

A rough estimate [7.13] for hydrogen gives $U_{eff} \approx 6$ eV, a significant reduction from its atomic value of 12.9 eV.

A full quantum mechanical treatment of the static image problem can, of course, be given in the LDF method, as described in Chap.3, but this method cannot address its dynamical aspects. A proper quantum mechanical treatment of these effects (retardation or energy dependence) can be given in terms of adatom self-energy corrections of the sort shown in Fig.7.7.

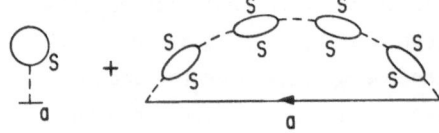

Fig.7.7. Dynamical image charge diagrams.
Lines marked s are electrons in the substrate;
those marked a, on the adatom

This problem has been treated in a simplified model where the substrate surface
density fluctuations are replaced by simple bosons (idealized surface plasmons)
[7.23]. A fully microscopic formulation has been given by KRANZ and GRIFFIN [7.24].
In both these formulations, one can study the nonlocal and energy-dependent deviations from the classical image potential. For example, one finds surface plasmon
satellites above the affinity level and below the ionization level.

There is an exchange counterpart of the image charge effect, the "image spin,"
which has been discussed by BRENIG [7.25], and KRANZ and GRIFFIN [7.24].

S electrons may, of course, also mix directly with the adsorbate orbital,
giving rise to a broadening of otherwise sharp bonding levels [7.26]. Another possible relevant effect in the presence of s electrons is the modification of the
s-d mixing of substrate orbitals in the vicinity of the adatom [7.17].

A further class of states which may be relevant to energetic questions in
chemisorption is the set of continuum states not bound by the self-consistent
potential of the solid [7.13]. These free-electron-like states thus lie at an
energy equal to or greater than the work function ($\sim$ 4 eV) above E_F. If we augment
the simple Anderson model (7.13), taking into account, the mixing between the adatom
orbital $|a>$ and these states $|k'>$ via matrix elements $V_{ak'}$, the adatom self-energy
Σ_v will acquire a new term:

$$|\text{Im}\{\Sigma_v(E)\}| = \pi \sum_k |V_{ak}|^2 \delta(E - \epsilon_k) + \pi \sum_{k'} |V_{ak'}|^2 \delta(E - \epsilon_{k'}) \quad . \tag{7.36}$$

[States labeled by k (without a prime) are taken to be the substrate d-level states
of the previous discussion.] Thus $\text{Im}\{\Sigma_v\}$ is nonzero both within the d-band and
above the vacuum level. This has several possible effects. First, it could heavily
damp an antibonding state that lie above the vacuum level; it could just decay
into the unbound continuum. Second, the part of Σ_v, $\text{Re}\{\Sigma_v\}$, that comes from the
Hilbert transform of the above-vacuum-level band in $\text{Im}\{\Sigma_v\}$ will be negative for
energies below the vacuum. To first order, this is equivalent to a downward shift
of ϵ_a.

However, this point brings us unavoidably to the question of nonorthogonal
basis functions [7.13,27]. The reason is that the whole set of metal states, bound
and unbound, is complete. Thus including the adatom orbital as well leads to an
overcomplete basis; the state $|a>$ is a not orthogonal to the $|k>$'s and $|k'>$'s,
but can be expanded in them.

Even if we decide to work in the (incomplete) basis of the bound d-band metal states and the adatom orbital, nonorthogonality is a problem because there will in general be a finite overlap between the adatom orbital and the eigenstates of the clean solid. (It will be very small if the binding is very weak, but this is not the limit of interest on metals). One can, of course, elude this difficulty by a redefinition: $|a>$ and the substrate orbitals $|i>$ can be defined to be the one-body Wannier (Löwdin) orbitals of the solid plus-adsorbate-core system [7.28]. They are thus orthogonal and one can proceed as we have previously. The trouble with this is that we lose whatever connection between atomic properties and the parameters of the model we had (such as ε_a = ionization level, U = affinity level — ionization level). The new localized orbitals are hard to solve for, and, furthermore, they have relatively slowly decaying oscillating tails (to make them mutually orthogonal). Hence the substrate orbitals may have appreciable amplitude on the adsorbate site, and if we want to keep correlation effects in a consistent way, we will have to generalize the model and include sizeable correction terms which look formally like correlations in the substrate (7.24). Thus, it may be preferable to retain non-orthogonality explicitly.

To see its effects, consider the Anderson model (7.13), in which the $|k>$'s are now taken to be the complete set of metal states, bound and unbound. Thus the adatom electron creation operator c_a^+ can be expanded in terms of the c_k^+

$$c_{a\sigma}^+ = \sum_k c_{k\sigma}^+ <k|a> \equiv \sum_k c_{k\sigma}^+ \Delta_{ka}.$$

(7.37)

If we express the adatom orbital operators in H terms of the $c_{k\sigma}$ or $c_{k\sigma}^+$ the adatom single-particle energy term

$$\sum_\sigma \varepsilon_a c_{a\sigma}^+ c_{a\sigma} = \sum_{kk'\sigma} c_{k\sigma}^+ \Delta_{ka} \varepsilon_a \Delta_{ak'} c_{k\sigma}$$

(7.38)

and the hopping term

$$\hat{V} = \sum_{k\sigma} (V_{k\sigma} c_{k\sigma}^+ c_{a\sigma} + h.c.) = \sum_{kk'\sigma} V_{ka} \Delta_{ak'} c_{k\sigma}^+ c_{k'\sigma}$$

(7.39)

appear as mixing terms in the orthogonal, complete basis of $|k>$'s. The correlation term, after Hartree-Fock factorization, just acts to shift ε_a to ε_a^σ so within this approximation, we just have an effective mixing of $|k>$'s given by

$$V_{kk',\sigma}^{eff} = \Delta_{ka} \varepsilon_a^\sigma \Delta_{ak'} + V_{ka} \Delta_{ak'}.$$

(7.40)

In this basis, it is simple to solve the Dyson equation

$$G = G^0 + G^0 V^{eff} G$$

(7.41)

because of the separability of V^{eff}. One finds [7.13]

$$G_{0a} \equiv \sum_{kk'} \Delta_{ak} G_{kk'} \Delta_{k'a} = G_{aa}^0 \left(1 - \epsilon_a^\sigma G_{aa}^0 - \sum_{k'} \frac{V_{k'a} \Delta_{ak'}}{E - \epsilon_{k'}} \right)^{-1} \tag{7.42}$$

where

$$G_{aa}^0 = \sum_k \frac{|\Delta_{ak}|^2}{E - \epsilon_k} \tag{7.43}$$

is the (a, a) matrix element of the Green's function with no adatom present (not to be confused with the noninteracting adatom Green's function $G_a^0 = (E - \epsilon_a)^{-1}$ of Sect.7.2.1. As before, ϵ_a^σ is determined self-consistently. The terms in the denominator of (7.43) have their spectral weight on the eigenvalue spectrum of the free metal; thus one can find the structure of G_{aa} in a way similar to what we have described for the orthogonal model, as shown in Fig.7.8. Notice, however, that the procedure is not quite the same; nonorthogonality and overcompleteness do have relevant qualitative consequences. In particular, to force (7.42) into the simple form obtained in the orthogonal Anderson model, one must make the mixing matrix elements V_{ka} energy dependent:

$$|V_{ak}^{eff}|^2 = (E - \epsilon_a) \Delta_{ak} (\epsilon^a \Delta_{ka} + V_{ka}) \quad . \tag{7.44}$$

Figure 7.9 shows the adatom density of states for a simple model of hydrogen on tungsten calculated by GOMER and LYO [7.13]. One sees bonding and mid-band resonances, but the antibonding resonance is wiped out because it lies in the continuum above the vacuum level.

In conclusion, the principal lesson of this subsection is that because of various omitted effects, the simple Anderson model is difficult to relate to any sort of first-principles starting point. The parameters in it depend on interactions and wave functions which are not simple to calculate and, indeed, may be functions of the self-consistent solution of the model itself.

On the other hand, the model is not useless in any sense if one is willing to allow its parameters to be fit by experiment. In this sense, it plays a role like that of Fermi liquid theory in the theory of metals, allowing calculations of experimental response functions in terms of a small set of parameters. This separation of the consequences of the model from the calculation of its parameters is useful because it allows meaningful calculations to be done without going all the way back to first principles. This point is particularly relevant when one wants to examine many-body effects, as we do in the next section.

This is also an opportune place to observe that theoretical work of this sort (including the original Edwards-Newns theory) parallels closely that on magnetism

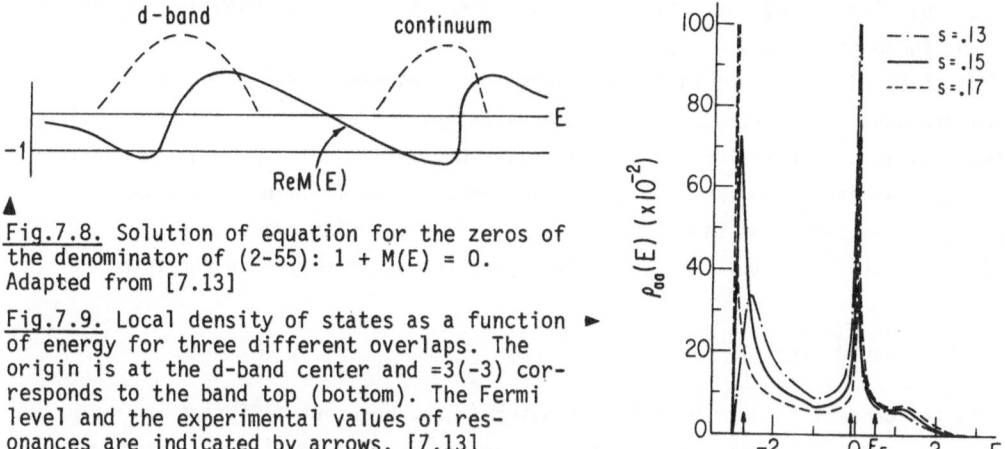

Fig.7.8. Solution of equation for the zeros of
the denominator of (2-55): 1 + M(E) = 0.
Adapted from [7.13]

Fig.7.9. Local density of states as a function ▶
of energy for three different overlaps. The
origin is at the d-band center and =3(-3) corresponds to the band top (bottom). The Fermi
level and the experimental values of resonances are indicated by arrows. [7.13]

in metals, particularly the magnetic impurity problem. The Hartree-Fock solution
of Anderson's model by Edwards and Newns was generalized directly from Anderson's
own treatment; his solution was in turn just an adaptation to the local-moment
problem of the Stoner Hartree-Fock theory of itinerant-electron magnetism. The
theoretical approaches we are about to examine also have their counterparts in
these older problems: The valence-bond theory corresponds to the Kondo limit
[7.29] of the impurity problem or to the Heisenberg exchange model of bulk magnetism. The local spin fluctuation description is carried over directly from a
successful phenomenological description of dilute magnetic alloys [7.30]. The
variational methods of Schönhammer and the interpolative decoupling schemes of
Brenig and Schönhammer, Schuck, and Bell and Madhukar also have closely related
predecessors [7.31-34] in both bulk and local moment magnetic problems. We now
survey these approximations.

7.3 Beyond the Hartree-Fock Theory

7.3.1 The Valence-Bond Picture

The principal defect of the Hartree-Fock approximation is the fact that it ignores
the Coulomb correlations between different electrons, thereby overemphasizing
ionic adatom configurations and underemphasizing neutral ones. For large U, we
expect the charged configurations to be quite suppressed. A way of dealing with
this situation, in the completely correlated limit where only neutral adatoms are
allowed, is the valence-bond theory of GOMER, SCHRIEFFER, and PAULSON [7.35]. It
is similar in spirit to the Heitler-London theory for molecules: In it one deals
with nonorthogonal states and obtains an exchange interaction explicitly dependent

on the overlap [7.36]. We outline the main ideas here. A more complete discussion can be found in [7.22].

We start with a set of single-particle energy eigenstates $|k\rangle$ for the relevant band or bands of the substrate (or, equavalently, its associated Wannier-Löwdin states $|i\rangle$), and an adatom orbital $|a\rangle$ which has finite overlap with the substrate states. ($|a\rangle$ cannot be expanded purely in terms of the $|k\rangle$ since the latter only span a finite number of bands of the solid). We define the overlaps $\Delta_{ka} = \langle k|a\rangle$ and $\Delta_{ia} = \langle i|a\rangle$. Now the neutral basis states for the theory have the form

$$|\bar{\alpha}\rangle = c^+_{a\sigma} \prod_{ks} c^+_{ks}|0\rangle \tag{7.45}$$

where the product is over N different pairs (k,s). For each different set of pairs we get a different state. (A more complete notation might be $|\bar{\alpha}\{k,s\}\rangle$, but we abbreviate this $|\bar{\alpha}\rangle$). Because of the nonorthogonality of the single-particle states, different states $|\bar{\alpha}\rangle$ are neither normalized nor orthogonal. For example, by applying the canonical Fermion algebra one can compute

$$\langle\bar{\alpha}|\bar{\alpha}\rangle = 1 - \sum_k |\Delta_{ka}|^2 \tag{7.46}$$

where the sum is over the occupied values of k. Similarly, for two such states $|\bar{\alpha}\rangle$ and $|\bar{\beta}\rangle$, the overlap is nonvanishing only when one $|ks\rangle$ state occupied in $|\bar{\beta}\rangle$ is replaced in $|\bar{\alpha}\rangle$ by $|k's'\rangle$. Then,

$$\langle\bar{\alpha}|\bar{\beta}\rangle = -(-1)^{\eta_{\alpha\beta}}\Delta_{k'a}\Delta_{ak}\delta_{s\sigma'}\delta_{s'\sigma} \tag{7.47}$$

where $\eta_{\alpha\beta}$ is the number of filled states separating $|k\sigma\rangle$ and $|k'\sigma'\rangle$ in some (any) enumeration of the single-particle states. In this basis Schrödinger's equation is

$$\sum_\beta (E\langle\bar{\alpha}|\bar{\beta}\rangle - \langle\bar{\alpha}|H|\bar{\beta}\rangle) a_\beta = 0 \qquad . \tag{7.48}$$

where a_β is the weighting coefficient of the groundstate wavefunction in terms of the basis states [7.35]. The overlap matrix elements can be tought of as giving rise to a pseudopotential $V_{\bar{\alpha}\bar{\beta}}(E) = -E\langle\bar{\alpha}|\bar{\beta}\rangle$ ($\bar{\alpha} \neq \bar{\beta}$).

To work with this, it is convenient to define an "overlap operator" D, an effective Hamiltonian H_{eff}, and a set of formally *orthogonal* states (without the bar), with the properties

$$\langle\alpha|(1 + D)|\beta\rangle = \langle\bar{\alpha}|\bar{\beta}\rangle \tag{7.49}$$

$$\langle\alpha|H_{eff}|\beta\rangle = \langle\bar{\alpha}|H|\bar{\beta}\rangle \qquad . \tag{7.50}$$

One can than construct D and H_{eff} from these requirements. For simplicity, consider the case where $|a>$ overlaps only one substrate Wannier state $|0>$. Then (7.46,47, and 49) can be satisfied by the choice

$$D_0 = -|\Delta|^2 \sum_{\sigma\sigma'} c^+_{0\sigma}c^+_{a\sigma'}c_{a\sigma}c_{0\sigma'} \quad . \tag{7.51}$$

This D gives the correct overlap (7.49) within the neutral manifold. If we wish as well to give correctly the overlap of neutral states with ionic states, such as

$$|a+> = \prod_{ks}^{N+1} c^+_{ks} |0> \tag{7.52}$$

and

$$|a-> = c^+_{a\uparrow}c^+_{a\downarrow} \prod_{ks}^{N-1} c^+_{ks} |0> \tag{7.53}$$

we can add to D_0 a term

$$D_1 = \sum_{\sigma} \Delta^* c^+_{0\sigma}c_{a\sigma} + h.c. \quad . \tag{7.54}$$

With the full overlap operator defined as $D = D_0 + D_1$, one can verify that the choice of effective Hamiltonian

$$H_{eff} = \sum_{ij\sigma} c^+_{i\sigma}(1 + D)<i|h|j>c_{j\sigma}$$

$$+ \frac{1}{2} \sum_{\substack{ijk\ell \\ \sigma\sigma'}} c^+_{i\sigma}c^+_{j\sigma'}(1 + D)<ij|U|k\ell>c_{\ell\sigma'}c_{k\sigma} + DE_{ion} \tag{7.55}$$

reproduces the desired matrix elements (7.50). In (7.55) i, j, etc. label either Wannier or adatom site and spin, $<i|h|j>$ and $<ij|U|k\ell>$ are one-body and Coulomb matrix elements, respectively, and E_{ion} is the electrostatic energy of all the ion cores.

In (7.55), we can identify several sources of exchange-like terms. The first, ferromagnetic in sign, is the direct Heisenberg exchange term in the Coulomb energy (to zeroth order in the overlap). The second, which is antiferromagnetic, comes from the terms where D_1 is sandwiched between the two field operators of the first term in (7.55). This part of the effective exchange is first order in the overlap. Finally, there is a ferromagnetic term from D_0 in the ionic term. Thus

$$J_{eff} = \Delta^* <a|V_a|0> + \Delta<a|V_M \; 0> + <0a|U_{Coulomb}|a0>$$

$$+ |\Delta|^2 e^2/R_{a1} \qquad\qquad (7.56)$$

where V_a is the adatom potential and V_M is the potential due to the atom "0". This effective exchange interaction generates the dominant physics of the problem in the valence-bond approach. The situation is different from the Heitler-London H_2 molecule because (in a one-body description of the substrate, at least) there is no well-defined substrate spin to couple to the adatom spin. The interaction (7.56) must *induce* a spin density near the adatom, and that induced density can couple to the adatom spin in a Heitler-London fashion. Hence the binding energy is second order in J_{eff} rather than first, as in the Heitler-London H_2 molecule, and the terminology "induced covalent bond" is used to characterize this description.

We distinguish two limiting cases, according to the size of J_{eff}. If J_{eff} is much less than the substrate bandwidth W, one can do perturbation theory in the overlap. To lowest order and for a half-filled band, the energy takes the form

$$\Delta E = 2 \sum_{kk'} \frac{f_k(1-f_{k'})}{\varepsilon_k - \varepsilon_{k'}} [|J_{eff}|^2 + (\varepsilon_{k'} - \varepsilon_k)|\Delta|^2 J_{eff}] - \frac{1}{2} J_0 \quad . \qquad (7.57)$$

The first term, quadratic in J_{eff}, is just proportional to the local spin susceptibility in the substrate, and corresponds to the lowering of the energy of the substrate spins in the field of the adatom spin. The term linear in J_{eff}, which is higher order in Δ, represents nonorthogonality corrections to the first term. The J^0 term is an exchange repulsion. The physical picture of this weak-coupling limit is as a Kondo problem [7.29]: the adatom spin induces a compensating cloud of opposite spin polarization in the substrate, yielding a singlet ground state. Above a characteristic temperature $T_K \approx W \exp(-W/J)$, thermal fluctuations decouple the adatom spin from its cloud.

In the opposite limit, $J \gg W$, one can first solve the problem of the surface molecule consisting of the adatom and substrate atom 0 (or, more generally, larger surface molecules, if the bonding is to several substrate orbitals), and then treat the rebonding of the surface molecule to the rest of the solid in perturbation theory. At $J/W = \infty$, the problem becomes trivial; a substrate electron is completely bound in the orbital 1 to form a singlet. At finite but large J/W, virtual excitation of the surface molecule singlet leads to an effective repulsive interaction between the electrons of the rest of the solid [7.37]. (Viewed on a long enough time scale $(t \gg T_K^{-1})$, the weak-coupling problem also looks like the surface molecule limit, since, as Wilson has shown, the renormalization process whereby short-time or high-frequency degrees of freedom are removed eventually converts any Kondo problem into a $J = \infty$ one [7.29].

7.3.2 Spin Fluctuation Model [7.38]

While the Hartree-Fock approximation makes sense for weak U and the valence-bond
picture is correct for very large U, hydrogen on transition metals seems to lie
between these extremes. To make this more precise, let us take the onset of magnetic
Hartree-Fock solutions [7.9,10] as the signal that the weak correlation theory is
beginning to break down in a serious way. For the wide substrate band case [7.9]
this happens when U is larger than the virtual level width $\text{Im}\{\Sigma_v(E_F)\} = \pi|V|^2\rho_s^0(E_F)$.
In the opposite (surface molecule) limit, magnetic solutions occur for $U \geq 4V$,
[7.38], so one can roughly define the intermediate coupling region by

$$U \approx \max\left[\frac{V^2}{W}, V\right] \quad . \tag{7.58}$$

For 3d series transition metals, V and W are both around 4-6 eV. while for the
heavier series W is larger. Taking into account the image corrections discussed in
Sect.7.3.3 gives a U of $\approx$ 6 eV, so that (7.58) is approximately satisfied. Thus
neither limiting theory discussed so far is likely to be satisfactory.

A semiphenomenological approach to the intermediate coupling situation is the
spin fluctuation or paramagnon model. It is based on the idea that since Hartree-
Fock theory indicates incipient adatom magnetism, it is important to include coupling
to the local collective spin degree of freedom. While the resulting theory cannot
be expressed as a rigorous perturbation theory in a small parameter, the hope is
that by adjusting the parametrization of the local spin fluctuations and their
coupling to electrons on the adatom, a reasonable phenomenological description of
the system can be achieved.

Such an approach is in the spirit of the Anderson model as discussed in Sect.
7.2.2. That is, it is perhaps better looked on as an alternative model (whose
parameters, like those of the Anderson model itself, have to be determined experimentally rather than calculated from first principles) than as an improved
solution of the Anderson model. However, it is simplest to introduce the theory
in the latter context.

Formally, this is achieved by including in the adatom self-energy, in addition
to the mixing (Σ_v) and Hartree terms, the family of diagrams

$$\Sigma_{sf}(E) = \quad\quad = \quad\quad$$

$$= \int \frac{d\omega}{2\pi i} [t(\omega) - U] G_a^H(E + \omega) \quad . \tag{7.59}$$

Here G^H is the Hartree-Fock adatom Green's function, and $t(\omega)$ is the electron-hole
t matrix, which is just proportional to the local spin susceptibility $\chi(\omega)$ on the
adatom:

$$t(\omega) = U + U^2\chi(\omega) \quad . \tag{7.60}$$

In the ladder (random-phase) approximation, the susceptibility is

$$\chi(\omega) = \frac{\chi_0(\omega)}{1-U\chi_0(\omega)} \tag{7.61}$$

where

$$\chi_0(\omega) = i \int \frac{dE}{2\pi} G_a^H(E)G_a^H(E + \omega) \tag{7.62}$$

is the local adatom susceptibility in the absence of correlation. [The latter is just a Pauli susceptibility; $\chi(0) = \rho_a(E_F)$.]

One can describe the physics of (7.59,60) in several ways. The diagrams suggest the following picture: In calculating the adatom energy in perturbation theory, we take into account virtual transitions to excited states with additional particle-hole pairs. The matrix element for these transitions is enhanced by the resonant final-state interactions between the original electron and the new hole. The latter are strong because their t matrix is proportional to the adatom susceptibility, which we know to be enhanced because of the tendency toward local moment formation. Another way to express it is to say that the hopping on and off the adatom by electrons generates a fluctuating magnetic moment, and the response of the electrons to this moment is given by Σ_{sf}. Finally, there is a field-theoretic language in which $\chi(\omega)$ is thought of as the propagator for a boson-like field, which is given the name "paramagnon" because it represents the collective magnetization fluctuations which, in a magnetic state of an extended system, become well-defined spin waves or magnons. Then (7.59) is the electron self-energy resulting from the lowest-order coupling to this boson field. Theories based on this kind of correction have been quite successful in describing bulk Fermi systems in both magnetic and nearly magnetic states [7.39,40], and as remarked above, in dilute magnetic alloys [7.30].

In the ladder approximation for χ(7.61), as U grows $\chi(0)$ grows and the spectral weight in $\chi(\omega)$ becomes concentrated at lower and lower frequencies. When $\chi(0)$ is very large, $\chi(\omega)$ can be well-approximated by a single-relaxation-time form

$$\chi(\omega) = \frac{1}{\omega_s-i\omega} \quad . \tag{7.63}$$

The parameter ω_s is the local spin fluctuation relaxation rate. In the ladder approximation, this happens when U is nearly at the Hartree-Fock threshold for magnetic solutions. However, we know that fluctuation corrections invalidate the magnetic solutions for any finite U, so the form (7.63) is actually valid for large U rather than $U\rho_a(E_F) \lesssim 1$.

This is the point at which we depart from strictly trying to make an improved solution to the original model and regard what we are doing more as a model in its own right. We adopt the hypothesis that the tendency toward local moment formation is strong, thus assuming χ to be given by (7.63), regardless of the value of U. Recognizing that other corrections will change the electron-paramagnon coupling vertex from its bare value of U, we then regard the corrected or effective value of U and ω_s as the parameters of the model. This is the spirit in which localized spin fluctuation theory has been applied to magnetic alloys [7.30].

We now consider qualitatively the effects of the spin fluctuation self-energy on the adatom spectrum, beginning with an examination of the spectral properties of $\Sigma_{sf}(E)$. It is straightforward to show from (7.59), using the spectral represen-tations of G_a and χ (or t), that Im{χ} can be written as a convolution of the spectral weight functions for G_a and

$$\text{Im}\{\Sigma_{sf}(E)\} = -U^2 \int dE' \rho_a(E') \text{Im}\{\chi(E - E')\}[\theta(E')\theta(E - E') - \theta(-E')\theta(E' - E)]$$

$$(7.64)$$

where θ is the unit step function. Under the assumption of small ω_s, Im{χ} is a narrow, peaked function (with width ω_s), so, crudely speaking, Im{Σ_{sf}} is just a smeared, slightly shifted image of the Hartree-Fock adatom density of states $N_a(E)$. For $E > E_F$ it is shifted to the right (larger E); for $E < E_F$, it is shifted to the left by roughly ω_s. The step functions in (7.64) (which arise because of the change in analytic structure of the Green's functions near the Fermi level) also insure that Im{Σ_{sf}} vanish at E_F and grow like $(E - E_F)^2$ for small $|E - E_F|$. (This behavior is thus a kind of Fermi liquid effect.) These qualitative features are shown in Fig.7.10a,b. We next consider (Fig.7.10c) the graphical solution of the Dyson equation $G^{-1} = G_0^{-1} - \Sigma$, with $\Sigma = \Sigma_v + U\langle n\rangle + \Sigma_{sf}$. Since Σ_{sf} varies so rapidly in regions where $\rho_a(E)$ has most of its spectral weight, it can lead to significant changes in the structure of the spectrum. The resulting adatom density of states is shown in Fig.7.10d. One possible spectral feature of note is a so-called Kondo resonance, pinned near E_F, of width $\approx\omega_s$, which comes about because of the Fermi-liquid behavior of Σ_{sf} near E_F. Figure 7.10 does not happen to exhibit this very well, since $\rho_a(E)$ already has a resonance there in the ab-sence of Σ_{sf} because of the structure of Σ_v. However, Fig.7.11 shows a case where the resonance comes from Σ_{sf}.

The paramagnon model can also be generalized to include the effects of corre-lation in the substrate. In addition to the obvious effect of altering $\rho_s^0(E)$, these correlations can be shown to affect the hopping processes on and off the adatom (Fig.7.12). Thus the magnitude of the Anderson model parameter V is changed, and it becomes a complex energy-dependent function.

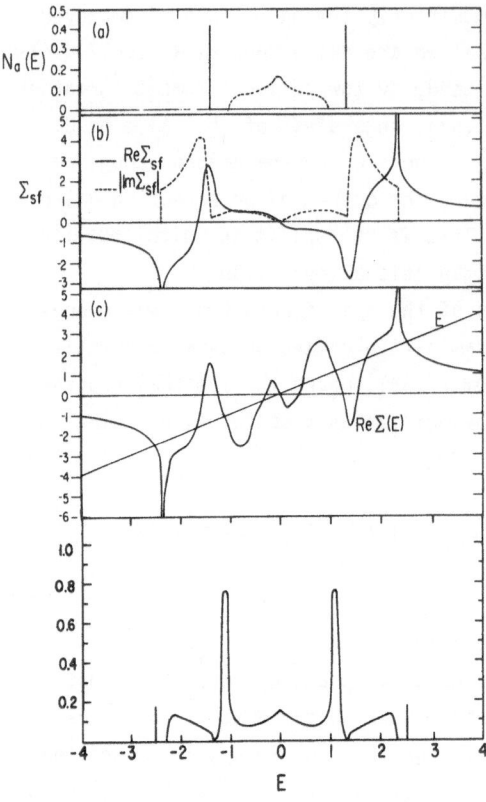

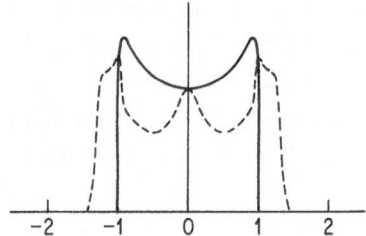

◀**Fig.7.10.** (a) Hartree-Fock $\rho_a(E)$; (b) spin fluctuation self-energy $\Sigma_{sf}(E)$; (c) graphical solution of $\varepsilon - \varepsilon_a^0 - \Sigma_v - \Sigma_{sf} = 0$; (d) resulting $\rho_a(E)$ (vertical lines at ± 2.5 indicate bonding/antibonding δ functions. [7.38]

Fig.7.11. Adatom density of states $\rho_a(E)$ without (solid line) and with (dotted line) spin fluctuation self-energy effects: $V = 0.6$ (in units of half the substrate bandwidth); semielliptic substrate surface model. [7.38]

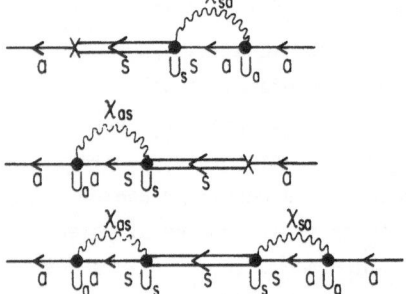

◀**Fig.7.12.** Adatom self-energy diagrams involving spin-fluctuation assisted hopping to and from the substrate (the double line stands for the full substrate propagator, including the effects of spin fluctuations). [7.38]

Models like this one, with different forms for the boson propagator and coupling constants if appropriate, can also be used to study the coupling of adatom electrons to other degrees of freedom, such as plasmons or phonons [7.23,41].

An approach complementary to that discussed here is just self-consistent perturbation theory to one higher order in U than Hartree-Fock [7.26]. This amounts to using χ_0 instead of the full χ in the t matrix (7.60), but using the correct χ_0 (7.62) rather than the phenomenological parametrization (7.63). This procedure appears to work reasonably well for H on Ni, suggesting that the local susceptibility is not too strongly enhanced.

7.3.3 Interpolative Decoupling Methods

If, instead of taking recourse in the phenomenology of the paramagnon model, we attempt to solve the Anderson model better in a well-defined way, we have to improve the qualitatively bad behavior of the Hartree-Fock adatom Green's function in the atomic limit $V = 0$. There the Hartree-Fock solution just has a single pole at ε_a^σ, while it is simple to show from the equation of motion of G_a that the correct form is [7.33,42]

$$G_a^a(E) = \frac{1-<n>}{E-\varepsilon_a} + \frac{<n>}{E-\varepsilon_a-U} = \frac{1}{E-\varepsilon_a-U<n>- \dfrac{U^2<n><1-n>}{E-\varepsilon_a-U<1-n>}} \quad . \tag{7.65}$$

The superscript a here stands for "atomic." This two-pole form has a simple interpretation. G_a^a describes the evolution of the system when an electron is added to the system. If the adatom is initially unoccupied, the final state will be an eigenstate with energy ε_a. If it is already occupied, the extra energy will be $\varepsilon_a + U$. The two configurations have probabilities $1 - <n>$ and $<n>$, respectively, so we obtain (7.65).

The schemes described in this subsection are attempts to calculate G_a for finite V in ways that respect this limit. One very simple scheme is the following: In the absence of correlation, we obtain an exact solution starting with the exact atomic limit Green's function and taking into account the mixing Hamiltonian through the self-energy correction Σ_V. Suppose we try to do the same thing here, starting with the exact Green's function (7.65) in the presence of finite correlation

$$G_a(E) = \frac{1}{E-\varepsilon_a-U<n>- \dfrac{U^2<n><1-n>}{E-\varepsilon_a-U<1-n>} -\Sigma_V(E)} \quad . \tag{7.66}$$

This expression is manifestly correct when $V = 0$ for any U, and correct to first order in U for any V. Such an approximation has been used in the magnetic impurity problem and in the Hubbard model by HEWSON [7.32], and HUBBARD [7.33], respectively although they did not derive it in this way. Our derivation is deceptively appealing because it appears to suggest, through the analogy with the uncorrelated problem, that it is exactly correct. Unfortunately, this is not true. Formally, the reason is that the usual diagrammatic perturbation theory has to be done around a *one-body* unperturbed Hamiltonian, since the latter is necessary to prove Wick's theorem, which is the basis of the diagrammatic representation. The atomic Hamiltonian with U present is not a one-body operator. Thus (7.66) is not even correct to order V^2. The schemes described below attempt to remedy this defect, getting things right at least to this order. To accomplish this, one decouples the chain of equations of motion of G_a and higher-order Green's functions by factorizing

certain higher-order ones into products of lower-order ones. By making sure that
the result is correct to lowest nontrivial order in U (for fixed V) and V (for
fixed U), one obtains a solution which interpolates reasonably between the two
soluble limits; thus, the title of this subsection. These methods are thus in
the spirit of Hubbard's improved decoupling for the extended problem [7.34] (equivalent to the coherent-potential-approximation (CPA) solution of the alloy analogy
[7.43]) or OGUCHI's approximation for the symmetric, wide-band Anderson model
[7.44].

There are several such schemes in the literature [7.45-47]. Here we discuss that
of BRENIG and SCHÖNHAMMER [7.45]; the others are nearly equivalent. In their
method one decomposes the adatom electron creation operators into parts which project onto states occupied and unoccupied by an electron of the opposite spin

$$c^+_{a\sigma} = c^+_{+,\sigma} + c^+_{-,\sigma}$$

$$= c^+_{a\sigma} n_{-\sigma} + c^+_{a\sigma} (1 - n_{-\sigma}) \tag{7.67}$$

and attempts to calculate the matrix Green's function

$$G_{\alpha\beta}(t) = -i <0|T(c_{\alpha,\sigma}(t)c^+_{\beta,\sigma}(0))|0>$$

$$\equiv \int \frac{dE}{2\pi} \exp(-iEt) \langle\!\langle c_{\alpha,\sigma} ; c^+_{\beta,\sigma} \rangle\!\rangle_E \tag{7.68}$$

where T is the time ordering operator and $|0>$ the ground state. Equivalently, one
may consider the associated self-energy, defined by the matrix Dyson equation

$$G_{\alpha\beta} = G^0_{\alpha\beta} + \sum_{\gamma\delta} G^0_{\alpha\beta} \Sigma_{\gamma\delta} G_{\delta\beta} \quad . \tag{7.69}$$

The unperturbed propagator $G^0_{\alpha\beta}$ is that for the atomic problem V = 0. Using the
easily verified anticommunication relations

$$\{c_{\alpha,\sigma}, c^+_{\beta,\sigma}\} = \delta_{\alpha\beta} n_{\alpha,\sigma} \tag{7.70}$$

where $n_{\alpha,\sigma}$ is defined by

$$n_{+,\sigma} = n_{-\sigma}$$

$$n_{-,\sigma} = 1 - n_{-\sigma} \tag{7.71}$$

and the Heisenberg equations of motion for the V = 0 case

$$i\dot{c}_{\alpha,\sigma} = [c_{\alpha,\sigma}, H_{atomic}] = \varepsilon_a c_{\alpha,\sigma} \tag{7.72}$$

where $\varepsilon_+ = \varepsilon_a + U$, $\varepsilon_- = \varepsilon_a$, we find

$$G_{\alpha\beta}^0(E) = \delta_{\alpha\beta} \frac{<n_{\alpha,\sigma}>}{E-\varepsilon_\alpha} \quad . \tag{7.73}$$

Then, from (7.67) the usual adatom Green's function is just $\sum_{\alpha\beta} G_{\alpha\beta}$, so one re-covers (7.65).

Since the algebra (7.70) of the projected field operators is not canonical, one cannot obtain a straightforward diagrammatic perturbation expansion. Instead, one tries to solve the equation of motion of G perturbatively. It turns out not to be too difficult to do this to order V^2. Consider the equation motion $G_{\alpha\beta}$

$$\frac{i\partial}{\partial t} G_{\alpha\beta}(t) = <n_{\alpha,\sigma}>\delta(t)\delta_{\alpha\beta} + \langle\!\langle \varphi_{\alpha,\sigma}; c_{\beta,\sigma}^+ \rangle\!\rangle \tag{7.74}$$

where $\varphi_{\alpha,\sigma}$ is defined by

$$\varphi_{\alpha,\sigma} = \left[c_{\alpha\sigma}, \sum_{k\sigma'} (V_{ak}c_{a\sigma'}^+ c_{k\sigma'} + h.c.) \right] \tag{7.75}$$

we find

$$\varphi_{\alpha,\sigma} = \sum_k \{V_{ak}c_{k\sigma}n_{\alpha,\sigma} + \alpha[V_{ak}c_{a\sigma}c_{a,-\sigma}^+ c_{k,-\sigma} + V_{ka}c_{a\sigma}c_{a,-\sigma}c_{k,-\sigma}^+]\} \quad . \tag{7.76}$$

In a matrix notation the Fourier transform of (7.74) can be written

$$(E - H_0)G = A + \langle\!\langle \varphi, c^+ \rangle\!\rangle_E \tag{7.77}$$

where H_0 is a diagonal 2×2 matrix with entries ε_+ and ε_- and A is diagonal with elements $<n_+>$ and $<n_->$. Similarly, differentiating (7.74) again with respect to time (from the right this time), we obtain

$$\{[(E - H_0)G - A](E - H_0)\}_{\alpha\beta} = \langle\!\langle \varphi_{\alpha,\sigma}; \varphi_{\beta,\sigma}^+ \rangle\!\rangle_E + <\{\varphi_{\alpha\sigma}, c_{\beta,\sigma}^+\}> \quad . \tag{7.78}$$

The Dyson equation (7.69) can then be used to write the left-hand side of (7.78) in terms of the self-energy instead of G, with the result

$$[A\Sigma(1 - G_0\Sigma)^{-1}A]_{\alpha\beta} = \langle\!\langle \varphi_{\alpha,\sigma}; \varphi_{\beta,\sigma}^+ \rangle\!\rangle_E + <\{\varphi_{\alpha,\beta}, c_{\beta,\sigma}^+\}> \quad . \tag{7.79}$$

Now the Green's function on the right-hand side is explicitly of order V^2, since each contains a factor of V. The anticommutator also turns out to be of order V^2, so to this order, we can identify the right-hand side as A Σ A.

It now remains to evaluate these quantities to this order. The operator (7.76) is composed of three terms, which we write

$$\varphi_{\alpha,\sigma} = \sum_{k,i=1}^{3} V_{ak}\varphi_{k\alpha,\sigma}^{(i)} \quad . \tag{7.80}$$

We have to consider the combinations $\left\langle\!\left\langle \varphi_{k\alpha,\sigma}^{(i)} ; \varphi_{k'\beta,\sigma}^{(j)^{+}} \right\rangle\!\right\rangle$. They all vanish to zeroth order in V, except for the cases i = j = 1 and i = j = 2. The first of these make a contribution

$$\sum_{kk'} V_{ak}\left\langle\!\left\langle \varphi_{k\alpha,\sigma}^{(1)} ; \varphi_{k'\beta,\sigma}^{(1)^{+}} \right\rangle\!\right\rangle_{E} V_{k'a} = \delta_{\alpha\beta}<n_{\alpha}>\Sigma_{V}(E) \tag{7.81}$$

to A Σ A in (7.79); the second gives

$$\left\langle\!\left\langle \varphi_{\alpha,\sigma}^{(2)} ; \varphi_{\beta,\sigma}^{(2)^{+}} \right\rangle\!\right\rangle_{E} = \frac{1}{2} \alpha\beta \ \Sigma_{V}(E) \quad . \tag{7.82}$$

[If we ignore the second contribution, we are led back to our naive result (7.66).] The anticommutator in (7.79) is also straightforwardly evaluated (see [7.45] for details) to order V^2:

$$<\{\varphi_{\alpha,\sigma};c_{\beta,\sigma}^{+}\}> = \alpha\beta \sum_{k} |V_{ak}|^{2} \left[\frac{f(\epsilon_{k})-f(\epsilon_{+})}{\epsilon_{k}-\epsilon_{+}} - (+ \rightarrow -) \right]$$

$$\equiv \frac{1}{2} \alpha\beta\Delta \quad . \tag{7.83}$$

Thus, combining (7.81-83), we find a self-energy

$$\Sigma(E) = A^{-1}[A\Sigma_{V}(E) + \frac{1}{2} (1 - \tau_{1})(\Sigma_{V}(E) + \Delta)]A^{-1} \quad . \tag{7.84}$$

(τ_1 is the first Pauli matrix in this two-dimensional space.) When (7.84) is used in (7.69), one finds that, to lowest order in V, G = $\sum_{\alpha\beta} G_{\alpha\beta}$ has resonances at ϵ_a and ϵ_a + U, but that their widths are twice the naive guess of $|\text{Im}\{\Sigma_{V}(\epsilon_a)\}|$ and $|\text{Im}\{\Sigma_{V}(\epsilon_a + U)\}|$. (If the poles lie outside the substrate band limits, the resonances become delta-function bonding or antibonding spikes). Figure 7.13 shows the adatom density of states for a case where the poles lie just outside the substrate band. The shape of the spectrum is quite similar to that obtained in the spin fluctuation model (Fig.7.11) (although the width of the resonance has to be put in ad hoc) except that this approximation misses the Kondo resonance at E_F. This is to be expected, since the approximation is only good to order V^2, and Kondo effects only come in at order $J^2 \propto V^4$ [7.48].

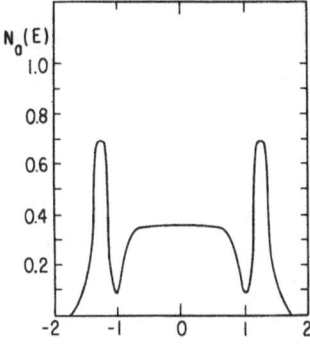

Fig.7.13. Adatom density of states calculated by Brenig-Schönhammer weak-bonding method (V = 0.6). Adapted from [7.45]

The foregoing sort of approximation is correct when V is the smallest energy in the problem. This is the classical exchange limit [7.48] in the limit of large substrate bandwidth W. However, as we have stressed, in chemisorption theory we often need to deal with strong-bonding situations where V is as large as or larger than W. BRENIG and SCHÖNHAMMER treat this case by evaluating expectation values like those which occur in (7.79), not in the V = 0 ground state, but in the Heitler-London surface molecule state

$$|\Phi_{sm}> = |\Phi_{is}> \frac{1}{\sqrt{2}} (c^+_{a\uparrow}c_{0\downarrow} + c^+_{0\uparrow}c_{a\downarrow})|0> \quad . \tag{7.85}$$

Here $|0>$ is the Wannier orbital (or group orbital) which bonds to the adatom, and $|\Phi_{is}>$ is the Fermi sea ground state of the "indented solid," that is, the rest of the substrate. (This procedure is exact in the symmetric ($E_F = \varepsilon_a + U/2$) surface molecule limit.) One finds then a self-energy of the form (7.84) except that the term proportional to $(1-\tau_1)$ is four times as large, and the form of Δ is modified

$$\Delta_{sm} = [-Re\{\Sigma_v(\varepsilon_a + U)\} -Re\{\Sigma_v(\varepsilon_a)\}]/2 \quad . \tag{7.86}$$

SCHUCK [7.49] has introduced an improved approximation procedure for the surface molecule limit which is correct in both the asymmetric and symmetric W→0 limits, for any value of U. Figure 7.14 shows adatom densities of states for a range of parameters, calculated by both the Brenig-Schönhammer and Schuck strong-coupling approximations.

The surface molecule limit is easy to understand qualitatively without doing any calculations. One finds *four* peaks (or spikes) in the adatom density of states: the bonding and antibonding peaks one would find in Hartree-Fock theory, and two satellite peaks. Thus the upper of the two peaks below E_F describes the removal of an electron from the bonding orbital of the surface molecule, while the (deeper) satellite corresponds to removing one electron from the bonding state and simultaneously raising the other one to the antibonding state.

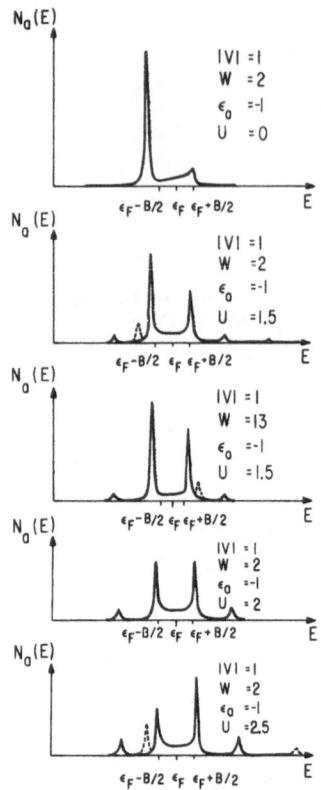

Fig.7.14. Adatom density of states: Brenig-Schönhammer
(dotted lines) and Schuck (solid lines) strong-binding
approximations. [7.49]

It is also possible to include spin-spin interactions between substrate electrons
in the present picture. These are relevant in a theory of chemisorption on ferromagnetic transition metals because they give rise to interesting temperature-depen-
dent effects near the Curie temperature T_c. There the substrate spin fluctuations
and thus the exchange-enhanced spin susceptibility of transition metals like Ni are
large. The effect of critical substrate spin fluctuations on the density of states
and chemisorption energy of hydrogen chemisorbed on such narrow-band transition
metals has recently been examined in a quantum mechanical calculation by KRANZ and
GRIFFIN [7.24].

These interpolative decoupling schemes represent the current state of the art in
solutions of the Anderson model for parameters in the ranges relevant to chemisorption on transition metals. Calculations within these approximations using realistic
substrate surface densities of states are lacking at present, although they involve
no technical obstacles. However, these methods appear capable of generating most
of the important observable structure in the adsorbate spectrum, except for the
possible Kondo resonance near the Fermi level. Improvements in this direction are
quite cumbersome in the decoupling formalism; progress may be possible using the
KEITER-KIMBALL perturbation theory [7.50], or the functional derivative formalism
[7.24].

7.3.4 Variational Methods

The final theoretical method we examine here is a variational technique. Such methods have provided insight into correlation effects in the Hubbard and related models [7.31,51]; their application to chemisorption theory is due to SCHÖNHAMMER [7.52,53].

The motivation for these calculations is the observation, mentioned in Sect. 7.3.1, that Hartree-Fock theory overemphasizes charged adatom configurations. One thus tries to make the relative weights of the different configurations variational parameters in a trial wave function. Formally, this procedure employs the projection operators P_i onto states with i electrons at the adatom site

$$P_0 = (1 - n_\uparrow)(1 - n_\downarrow)$$

$$P_1 = n_\uparrow(1 - n_\downarrow) + n_\downarrow(1 - n_\uparrow)$$

$$P_2 = n_\uparrow n_\downarrow \quad . \tag{7.87}$$

The simplest thing one can try is then to write the variational state as [7.52]

$$|\Phi_0\rangle = \sum_{i=0}^{2} \lambda_i P_i |HF\rangle \tag{7.88}$$

where $|HF\rangle$ is the Hartree-Fock state. The requirement of stationarity of the ground state energy with respect to variation of the λ_i then leads to an eigenvalue problem which is equivalent to the diagonalization of the Hamiltonian in the space spanned by the three states $P_i|HF\rangle$. (These states are orthogonal but not normalized.) The elements of the Hamiltonian matrix involve expectation values of the form $\langle c_{a\sigma}^+ c_{a\sigma}\rangle$, $V\langle c_{a\sigma}^+ c_{0\sigma}\rangle = \sum_k V_{ka}\langle c_{a\sigma}^+ c_{k\sigma}\rangle$, and so forth. All of these can be expressed in terms of moments of the occupied or unoccupied adatom density of states in the Hartree-Fock theory. In the case of $\langle c_{a\sigma}^+ c_{a\sigma}\rangle$ this is trivial; for the expressions involving the bond charges, we have, for example see Fig.7.15,

$$\sum_k V_{ka}\langle c_{a\sigma}^+ c_{k\sigma}\rangle = \sum_k V_{ka} \int \frac{dE}{2\pi} G_a V_{ak} G_k^0$$

$$= \int \frac{dE}{2\pi i} \frac{\Sigma_v(E)}{E - \epsilon_a^\sigma - \Sigma_v(E)} = -\frac{1}{\pi} \int_{-\infty}^{\infty} dE(E - \epsilon_a^\sigma) \, \text{Im}\{G_a(E)\}f(E) \quad . \tag{7.89}$$

Thus, all relevant averages can be expressed in terms of zeroth and first moments of $f(E) \, \text{Im}\{G_a(E)\}$ or $[1 - f(E)] \, \text{Im}\{G_a(E)\}$. Explicitly, the Hamiltonian is

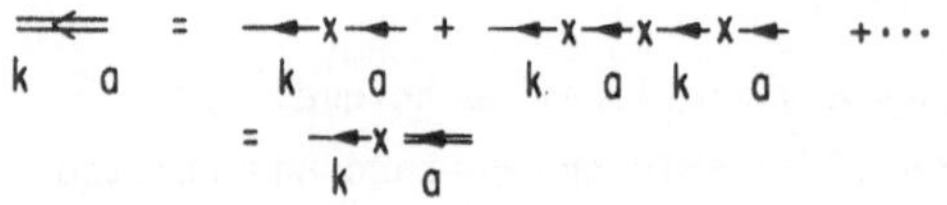

Fig.7.15. Dyson Equation for G_{ka}

$$H = E_{HF}\mathbf{1} + \begin{pmatrix} Un^2 + 2nA & -\sqrt{2n(1-n)}A & 0 \\ -\sqrt{2n(1-n)}A & -Un(1-n) + A & -\sqrt{2n(1-n)}A \\ 0 & -\sqrt{2n(1-n)}A & U(1-n)^2 + 2(1-n)A \end{pmatrix} \quad (7.90)$$

where

$$n = \frac{1}{\pi} \int_{-\infty}^{\infty} Im\{G_a(E)\}f(E)dE \quad (7.91)$$

and

$$A = -\frac{1}{\pi} \int_{-\infty}^{\infty} dE(E - E_F) \left[\frac{f(E)}{n} + \frac{1-f(E)}{1-n}\right] Im\{G_a(E)\} \quad . \quad (7.92)$$

In the foregoing, the state |HF> was taken to be the (nonmagnetic) Hartree-Fock state described in Sect.7.3.1. This can lead to some inaccuracy in treating problems with small V, where the nonmagnetic HF state corresponds to a local maximum of the energy, rather than a minimum. One might try to use the magnetic HF states in (7.88) in place of the nonmagnetic one, but this raises worries of the sort we discussed earlier about the instability of magnetic solutions. As an alternative, SCHÖNHAMMER proposed to let |HF> be the ground state of an effective one-electron Hamiltonian [7.53]

$$H_1 = \sum_{k\sigma} \varepsilon_k n_{k\sigma} + \sum_\sigma (\varepsilon_a + Un_0) + \tilde{V} \sum_\sigma (c_{a\sigma}^+ c_{0\sigma} + c_{0\sigma}^+ c_{b\sigma}) \quad (7.93)$$

in which n_0 and $\tilde{V}$ are variational parameters. The trial state (7.88) is also generalized

$$|\Phi_0> = \sum_{i=0}^{2} \lambda_i P_i |HF> + (\lambda_3 P_0 + \lambda_4 P_2)\hat{V}P_1|HF> \quad (7.94)$$

where $\hat{V}$ is the electron transfer part of the Hamiltonian

$$\hat{V} = \sum_{k\sigma} V_{ak} c_{a\sigma}^+ c_{k\sigma} + h.c. = V \sum_\sigma (c_{a\sigma}^+ c_{0\sigma} + c_{0\sigma}^+ c_{a\sigma}) \quad . \quad (7.95)$$

There are thus, in all, seven variational parameters in this improved scheme, rather than three. (In the symmetric case, this number is reduced to four; only λ_0, λ_1, λ_3, and V are independent.)

SCHÖNHAMMER obtained quite accurate values of the binding energy in this way, as indicated by comparison with the exact results (obtained by EINSTEIN) [7.54] for substrates consisting of finite chains of atoms. For a three-atom chain, the fractional error is less than 10^{-3} over a wide range of V (for U = 5W/4.). Figure 7.16 shows the calculated binding energy as a function of V for the two variational ansätze and the Hartree-Fock solution. In view of this noteworthy success, it would be very interesting to see calculations of G_a or the adsorbate density of states using these ideas. (Indeed, we argued earlier that the principal rationale for studying model Hamiltonians rather than proceeding in the density functional formalism was the possibility of doing such calculations.) Unfortunately, none have been done so far.

The short chain model also permits comparison of the spin-polarized (unrestricted) and unpolarized (restricted) Hartree-Fock approximations with two other simpler approximations. For weak V, the adatom-chain coupling can be treated by second-order perturbation theory. For strong V, the last atom of the chain forms a surface molecule with the adatom; this dimer then is rebonded to the indented chain by second-order perturbation in the chain nearest-neighbor hopping. As shown in Fig.7.17, Einstein found that 1) it is easy to interpolate between these two limits; 2) this interpolation does far better than Hartree-Fock in matching exact results; and 3) the interpolation seems to overestimate the binding energy, so that in conjunction with a variational calculation it brackets the exact results.

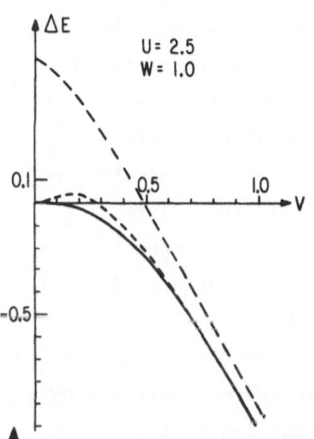

Fig.7.16. Binding energy for absorption on an infinite chain (semielliptic $\rho^0(E)$): long-dashed curve: spin-polarized Hartree-Fock; short-dashed curve: 3-parameter ansatz (7.88); solid curve: 7-parameter ansatz (7.94). [7.53]

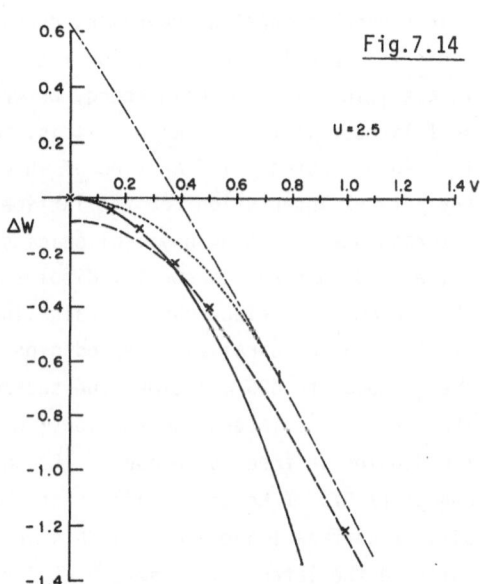

Fig.7.14

Fig.7.17. Interaction energy (negative of binding energy) vs hopping parameter V for intra-adatomic Coulomb U = 2.5. All energies are in units of 2T, where -T is the interatomic hopping parameter of the three-atom "bulk" chain. Shown are exact points derived by computer (x's), weak limit results (solid curve), rebond surface molecule (long-dashed curve), and unrestricted (short-dashed curve) and restricted (dash-dotted curve) Hartree-Fock. The small vertical line indicates where unrestricted solution reduces to the restricted one

The computational solution of short-chain models has also not been fully exploited. Some efforts were directed toward explicitly including overlap in the model [7.55]. In this way, one could examine the valence-bond picture more closely. Moreover, diagonalization routines typically produce not just eigenvalues but also the associated eigenvectors in terms of the basis states. One can then write down the spectral density function for G_a (or any other component of G) in terms of the eigenfunctions of the ground state and of each state in the singly ionized manifold [7.55].

7.4 Adatom-Adatom Interactions [7.55]

While theorists have typically focused on single adatom situations, real systems involve large numbers of adsorbed species. In this section, we give a brief discussion of adatom-adatom interaction effects, beginning with low coverage and working up to a full monolayer.

At low coverages, we find one of nature's best realizations of a two-dimensional lattice gas. The binding energy per adatom ranges from 1 to several eV; it is the largest energy in the problem. For moderate temperatures, we can be fairly sure of constant coverage, provided the adatoms do not sorb into the bulk. (This is a problem, however, for the hydrogen and for less cohesive transition metals.) The difference in energy between various possible symmetry sites is the next largest in the hierarchy; it is, roughly, the diffusion barrier, and is thus some substantial franction of an eV [7.56]. This energy must be greater than any lateral interaction for the lattice gas model to be valid. These lateral interactions have many physical origins. If there is substantial charge transfer between adatoms and substrate, dipoles will arise normal to the surface. (Usually, the adatom gains electrons; for alkaline and inert gas adsorbates, it loses electrons.) Two parallel-oriented dipoles will repel each other with an interaction energy equal to 625 meV times the squared dipole moment (in debyes), divided by the cube of the lateral separation (in angstroms). This repulsion is twice that for dipoles in free space due to the substrate screening in the form of the image charge [7.57]. (The second adatom sees both the adatom-surface dipole and the similar surface-image dipole.) This interaction is usually negligible, but does dominate the lateral interaction at very large separations.

Direct interactions between adatoms, i.e, those involving overlapping adatom orbitals are ipso facto comparable to chemical bond energies. For example, as a diatomic molecule approaches a surface, the adatom-substrate attraction may dissociate the dimer. Depending on how the dimer bond weakens with increasing adsorption bond strength, this dissociation may or may not have an activation barrier, as has been shown (schematically) by DAVENPORT et al. [7.58]. The details of this process have been considered in various numerical schemes [7.59] involving cluster calculations. The role of the semi-infinite substrate is perturbative in this problem; HYMAN [7.60] has showed formally how one might treat the important correlation effects thoroughly within a small cluster and then reembed this cluster into an (indented) substrate. GRIMLEY and PISANI [7.19] have focused on such clusters containing single adatoms with a self-consistent-field LCAO-MO scheme. Progress on this problem is obviously of vast importance to the understanding of catalysis. This approach may also have value in describing physisorbed overlayers. Here the direct lateral interactions determine the two-dimensional net, with the substrate reducing the gas phase van der Waals attraction by roughly 20% [7.61].

In the low coverage lattice gas regime, the adatoms are sufficiently separated that there is negligible adatom-adatom overlap. Their principal mode of interaction, when the adsorption-induced dipole is small and hence the adsorption band is pre-dominantly covalent, is an indirect coupling via the substrate electronic states. In many ways this indirect interaction is analogous to the RKKY interaction between dilute bulk magnetic impurities [7.62]: the coupling can be attractive or repulsive depending on whether the two impurity states are in phase or out of phase (bonding or antibonding), respectively, after matching onto the intermediate substrate eigenstates. Both interactions fall off rapidly with separation. In the RKKY case the interaction goes as $r^{-3} \cos(2k_F r)$ (this is the asymptotic form at large sep-arations for a free-electron host). For a pair of adatoms on jellium, the correspond-ing expression has r^{-5} as the leading power, at large r, and a directional depend-ence related to the parallel Fermi velocity, as found long ago for the bulk [7.63]. However, for the small separations at which it is significant this pair inter-action is not merely carried by the highest-lying state but by all the occupied states. With increased separation, the matching of adatom wave functions to the propagating substrate states involves increasing interference. The strength con-sequently falls off much faster than an inverse power, more nearly exponentially. At separations where the r^{-5} asymptotic behavior is reached, the interaction is negligibly small.

Formally, one can straightforwardly derive the change in energy for two identical adsorbed atoms adsorbed above sites 0 and n [7.15]. One begins by adding to (7.14) a similar Hamiltonian with 0 replaced by n

$$\hat{V} = V[c_0^+ c_a + c_a^+ c_0 + c_n^+ c_b + c_b^+ c_n] \quad . \tag{7.96}$$

One can then begin with the general formal expression for the change in density of states

$$\Delta \rho(E) = \sum_j [\delta(E - E_j) - \delta(E - \varepsilon_j)]$$

$$= \frac{1}{\pi} \, \text{Im} \left\{ \sum_j [G_{jj}(E - i\delta) - G_{jj}^0(E - i\delta)] \right\} \tag{7.97}$$

where ε_i are the eigenvalues of the unperturbed Hamiltonian H_0 and E_i are those of the full Hamiltonian $H_0 + \hat{V}$. Equation (7.97) can be manipulated into the form [7.15,64]

$$\Delta\rho(E) = \frac{1}{\pi} \, \text{Im} \left\{ \frac{\partial}{\partial E} \ln \det \left(\frac{E - H - i\delta}{E - H_0 - i\delta} \right) \right\}$$

$$= \frac{1}{\pi} \, \text{Im} \left\{ \frac{\partial}{\partial E} \ln \det[1 - G_0(E - i\delta)V] \right\} \tag{7.98}$$

so that (with the help of an integration by parts)

$$\Delta E = - \frac{2}{\pi} \int_{-\infty}^{E_F} \text{Im}\{\ln \det[1 - G_0(E - i\delta)\hat{V}]\} \quad . \tag{7.99}$$

For the perturbing Hamiltonian (7.96) the resulting determinant (which is effectively only 4×4) is easily evaluated

$$\det(1 - G_0\hat{V}) = (1 - V^2 G_a^0 G_{00}^0)^2 (1 - V^4 G_{0n}^{02} G_a^2) \quad , \tag{7.100}$$

where G_a is the adsorbate Green's function for the single-adatom problem. $(G_{nn}^0 = G_{00}^0$, and $G_{0n}^0 = G_{n0}^0$ by virtue of the inversion symmetry of the surface Brillouin zone.) The factorization naturally separates the pair interaction from single-adatom effects, eliminating the need for a subtraction of two large energies. Thus, this energy can be written simply as

$$\Delta E_{pair} = \frac{2}{\pi} \int_{-\infty}^{E_F} dE \, \text{Im}\{\ln[1 - V^4 G_{0n}^{02}(E - i\delta)G_a^2(E - i\delta)]\} \quad . \tag{7.101}$$

By taking the lowest order term in the expansion of the logarithm (which is in general a good approximation) and ignoring G_a (which is not) we recover the equivalent of the RKKY result. Even in this simple limit we find that the pair interaction is isotropic only when the substrate band structure itself is, which is rarely the case for metals active in adsorption.

This treatment implicitly positioned the adatoms in the "atop" position, directly above substrate atoms. The procedure to generalize to bridged or to centered sites was outlined above [see, e.g., (7.21,22)]. Calculations implementing this procedure show that with all other parameters held fixed the pair interaction energy depends sensitively on the binding site symmetry, not only in strength but even in sign! This result confirms the assertion that the adsorption system should be viewed as a lattice gas rather than a two-dimensional liquid. The values of the interaction energy can vary over roughly an order of magnitude for each separation. Typical numbers for the magnitudes of nearest, next-nearest, and third nearest pair energies are 0.6, 0.12, and 0.05, respectively, in units of the substrate bandwidths (typically 6 to 12 eV).

While the above calculations were performed for the case of the (100) face of a single-band, simple cubic crystal, BURKE [7.65] obtained similar answers in a model study of W on the (100) and (110) faces of W. His approach considers only the d states, and uses SLATER-KOSTER [7.66] matrix elements between nearest and next-nearest sites (i.e., 6 parameters) which are simply scaled up from narrow-band values. Unfortunately, his attempt to obtain the experimental result (0.3 eV) for a nearest-neighbor W dimer on W (110) does not succeed; his answer is five times too large.

A remarkable feature of the pair interaction is that the self-consistency and correlation corrections to this simple treatment, which were seen to have such impact in the one-adatom problem, play a minor role here.

Self-consistency can only be pursued in a crude way in the LCAO models. Since the electron orbitals are fixed a priori, Poisson's equation is not suitable; instead, the Friedel sum rule is invoked to demand charge neutrality within some finite range of the adatom, often on each site [7.67,68] or within a small surface molecule [7.67,69], and thus artificially short. The essence of these schemes is that the local levels will tend to shift by $I_i|n_{i,-\sigma}>$ to impede charge transfer to or from the site they represent. Another tack treats small clusters carefully with the background ignored [7.70] or embedded in some simple background [7.19]. While useful in studying direct interactions, the clusters become unwieldy for moderately separated adatom pairs, since the distance from each adatom to the edge of the cluster should at least equal the interaction distance. GRIMLEY and WALKER [7.71] observed that while sizeable charge transfer might occur during the adsorption of an atom, little more should occur as a function of the relative orientation of two of them. This feature arises from the weak nature of the pair interaction. If the local one-electron adatom levels are reasonably parametrized (as by PANDEY's prescription [7.72] of fitting values to achieve known results for small analogous molecular clusters), the pair interaction should work out adequately even when single adatom results are somewhat unsatisfying. Moreover, explicit calculations show the pair interaction to depend only weakly on these levels. Correspondingly, in an extension of the single-adatom theory described in Sect. 7.3.3, SCHÖNHAMMER et al. [7.73] showed that correlation had hardly any effect on the pair interaction energies. Specifically, in their density of states calculation, they plotted the local density of a states on an adatom (Fig.7.18). The correlation-induced peaks (corresponding to atomic levels in the weak V limit and shake-up processes more generally) are apparently unaffected by pair interactions. Both the bonding and antibonding (molecular orbital or Hartree-Fock-like) peaks are slightly

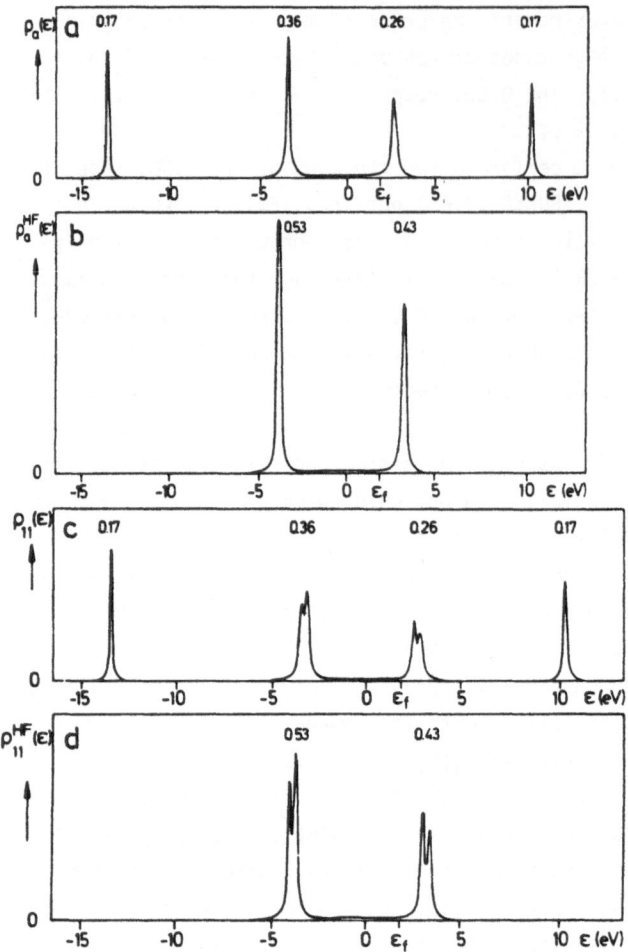

Fig.7.17a-d. Adatom densities of states calculated by SCHÖNHAMMER et al. (a) isolated adatom, BS approximation; (b) isolated adatom, HF; (c) two-adjacent adatoms atop nearest neighbor sites, BS approximation; (d) same, HF approximation. [7.73]

split, with no change in overall weight. The larger subpeak lies nearer the center of the substrate band. While insufficient data are presented for quantitative statements it is apparent that Hartree-Fock theory does a respectable job of indicating the qualitative *changes* in the density of states due to pair effects.

With the notable exception of field ion microscopy, experimental observations usually require large numbers of adsorbates to produce observable signals. If there are only dipolar repulsions between (partially ionized) atoms (as in the case for alkali adsorbates) or if the significantly strong indirect interactions are all repulsive, adatoms will stay as far apart as the lattice model permits, and ordered patterns will arise for coverages near the saturation. For instance, on a square face, one would expect (2×2) and centered (2×2) structures only near quarter and

half monolayer coverage, respectively. Attractive interactions as well as repulsive
ones can be involved in the formation of ordered overlayers. For example, a centered
(2 × 2) adlayer might be the product of a repulsive nearest-neighbor and attractive
next-neighbor interaction. When attractions are important, islands of the ordered
phase, with a dilute "gas" of adatoms in the rest of the plane, will form at low
coverages. In the presence of islands, even at low coverages spectroscopies should
show densities of states, etc., characteristic of saturated coverage rather than
of isolated adatoms.

The general problem of phase transitions of adatoms has recently attracted great
interest. Monte Carlo computations have generated phase diagrams for lattice gases
with one or two nearest-neighbor interactions [7.74]. Moreover, the associated
Landau-Ginzburg-Wilson free energy expression is often not that of the Ising model,
but rather the XY or the Potts model, depending on the substrate symmetry [7.75].
There are, however, many complications in real systems. For example, configurations
at low coverage should be one huge island surrounded by a dilute gas, while experiments on oxygen on the (110) face of tungsten [7.76] suggest a maximum island
size of 60 Å. Surface heterogeneity or diffusion limitations may be responsible.
Also, since "occupied" and "unoccupied" in a lattice model correspond to "spin-up"
and "spin-down" in an Ising model, such models predict phase diagrams to be symmetric about half monolayer coverage. Significant three-adatom interactions could
account for deviations from this behavior.

From a theoretical standpoint, the problem of a complete monolayer is easier —
particularly if adatoms do not overlap significantly — than the single-adatom case
since the adsorbed system then has the full symmetry of the substrate, and $k_{\parallel}$
(the wave number in the surface plane) is a good quantum number. In the LCAO formalism used above, we find the total change in density of states per adatom for a
1 × 1 overlayer to be

$$\Delta\rho_{1 \times 1}(E) = -\frac{1}{\pi N} \text{Im} \left\{ \sum_{k_{\parallel}} \ln[1 - V^2 G_a^0 G_s^0(k_{\parallel})] \right\}$$

$$= \Delta\rho_{single}(E) - \frac{1}{\pi N} \text{Im} \left\{ \sum_{k_{\parallel}} \ln[1 - V^2 G_a (G_s^0(k_{\parallel}) - G_{00}^0)] \right\} \qquad (7.102)$$

where N is the number of adatoms (or, equivalently, the number of surface atoms
and the number of $k_{\parallel}$ vectors) and $G_s^0(k_{\parallel})$ the two-dimensional Fourier transform of
the clean surface Green's function G_{0n}^0. The second form of (7.102) shows that it
is readily possible to separate formally the single adsorption process from the
lateral interaction effects. No further simplification is possible, except in a
perturbation series in V, which misemphasizes the relevant physics. As suggested in
Fig.7.19, the monolayer tends to sharpen the bonding and antibonding peaks, introducing van Hove-like singularities. The depleted region in the center of the band

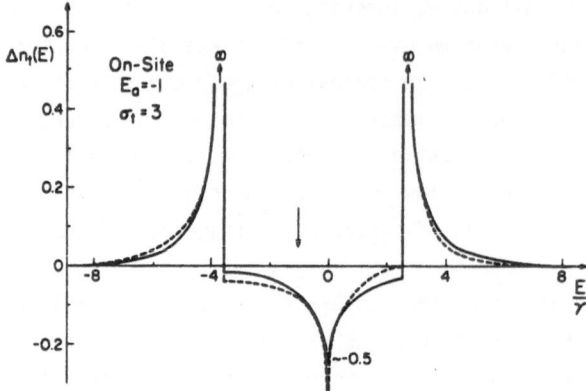

Fig.7.19. Total change in density of states for a monolayer in the "atop" position on the (100) face of bcc lattice. (Energy unit = bandwidth/16) Solid curve: non-self-consistent calculation with $\varepsilon_a = 1$; dashed curve: $\varepsilon_a = 0$ (trivially self-consistent). [7.69]

is characteristic of a bulk rather than a surface site. Physically we are now re-moving the whole top layer of orbitals in forming surface molecules. The surface of the crystal is essentially moved one layer down, which is equivalent to removing a bulk layer. For transition metals on transition metals, direct interactions will also enter and should play a major role. For light gas adsorbates (e.g., H, O, N) the dimer internuclear separation is of order half the lattice spacing, so that direct effects may sometimes be negligible, especially on the less dense surfaces. On the other hand, for a two-dimensional net of O atoms with spacings appropriate to a (1×1) overlayer on Ni(100), LIEBSCH [7.77] found that the p levels broadened to a band of width nearly 5 eV. When the spacing is increased (by $\sqrt{2}$) to that of a centered (2×2) overlayer, the bandwidth decreases to about 1.3 eV, a relatively modest fraction of a typical transition metal substrate.

Since in the case of a centered (2×2) overlayer the real space primitive cell is doubled in area, the corresponding surface Brillouin zone is halved. For each $k_\shortparallel$ in the new surface Brillouin zone, there is another $k_\shortparallel$ outside the new zone (but inside the old one) related to it by a new reciprocal lattice vector. We must fold these excluded $k_\shortparallel$'s back into the smaller zone in a manner reminiscent of the formation of optical phonon modes. The upshot is to replace $G_s^0(k_\shortparallel)$ in (7.102) by $\{G_s^0(k_\shortparallel) + G_s^0[k_\shortparallel - \pi(1,1)]\}/2$. Now N must be interpreted as the number of adatoms, i.e., half the number of substrate sites. The resulting lateral inter-action energy per adatom is well described, at least qualitatively and to a large extent semi-quantitatively, by twice the second neighbor pair interaction. The as-sociated total change in density of states per adatom looks rather like that for a monolayer, but with the new feature of a sharp narrow peak near the center of the band. This peak can be interpreted if we view the centered (2×2) overlayer as being formed by desorbing half of a full monolayer. Then the surface orbital associated with the desorbed adatom must rebond to the substrate (indented by a whole layer in the monolayer case), producing the narrow peak at its atomic energy. More detailed analysis of these effects can be achieved using a scattered-wave formalism in a layer format, with two-dimensional periodicity [7.77,78].

7.5 Photoemission

Much of the impetus to theorize about chemisorption-induced changes in electronic properties has come from dramatic progress in experimental probes of these properties, rooted in advances in ultra-high-vacuum technology. We therefore will comment on how the formal quantities discussed here relate to the curves measured in the laboratory. On the other hand, because of the availability of a thorough review on various spectroscopies, [7.79] we limit our specifics to the simple LCAO models. The usual goal is some measure of the one-electron density of states. To probe the *filled* states in a direct way, one must somehow get the electron in question out of the material, either by exciting it to an unbound state with a photon (photoemission) or by electrostatically lowering the barrier which holds it inside (field emission). Other techniques, which require deconvoluting a two-electron process, include Auger and ion neutralization spectroscopies. Gauging the *unfilled* densities of states requires the analogous process of appearance potential spectroscopy, in which a core electron is electronically excited to some vacant state. These two-electron processes are viable probes of density of states because, due to the sharp nature of the Fermi surface, differentiated spectra are dominated by one electron pinned there and the second monitoring the density of states at the energy-conserving energy. We will restrict our discussion to photoemission as the most widely used, flexible, and direct technique.

Photoemission involves the absorption of a photon by an electron in an occupied state of the sample. This electron is excited to an energy-conserving unoccupied state via which it may escape from the sample and be collected. To gain insight into this process, we invoke a number of simplifying approximations whose validity we will subsequently discuss. First we assume the photon can be represented by a spatially uniform vector potential $\underline{A}$, which we choose in the Coulomb gauge $\underline{\triangledown} \cdot \underline{A} = 0$ to avoid symmetrization problems. The perturbing Hamiltonian has the simple dipole form $(e/mc) \, \underline{A} \cdot \underline{p}$. Second we represent the final-state electron by a plane wave with wave vector $\underline{k}$. Third, we model the initial-state electrons by single-band tight-binding wave functions. We neglect all explicit many-body effects.

According to the Golden Rule, the photoemission rate is proportional to

$$\sum_\kappa \, |<k|\underline{A} \cdot \underline{p}|\kappa>|^2 \delta(\varepsilon_k - \omega - \varepsilon_\kappa) \quad . \tag{7.103}$$

Since $|k>$ is imagined a plane wave at this stage, the matrix element simplifies
to $\underline{A} \cdot \underline{k} < \underline{k} | \kappa >$. In terms of the (generalized) Wannier orbitals $|R_i>$ centered on
atomic sites,

$$|R_i> = \sum_\kappa |\kappa><\kappa|R_i> \tag{7.104}$$

(7.103) becomes

$$\sum_{ij} |\underline{A} \cdot \underline{k}|^2 <\underline{k}|R_i><R_j|\underline{k}> \sum_k <R_i|\kappa><\kappa|R_j>\delta(\epsilon_{\underline{k}} - \omega - \epsilon_\kappa)$$

$$= \pi^{-1}|\underline{A} \cdot \underline{k}\phi(\underline{k})|^2 \sum_{ij} \exp[-i\underline{k} \cdot (\underline{R}_i - \underline{R}_j)] |Im\{G_{ij}(\epsilon_k - \omega)\}| \tag{7.105}$$

where $\phi(\underline{k})$ is the Fourier transform of the Wannier orbital (assuming them all identical). (Clearly for chemisorption systems, this might be valid only in the case
of "self-adsorption" [adatom same as substrate element].) In general surface orbitals will also distort. The orbital factors can then be attached separately to
each summation within a layer.

The angle over which photoelectrons are collected substantially influences the
interpretation of the intensity profiles as a function of ϵ_k. Traditionally one
had to collect over a large spherical angle to obtain adequate intensity. Such
angular-averaged data give some feel for the local DOS. More importantly, since
this picture is quite clouded by the short mean free path of the photoelectron,
difference spectra [adsorbed surface spectrum minus clean surface spectrum] give
a better feel for $\Delta\tilde{\rho}$, which, as we have seen, is concentrated near the surface.
Angular-resolved measurements bring out structure that is more band-like, i.e.,
localized in k space rather than real space. This experimental scheme provides a
more detailed probe of adsorption properties and, when used in conjunction with
suspected symmetries of the system, can provide a powerful means of identifying
electronic orbitals and molecular orientations [7.79]. When only the photoelectrons
emitted normal to the surface are collected, the data should resemble field
emission, which samples only the center of the surface Brillouin zone. Actual
comparisons emphasize that the collection angle in the angular-resolved mode must
be quite small in order to pick out all resonances [7.80].

To illustrate these general remarks, we first suppress the factor preceding
the double summation in (7.105). In the angular-averaged case, we imagine that
electrons of constant $|\underline{k}|$ are collected over the entire hemisphere above the
sample. Since $G_{ji} = G_{ij}$, we can write the summand as $(2\pi)^{-1} \cos[k \cdot (R_i - R_j)]$
$Im\{G_{jj}\}$; performing the hemispherical average changes the cosine to a factor of
$x_{ij}^{-1}\sin x_{ij}$, where $x_{ij} \equiv |\underline{k}||\underline{R}_i - \underline{R}_j|$. For uv photons and realistic interatomic
distances, this factor produces roughly an order of magnitude reduction of a term

with i and j nearest neighbors compared with a diagonal, j = i, term. Thus, upon angular averaging the summation in (7.105) reduces to approximately a sum over the local density of states.

As alluded to above, in uv photoemission the ejected electron lies in the moderately low energy range (10-100 eV) in which the electron scatters strongly off core electrons and has a short mean free path due to a great tendency to excite electron-gas loss modes. The plane-wave approximation is really justified only for higher energy electrons (X-ray regime). (For very low energy electrons the mean free path again grows, but the strong interaction with the crystal produces Bloch-like states.) The final-state wave function should properly be described by the same complicated wave function used in analyzing LEED experiments, as has been discussed in detail by FEIBELMAN and EASTMAN [7.81]. In essence, the simple decomposition of (7.105) must be replaced by the matrix element of $\underline{A} \cdot \underline{p}$ between the appropriate LEED function and the initial-state orbital (here a tight-binding combination of atomic orbitals). Since the angular averaging will integrate away many of the k-dependent wiggles, these effects are relatively less noticeable than the short mean free path. As a gauge of these effects, one can sum the local density of states layer by layer and apply an exponential damping factor; the summation in (7.104) becomes, *in the case of a difference spectrum*, due to adsorption of a single atom

$$\rho_a(E) + \sum_{m=1}^{\infty} \Delta\rho_m(E)\exp[-\lambda(m-1)] \quad , \tag{7.106}$$

where $\Delta\rho_m$ is the change in density of states in the mth layer. This expression will be dominated by ρ_a and $\Delta\rho_1$ anyway, so the damping factor is not crucial [7.12, 82]. [The λ here is greater than that deduced from the mean free path since an (implicit) angular average has been performed. Electrons from some layer emerging at some near grazing angle travel a far longer path than those from the same layer propagating normal to the surface; hence they should be more heavily damped. The λ gives some average value.]

Correlations, even strong ones, can be incorporated into this description quite simply for strict Anderson models (those containing only the adatom intra-atomic Coulomb interaction $Un_{a\uparrow}n_{a\downarrow}$). Then the ρ_a and $\Delta\rho_m$ of (7.106) must be calculated taking these correlations into account (as, e.g., in Sects.7.3.2 or 3), but the expression for the measured spectrum in terms of these densities of states is unchanged.

In the angular-resolved mode, just one value of $\underline{k}$ is measured. To illustrate the different physics involved, we comment here on the easiest case of the clean substrate. (See [7.12] for comments on the more general case.) Returning to the plane-wave approximation and (7.104), we note first that the "Im" operation can be pulled outside the double summation, since only the cosine part of the exponentials contributes (by the i↔j symmetry used in the angular averaged derivation). The work of KALKSTEIN and SOVEN [7.14] allows us to express the Green's function for the semi-infinite system in terms of that for the infinite lattice, with the result

$$-(N_{\shortparallel}/\pi) \; \text{Im} \sum_{n_i=1}^{\infty} \sum_{n_j=1}^{\infty} \sum_{g_{\shortparallel}} \exp[-ik_{\perp}d(n_i - n_j)]\{G(|n_i - n_j|, \underline{k}_{\shortparallel} + g_{\shortparallel};\varepsilon_k - \omega)$$

$$-G(n_i + n_j,\underline{k}_{\shortparallel} + g_{\shortparallel};\varepsilon_k - \omega)\} \tag{7.107}$$

where $N_{\shortparallel}$ is the number of surface atoms, d is the lattice spacing, and $g_{\shortparallel}$ are the surface reciprocal lattice vectors. The sum over $g_{\shortparallel}$ simply gives the possibility of more than one emergent direction; we neglect such surface Umklapp processes in the following. The contribution due to $G(|n_i - n_j|,k_{\shortparallel};)$ can be broken into two parts: 1) $N_{\perp}$ times its Fourier inversion in $k_{\perp}$, producing a final result going like $N_{\shortparallel}N_{\perp}\delta(\varepsilon - \varepsilon_k)$ (i.e., the bulk band structure times the number of sites) and 2) a correction term of order 1 (rather than $N_{\perp}$) due to the fact that the layer summations are semi-infinite rather than infinite. For the single band model one can exactly calculate this term as well as the contribution from the reflected wave [7.12], also of order 1. In the no-damping limit we get behavior character-istic of the bulk bands. This viewpoint is embodied in the old "three-step model" of photoemission [7.83], in which a bulk-like optical excitation within an in-elastic mean-free-path length of the surface is followed by propagation to the surface and escape into the vacuum. For short mean free paths only the top few layers contribute, so that the correction terms to the bulk term become comparably important. A similar analysis can be applied to the adsorption difference spec-trum [7.12]. The photocurrent breaks naturally into three contributions: 1) an adatom term, 2) a change in the substrate term, and 3) an adatom-substrate inter-ference term.

To the extent that the mean free path $\lambda(\varepsilon)$ is so long that bulk processes do dominate, the three-step picture suggests that the excitation process is direct and hence, as in optical studies, probes the joint density of states rather than the initial (or "single-particle") density of states. In the low-energy regime (say less than 10 eV photoelectrons), this feature can complicate the inter-pretation. For high energies (100's of eV), λ again increases, but here the Fourier components of the lattice potential are weak and the band structure consequently nearly free-electron-like and smoothly varying (on the scale of λ^{-1}). X-ray photo-emission is thus known to measure the initial density of states. FEIBELMAN and EASTMAN [7.81] have formalized this criterion for initial rather than joint den-sity of states as that the separation in k between available states at energy be smaller than the imaginary part of k, produced be inelastic effects [7.84].

With the advent of dedicated facilities producing high intensity synchrotron radiation, it has become possible in recent years to go beyond the handful of energies available from gas lamps to generate spectra over a continuum of uv photon energies ω. One then plots these spectra simultaneously by versus initial state energies (i.e., by putting the spectrum threshold near the zero of the abscissa). Features which remain relatively unchanged from curve to curve are attributed to initial-state effects. Those which shift linearly in ω from curve to curve are called final-state effects. Other variations can be ascribed to such complications as variations of initial-state wave functions in energy [i.e., effects due to $\phi(\underline{k})$ in (7.105)] and modifications of the vector potential $\underline{A}$ due to the dielectric response of the sample (especially at lower photon energies).

LEED theorists have led the way in studying final-state effects [7.85-87]. Their analyses are based on a scattering theory formulation of photoemission begun by ADAWI [7.88], greatly expanded by MAHAN [7.89] for nearly free electron substrates, and generalized somewhat by FEIBELMAN and EASTMAN [7.81]. The results often show strong dependence on photoelectron energy and direction. LIEBSCH [7.85] found in fact that the positions of intensity extrema for emission from localized adsorbate levels are determined solely by the substrate symmetry (linked to crossing of free-electron bands); the adatom position produces phase factors determining the relative intensity of these extrema. The fraction of the photocurrent which has been back-scattered is in the range 1/6 to 1/3 of the total at the detector. In contrast to the case of LEED, however, multiple scattering produces little change form the single scattering result, perhaps because the wave source is localized here.

When the initial state is very localized, the independent-particle formulation espoused above is unreliable since there is no natural means to include relaxation effects of the final-state hole. Moreover, the corrections discussed above to simple treatments form an unsatisfying, patchwork picture. Much theoretical effort has been recently devoted to developing a systematic many-body approach to photoemission. The starting point is SCHAICH and ASHCROFT's description [7.84] of photoemission in terms of a three-current correlation function. The idea is to evaluate the expectation value of the photoemission current operator outside the sample (at the detector) using the interaction picture. As in the standard Kubo theory, one expands the exponential of the time evolution operator. Since the $\underline{p}$ of the interaction Hamiltonian can be rewritten as a [paramagnetic] current, each $\underline{A}$ carries along a $\underline{J}$. As the photoemission current is proportional to the intensity of the radiation, i.e., A^2, one must use quadratic instead of just linear response, yielding two currents plus the original. CAROLI et al. [7.90] used KELDYSH's perturbation theory [7.91] for nonequilibrium many-body problems to produce a formal diagrammatic expansion. FEIBELMAN and EASTMAN [7.81] showed how the lowest order diagram reduces to the several independent-particle schemes discussed above. KEITER [7.92] has very recently formulated a diagrammatic perturbation expansion for photoemission from the standpoint of conventional finite-temperature many-body theory.

On the other hand, in many cases the qualitative changes in spectra due to many-body effects may be too small to be observed experimentally. For example, for hydrogen on Ni, SCHÖNHAMMER found this result using an Anderson model parametrized from cluster calculations [7.26].

7.6 Concluding Remarks

It should be apparent to the reader at this point that model Hamiltonian approaches to chemisorption theory can be of great value in understanding important problems inaccessible to local density methods in their present form, but that they have not yet been exploited fully in applications to real experimental data. The basic theoretical issues raised in Sect.7.1 have been settled reasonably well in the context of very simple models (semicircular density of states, etc.) but meaning-ful comparisons with, e.g., photoemission spectra are lacking at present.

In our view, the machinery with which to fill this gap is now available. The initial step is the proper parametrization of the Anderson or generalized Anderson model to be used. It seems that a profitable approach might be to use the results of local density calculations to determine the parameters of the model, as already tried with some success by SCHÖNHAMMER [7.26] and in a different context, by PANDEY [7.72]. In this sort of approach, one absorbs some of the correlation effects (those judged weaker or less important) into the parameters of the model via the local density calculation, leaving the part of the correlation problem judged to be too large for such approximations to be treated specially in an approximate solution of the resulting model. Only through such a procedure does the general correlation problem become amenable to insightful theoretical analysis. Once this step is accomplished, there is the task of calculating the spectra measured in experiments. While it is apparent from Sect.7.5 that progress in this direction, thus far, is quite limited because of the complexity of the photoemission process, some relevant calculations (e.g., difference spectra) seem possible now within the sorts of models we have discussed. We hope that in the near future such calculations will be performed and that they will stimulate further useful interactions between theoreticians and experimentalists.

References

7.1   J.C. Slater: Phys. Rev. *81*, 385 (1951)
        P.O. Löwdin (ed.): *Advances in Quantum Chemistry*, Vol.6 (Academic Press,
        New York 1971)
7.2   W. Kohn, L.J. Sham: Phys. Rev. *140*, 1133 (1965); *145*, 561 (1966)
        P. Hohenberg, W. Kohn: Phys. Rev. *136*, B 864 (1964)
7.3   O. Gunnarsson, B.I. Lundqvist, J.W. Wilkins: Phys. Rev. B *10*, 1319 (1974)
7.4   J.C. Stoddard, N.H. March: Ann. Phys. *64*, 174 (1971)
        N. von Barth, L. Hedin: J. Phys. C *5*, 1629 (1972)
7.5   A.K. Rajagopal, J. Callaway: Phys. Rev. B *7*, 1912 (1973) see ref. (A)
        S.K. Ma, K.A. Brueckner: Phys. Rev. *165*, 18 (1968)
        F. Herman, J.P. Van Dyke, I.B. Ortenburger: Phys. Rev. Lett. *22*, 807 (1969)
        D.J.W. Geldart, M. Rasolt, R. Taylor: Solid State Commun. *10*, 279 (1972)
        D.J.W. Geldart, M. Rasolt: Phys. Rev. B *13*, 1477 (1976)
7.6   O. Gunnarson, M. Jonson, B.I. Lundqvist: Phys. Lett. A *59*, 177 (1976)
7.7   D.C. Langreth, J.P. Perdew: Solid State Commun. *17*, 1425 (1975)
7.8   N.D. Lang, A.R. Williams: Phys. Rev. Lett. *37*, 212 (1976)
7.9   P.W. Anderson: Phys. Rev. *124*, 41 (1961)
7.10   D.M. Edwards, D.M. Newns: Phys. Lett. *24A*, 236 (1967)
        D.M. Newns: Phys. Rev. *178*, 1123 (1969)
7.11   T.B. Grimley: J. Phys. C *3*, 1934 (1970)
7.12   T.L. Einstein: Phys. Rev. B *12*, 1262 (1975)
7.13   S.K. Lyo, R. Gomer: Phys. Rev. B *10*, 4161 (1974); and in *Interactions on
        Metal Surfaces*, ed. by R. Gomer, Topics in Applied Physics, Vol.4 (Springer,
        Berlin, Heidelberg, New York 1975) Chap.2
7.14   D. Kalkstein, P. Soven: Surf. Sci. *26*, 85 (1971)
7.15   T.L. Einstein, J.R. Schrieffer: Phys. Rev. B *7*, 3629 (1973)
7.16   J.W. Gadzuk: In *Surface Physics of Materials*, ed. by J.M. Blakely (Academic
        Press, New York 1975); Surf. Sci. *43*, 44 (1974); Phys. Rev. B *10*, 5030
        (1974)
7.17   These effects have been discussed from a point of view similar to the one
        we take here by A. Madhukar: IBM Research Dept. RC5413 (unpublished)
7.18   See, e.g., R.T.H. Voorhoeve: AIP Conf. Proc. *18*, 19 (1974)
7.19   T.B. Grimley, C. Pisani: J. Phys. C.*7*, 2831 (1974)
7.20   T.B. Grimley, E.E. Mola: J. Phys. C *9*, 3437 (1976)
7.21   R. Gomer: Accounts of Chemical Research *8*, 420 (1975)
7.22   J.R. Schrieffer: In *Dynamic Aspects of Surface Physics*, ed. by F.O. Goodman,
        Proc. Enrico Femi Summer School, Varenna, 1973 (Italian Physical Society,
        1974)
7.23   B. Gumhalter, D. Newns: Phys. Lett. *57A*, 423 (1976)
7.24   H. Kranz, A. Griffin: To be published
        H. Kranz: thesis, University of Toronto (1977)
7.25   W. Brenig: Z. Phys. B *20*, 55 (1975)
7.26   K. Schönhammer: Solid State Commun. *22*, 51 (1977)
7.27   A. Bagchi, M.H. Cohen: Phys. Rev. B *9*, 4103 (1974)
7.28   P.O. Löwdin: J. Chem. Phys. *18*, 365 (1950)
        J.J. Rehr, W. Kohn: Phys. Rev. B *10*, 448 (1975)
        J.G. Gay, J.R. Smith: Phys. Rev. B *11*, 907 (1975)
7.29   For reviews see P.W. Anderson, G. Yuval: In *Magnetism*, Vol.5, ed. by H. Suhl
        (Academic Press, New York 1973) Chap.7
        K.G. Wilson: Rev. Mod. Phys. *47*, 773 (1975)
        G. Grüner, A. Zawadowski: Repts. Progr. Phys. *37*, 1497 (1974)
7.30   A useful review is given by D.L. Mills, M.T. Béal-Monod, P. Lederer: In
        Magnetism, Vol.5, ed. by H. Suhl (Academic Press, New York 1973) Chap.3
7.31   M.C. Gutzwiller: Phys. Rev. Lett. *10*, 159 (1963); Phys. Rev. *134*, A923
        (1964); *137*, A1726 (1975)
7.32   A.C. Hewson: Phys. Rev. *144*, 420 (1966)
7.33   J. Hubbard: Proc. Roy. Soc. A *276*, 238 (1963)
7.34   J. Hubbard: Proc. Roy. Soc. A *281*, 401 (1964)

7.35 J.R. Schrieffer, R. Gomer: Surf. Sci. *25*, 315 (1971)
      R.H. Paulson, J.R. Schrieffer: Surf. Sci. *48*, 329 (1975)
      see also A. Bagchi, M. Cohen: Phys. Rev. B *13*, 5351 (1975)
7.36 See, e.g., D. Mattis: *The Theory of Magnetism* (Harper and Row, New York 1965)
      Chap.2
7.37 P. Nozieres: J. Low Temp. Phys. *17*, 31 (1974)
7.38 J.A. Hertz, J. Handler: Phys. Rev. B *15*, 4667 (1977)
7.39 J.A. Hertz, D.M. Edwards: Phys. Rev. Lett. *28*, 1334 (1972); J. Phys. F *3*,
      2174, 2191 (1973)
7.40 S. Doniach, S. Engelsberg: Phys. Rev. Lett. *17*, 750 (1966)
      W. Brinkman, S. Engelsberg: Phys. Rev. *169*, 417 (1968)
7.41 Such an approach has been used in the related problem of valence fluctuations
      by F.D.M. Haldane: Phys. Rev. B *15*, 281, 2477 (1977)
7.42 See, e.g., S. Doniach: Adv. Phys. *18*, 819 (1969)
7.43 B. Velicky, S. Kirkpatrick, H. Ehrenreich: Phys. Rev.*175*, 747 (1968)
      K. Levin, K. Benneman: Phys. Rev. B *5*, 3770 (1972)
7.44 A. Oguchi: Progr. Theor. Phys. *43*, 257 (1970)
7.45 W. Brenig, K. Schönhammer: Z. Phys. *267*, 201 (1974)
7.46 A. Madhukar, B. Bell: Phys. Rev. Lett. *34*, 1631 (1975); Phys. Rev. B *14*,
      4281 (1976)
7.47 E. Anda, N. Majlis, D. Grenupel: J. Phys. C *10*, 2365 (1977)
7.48 J.R. Schrieffer, P.A. Wolff: Phys. Rev. *149*, 491 (1966)
7.49 P. Schuck: Phys. Rev. B *13*, 5225 (1976)
7.50 H. Keiter, J. Kimball: Int. J. Magnetism *1*, 233 (1971)
7.51 W. Brinkman, T.M. Rice: Phys. Rev. B *2*, 4302 (1970)
7.52 K. Schönhammer: Z. Phys. B *21*, 389 (1975)
7.53 K. Schönhammer: Phys. Rev. B *13*, 4336 (1976)
7.54 T.L. Einstein: Phys. Rev. B *11*, 577 (1975)
7.55 A more complete discussion is given in T.L. Einstein: CRC Crit. Rev. in
      Solid State and Mat. Sci. *7*, 260 (1978); Ph.D. dissertation, Univ. of
      Pennsylvania (1973), unpublished
7.56 R. Gomer: Solid State Phys. *30*, 93 (1975)
7.57 W. Kohn, K.H. Lau: Solid State Commun. *18*, 553 (1976)
7.58 J.W. Davenport, T.L. Einstein, J.R. Schrieffer, P. Soven: In *The Physical
      Basis for Heterogeneous Catalysis*, ed. by E. Drauglis, R.I. Jaffee (Plenum
      Press, New York 1975) p.295
7.59 C.F. Melius, J.W. Moskowitz, A.P. Mortola, M.B. Baillie, M.A. Ratner: Surf.
      Sci. *59*, 279 (1976)
      J.R. Schrieffer: J. Vac. Sci. Technol. *13*, 335 (1976)
7.60 E.A. Hyman: Phys. Rev. B *11*, 3739 (1975)
7.61 J.O. Hirschfelder, C.F. Curtiss, R.B. Bird: *Molecular Theory of Gases and
      Liquids* (John Wiley, New York 1954)
      Henry Margenau: J. Chem. Phys. *9*, 896 (1938)
      J.A. Barker, R.O. Watts, J.K. Lee, T.P. Schafer, Y.T. Lee: J. Chem. Phys.
      61, 3081 (1974)
      B.M. Axilrod, E. Teller: J. Chem. Phys. *11*, 299 (1943)
      Y. Muto: Proc. Phys. Math. Soc. Jpn *17*, 629 (1943)
      O. Sinanoglu, K.S. Pitzer: J. Chem. Phys. *32*, 1279 (1960)
      A.D. McLachlan: Mol. Phys. *7*, 381 (1964)
      T.B. MacRury, B. Linder: J. Chem. Phys. *54*, 2056 (1971)
      D.L. Freeman: J. Chem. Phys. *62*, 4300 (1975)
      L.W. Bruch, P.I. Cohen, M.B. Webb: Surf. Sci. *59*, 1 (1976)
      J.J. Rehr, E. Zaremba, W. Kohn: Phys. Rev. B *12*, 2062 (1975)
7.62 M.A. Ruderman, C. Kittel: Phys. Rev. *96*, 99 (1954)
      T. Kasuya: Prog. Theor. Phys. (Kyoto) *16*, 45 (1956)
      K. Yosida: Phys. Rev. *106*, 893 (1957)
      J.H. Van Vleck: Rev. Mod. Phys. *34*, 681 (1962)
7.63 G.F. Koster: Phys. Rev. *94*, 1498 (1954)
7.64 K. Gottfried: *Quantum Mechanics I* (W.A. Benjamin, New York 1966) p.382
7.65 N.R. Burke: Surf. Sci. *58*, 349 (1976)
7.66 J.C. Slater, G.F. Koster: Phys. Rev. *94*, 1498 (1954)

7.67   G. Allan: Ann. Phys. (Paris) 5, 169 (1970)
        C. Brunel, G. Allan: Surf. Sci. 39, 385 (1973)
7.68   M. Leynaud, G. Allan: Surf. Sci. 53, 359 (1975)
7.69   W. Ho, S.L. Cunningham, W.H. Weinberg: Surf. Sci. 62, 662 (1977)
7.70   T.B. Grimley, M. Torrini: J. Phys. C 6, 868 (1973)
        T.B. Grimley: Ber. Bunsenges. Phys. Chem. 75, 1003 (1971)
7.71   T.B. Grimley, S.M. Walker: Surf. Sci. 14, 395 (1969)
7.72   K.C. Pandey: Phys. Rev. B 14, 1557 (1976)
7.73   K. Schönhammer, V. Hartung, W. Brenig: Z. Phys. B 22, 143 (1975)
7.74   K. Binder, D.P. Landau: Surf. Sci. 61, 577 (1976)
        B. Mihura, D.P. Landau: Phys. Rev. Lett. 38, 977 (1977)
        K. Binder, (ed.): Monte Carlo Methods in Statistical Physics, Topics in
        Current Physics, Vol.7 (Springer, Berlin, Heidelberg, New York 1979)
7.75   E. Domany, M. Schick, J.S. Walker: Phys. Rev. Lett. 38, 1148 (1977)
        A.N. Berker, S. Ostland, F.A. Putnam: Phys. Rev. B 17, 3650 (1978)
        S. Krinsky, D. Mukamel: Phys. Rev. B 16, 2313 (1977)
        E. Domany, E.K. Riedel: Phys. Rev. Lett. 40, 561 (1978)
7.76   T.M. Lu, G.C. Wang, M.G. Lagally: Phys. Rev. Lett. 39, 411 (1977)
        M.G. Lagally: CRC Crit. Rev. in Solid State and Mat. Sci. 7, 233 (1978)
        also see W.Y. Ching, D.C. Huber, M.G. Lagally, G.C. Wang: Surf. Sci. 77,
        550 (1978)
7.77   A. Liebsch: Phys. Rev. B 11, 577 (1975)
7.78   N. Kar, P. Soven: Solid State Commun. 20, 977 (1976)
7.79   E.W. Plummer: In Interactions on Metal Surfaces, ed. by R. Gomer, Topics
        in Applied Physics, Vol.4 (Springer, Berlin, Heidelberg, New York 1975)
        Chap.5
7.80   S.L. Weng, E.W. Plummer, T. Gustafsson: Phys. Rev. B 18, 1718 (1978)
7.81   P.J. Feibelman, D.E. Eastman: Phys. Rev. B 10, 4932 (1974)
7.82   T.L. Einstein: Surf. Sci. 45, 713 (1974)
7.83   C.N. Berglund, W.E. Spicer: Phys. Rev. 136, A 1030 (1964); 136, A 1044 (1964)
7.84   W.L. Schaich, N.W. Ashcroft: Solid State Commun. 8, 1959 (1970); Phys. Rev.
        B 3, 2452 (1971)
7.85   A. Liebsch: Phys. Rev. Lett. 32, 1203 (1974); Phys. Rev. B 13, 544 (1976);
        Solid State Commun. 19, 1193 (1976)
7.86   J.B. Pendry: Surf. Sci. 57, 679 (1976)
7.87   S.Y. Tong, C.M. Li, A.R. Lubinsky: Phys. Rev. Lett. 39, 498 (1977)
7.88   I. Adawi: Phys. Rev. 134, A 788 (1964)
7.89   G.D. Mahan: Phys. Rev. B 2, 4334 (1970)
7.90   C. Caroli, D. Lederer-Rozenblatt, B. Roulet, D. Saint-James: Phys. Rev.
        B 8, 4552 (1973)
7.91   L.V. Keldysh: Zh. Eksp. Teor. Fiz. 47, 1515 (1964) [English transl.:
        Sov. Phys.-JETP 20, 1018 (1965)]
7.92   H. Keiter: 7. Phys. B 30, 167 (1978)

Subject Index

Applied Physics

A monthly journal

Board of Editors

S. Amelinckx, Mol; **V. P. Chebotayev**, Novosibirsk;
R. Gomer, Chicago, IL; **P. Hautojärvi**, Espoo;
H. Ibach, Jülich; **K.-L. Kompa**, Garching;
V. S. Letokhov, Moskau; **H. K. V. Lotsch**, Heidelberg;;
H. J. Queisser, Stuttgart; **F. P. Schäfer**, Göttingen;
K. Shimoda, Tokyo; **R. Ulrich**, Stuttgart;
W. T. Welford, London; **H. P. J. Wijn**, Endhoven

Coverage

application-oriented experimental and theoretical
physics

Solid-State Physics	*Quantum Electronics*
Surface Science	*Laser Spectroscopy*
Solar Energy Physics	*Photophysical Chemistry*
Microwave Acoustics	*Optical Physics*
Electrophysics	*Optical Communications*

Special Features

rapid publication (3–4 months)
no page charges for concise reports
microform edition available

Languages
mostly English

Articles

original reports, and short communications
review and/or tutorial papers

Manuscripts

to Springer-Verlag (Attn. H. Lotsch), P.O.Box 105 280
D-6900 Heidelberg 1, FRG

Place North-America orders with:
Springer-Verlag New York Inc., 175 Fifth Avenue,
New York, N.Y. 10010, USA

Springer-Verlag
Berlin
Heidelberg
New York

M. A. Van Hove, S. Y. Tong

Surface Crystallography by LEED

Theory, Computation and Structural Results
1979. 19 figures, 2 tables. IX, 286 pages
(Springer Series in Chemical Physics, Volume 2)
ISBN 3-540-09194-7

Contents: Introduction. – The Physics of LEED. – Basic Aspects of
the Programs. – Symmetry and Its Use. – Calculation of Diffraction
Matrices for Single Bravais-Lattice Layers. – The Combined Space
Method for Composite Layers: by Matrix Inversion. – The Combined
Space Method for Composite Layers: by Reverse Scattering Pertur-
bation. – Stacking Layers by Layer Doubling. – Stacking Layers by
Renormalized Forward Scattering (RFS) Pertubation. – Assembling a
Program: The Main Program and the Input. – Subroutine Listings. –
Structural Results of LEED Crystallography. – Appendices. – Refe-
rences. – Subject Index.

Inelastic Electron Tunneling Spectroscopy

Proceedings of the International Conference, and Symposium on
Electron Tunneling, University of Missouri-Columbia, USA,
May 25–27, 1977
Editor: T. Wolfram
1978. 126 figures, 7 tables. VIII, 242 pages
(Springer Series in Solid-State Sciences, Volume 4)
ISBN 3-540-08691-9

Contents: Review of Inelastic Electron Tunneling. – Applications of
Inelastic Electron Tunneling. – Theoretical Aspects of Electron
Tunneling. – Discussions and Comments. – Molecular Adsorption
on Non-Metallic Surfaces. – New Applications of IETS. – Elastic
Tunneling.

Interactions on Metal Surfaces

Editor: R. Gomer
1975. 112 figures. XI, 310 pages
(Topics in Applied Physics, Volume 4)
ISBN 3-540-07094-X

Contents: *J. R. Smith:* Theory of Electronic Properties of Surfaces. –
S. K. Lyo, R. Gomer: Theory of Chemisorption. – L. D. Schmidt: Chem-
isorption: Aspects of the Experimental Situation. – *D. Menzel:*
Desorption Phenomena. – *E. W. Plummer:* Photoemission and Field
Emission Spectroscopy. – *E. Bauer:* Low Energy Electron Diffraction
(LEED) and Auger Methods. – *M. Boudart:* Concepts in
Heterogeneous Catalysis.

Secondary Ion Mass Spectrometry SIMS-II

Proceedings of the Second International Conference on
Secondary Ion Mass Spectroscopy
Stanford University, Stanford, California, USA, August 27–31, 1979
Editors: A. Benninghoven, C. A. Evans jr., R. A. Powell, R. Shimizu,
H. A. Storms
1980. 234 figures. XIII, 298 pages
(Springer Series in Chemical Physics, Volume 9)
ISBN 3-540-09843-7

Contents: Fundamentals. – Quantitation. – Semiconductors. – Static
SIMS. – Metallurgy. – Instrumentation. – Geology. – Panel Discussion. – Biology. – Combined Techniques. – Postdeadline Papers.

Springer-Verlag
Berlin
Heidelberg
New York